INSECT
IMMUNOLOGY

INSECT IMMUNOLOGY

Nancy E. Beckage

AMSTERDAM • BOSTON • HEIDELBERG • LONDON • NEW YORK • OXFORD
PARIS • SAN DIEGO • SAN FRANCISCO • SINGAPORE • SYDNEY • TOKYO

Academic Press is an imprint of Elsevier

Academic Press is an imprint of Elsevier
525 B Street, Suite 1900, San Diego, CA 92101-4495, USA
30 Corporate Drive, Suite 400, Burlington, MA 01803, USA
84 Theobald's Road, London WC1X 8RR, UK
Radarweg 29, PO Box 211, 1000 AE Amsterdam, The Netherlands
Linacre House, Jordan Hill, Oxford OX2 8DP, UK

First edition 2008

Library of Congress Cataloging in Publication Data
A catalog record for this book is available from the Library of Congress

British Library Cataloguing in Publication Data
A catalogue record for this book is available from the British Library

ISBN: 978-0-12-373976-6

For information on all Academic Press publications
visit our website at books.elsevier.com

Typeset by Charon Tec Ltd (A Macmillan Company), Chennai, India
www.charontec.com

Printed and bound by CPI Group (UK) Ltd, Croydon, CR0 4YY

Transferred to Digital Print 2011

Working together to grow
libraries in developing countries

www.elsevier.com | www.bookaid.org | www.sabre.org

ELSEVIER   BOOK AID
            International   Sabre Foundation

CONTENTS

PREFACE

AN INSECT'S PERSPECTIVE ON IMMUNITY

The subject of insect immunology is critically important to both basic insect biology, in clarifying immune mechanisms and disease etiology, and applied entomology, as immunological factors influence the success of parasitoids in biological control. As insects transmit many highly virulent diseases of animals and plants, and play critical roles as agricultural pests, the study of insect immunology has major impacts on global health, economy, and human welfare.

Insect immune responses form the focus of this book. This volume offers current perspectives on both novel and well-established lines of insect immunological research, with emphasis on physiological, biochemical, and molecular aspects of immunity. Many of its chapters including the Epilogue (chapter 13) suggest avenues for future research that are likely to prove especially fruitful.

The insect immune system is comprised of both humoral plasma-borne factors and cellular or hemocyte-associated molecules that are mobilized in response to parasitic or pathogenic infection. Many of these elements act cooperatively during an immune response, revealing a complex level of interaction between cell-based and humoral factors. Examples of such interactions are outlined in several chapters of this book (e.g. chapters 2, 3, 4 and 9), generating an appreciation of using integrative approaches to study insect immunity.

The concept of 'immunity' is often inferred as representing both 'innate' constitutive and 'acquired' or 'adaptive' immunity based on the vertebrate literature. While insects have long been recognized as having innate immunity, evidence for the existence of acquired immunity, the basis of a 'memory' function, in insects has been slow to materialize. However, as described in chapters 1 and 5, we are rapidly developing an awareness of certain aspects of 'acquired immunity' in the insect realm.

Insects are susceptible to parasites and pathogens during many stages of their life cycle, and many of their attributes contribute to their infection status. For example, insects are particularly vulnerable to pathogenic infections during the critical periods of molting, when the lining of the foregut and hindgut, as well as the external cuticle, are sloughed during the ecdysis process. At this time, microbes that normally reside in the gut lumen can cross the gut epithelium and move into the hemocoel potentially causing infection. After a molt but before tanning of the new external cuticle has occurred, an insect's soft integument also provides opportunities for invasion of the hemocoel.

The surrounding environment can also impact an insect's susceptibility to infection. Many insects live in microbe-rich environments (e.g. soil, decaying wood), which pose additional immunological challenges. The maintenance of high levels of immunocompetence is critical to survival when the insect lives in close contact with many potential pathogens. On the other hand, insect microbial symbionts must evade immunological recognition in the insects carrying them, and the underlying mechanism(s) that allow symbionts to complete their life cycle successfully is an open question. As we delve deeper into identifying molecular signatures that trigger or suppress recognition of non-self in insects, as described in chapters 1, 2, and 11, some of these questions will undoubtedly be answered.

The disease susceptibility of insects (as well as vertebrates) is also influenced by stress. Stress-inducing factors can take the form of reduced food availability, starvation, or crowding as examples. During insect outbreaks, many latent or cryptic viral pathogen infections can erupt as epizootics, causing huge declines in insect population levels that cycle over time. Under crowded conditions, rapid insect-to-insect disease transmission occurs.

Parasitoids are unusual parasites in that they invariably cause death of their insect host at the conclusion of the host–parasite relationship. Hence, they are used in biological control programs to kill insect pests. The physiological attributes of an insect host that render it refractory or permissive to a particular wasp parasitoid species are becoming clearer as the immunosuppressive roles of parasitoid polydnaviruses in preventing encapsulation of parasitoid eggs and larvae are being analyzed on a functional level. This subject is addressed in chapters 10 and 11.

As a field, insect immunology is challenging a growing cadre of researchers in genomics and proteomics to complement others focusing on physiological and biochemical approaches in studying how insects counter parasitic and pathogenic infection. The rapid pace with which the field of insect immunology has matured in the past 5–10 years in the genomic era is phenomenal. The wealth of genomic and genetic information available for the fruitfly insect model *Drosophila melanogaster* has facilitated elucidation of mechanisms of insect antimicrobial immunity. Recent development of new technologies in gene silencing afford us ample opportunity for studying genes and gene products that play seminal roles in insect anti-pathogen and anti-parasite immunity (chapter 12). Major breakthroughs in developing novel strategies to reduce insect vector-borne disease transmission (chapters 7 and 8) and prevent insect damage to agricultural crops (chapters 9 and 10) will be generated by enhanced focus in this area, as we continue to unravel previously understudied host–pathogen and host–parasite immunological interactions.

Two key biochemical mediators of insect immune responses are eicosanoids (chapter 3) and phenoloxidase enzymes (chapter 4). These molecules act in insect defenses against bacteria, viruses, fungi, and parasitoids. The recently recognized roles of lipids in insect immunity are discussed in chapters 1 and 11. As new immunomodulatory molecules and pathways are identified, our appreciation of the complexity of the roles they play will continue to grow.

New insights include recognition that the immune system and nervous system are functionally linked in insects as in vertebrates, as explored in chapter 6. These interactions are bidirectional, and ultimately impact the neural regulation of behavior of infected hosts. The emerging field of insect psychoneuroimmunology offers considerable promise for the future (chapter 13).

The delicate interplay between resistance traits of the insect host and virulence characteristics of the pathogen or parasite can now be dissected at genetic and molecular levels to provide deeper insight into the co-evolutionary 'arms race' between the host and its invader. Host immune competence is critical to the success of, or defense against, pathogens and parasites that play varied roles as regulators of host development and survival. Molecular strategies of 'parasite offense' and 'host defense' offer novel opportunities for manipulating those interactions in a beneficial manner, including development of biopesticides with enhanced virulence to control insect pests, and implementation of novel strategies to stop the movement of parasites and pathogens through insect vectors to halt cycles of disease transmission.

A major goal of this book is to provide an insect-oriented complement to the large library of books focusing intensively on mammalian immunity (e.g. 'The Immune Response: Basic and Clinical Principles' by T. Y. Mak and M. E. Saunders (2006; Elsevier/Academic Press)). Vertebrate immunologists will acquire a comparative perspective in learning about the insect systems described in this book. As many immunological signaling pathways are highly conserved between vertebrates and invertebrates, many similarities as well as differences in immunological molecules and processes exist in these two diverse animal groups (chapter 1).

Earlier books focusing on invertebrate or insect immunology include: 'Molecular Mechanisms of Immune Responses in Insects' edited by P. T. Brey and D. Hultmark (1998; Chapman and Hall); 'Techniques in Insect Immunology' edited by A. Wiesner, G. B. Dunphy, V. J. Marmaras, I. Morishima, M. Sugumaran, and M. Yamakawa (1998; SOS Publications); 'Parasites and Pathogens of Insects: Vol. 1: Parasites, Vol. 2: Pathogens' edited by N. E. Beckage, S. N. Thompson, and B. A. Federici (1993; Academic Press); 'Insect Immunity' edited by J. P. N. Pathak (1993; Kluwer Academic Publishers); 'Immunology of Insects and Other Arthropods' edited by A. P. Gupta (1991; CRC Press); 'Hemocytic and Humoral Immunity in Arthropods' edited by A. P. Gupta (1986; John Wiley & Sons); 'Immune Mechanisms in Invertebrate Vectors' edited by A. M. Lackie (1986; Clarendon Press); and 'Invertebrate Immunity' edited by K. Maramorosch and R. E. Shope (1975; Academic Press).

Our intended audience encompasses students at the upper undergraduate and graduate levels and professional researchers in many subfields of insect science (immunology, pathology, parasitology, physiology, biochemistry, molecular biology, genomics, biotechnology, and biological control). This book is a classroom teaching tool and a research reference work to transmit current knowledge about the insect immune response to students and researchers in both basic and applied sciences. We encourage its readers to use this information to devise new technologies for manipulation of insects that play significant roles in agriculture, forestry, and disease transmission. Finally, insects serve as excellent models of many immunologically based host–parasite and host–pathogen interactions that are relevant to human biology and medicine.

Nancy E. Beckage
Departments of Entomology and Cell Biology and
    Neuroscience and Center for Disease Vector Research
University of California-Riverside
Riverside, CA 92521, USA
E-mail: nancy.beckage@ucr.edu
Phone: 951-827-3521.

ACKNOWLEDGMENTS

The chapter authors merit special praise for the excellence of their respective written contributions as well as valued discussions during the preparation of this book. I especially thank Shelley Adamo, Bryony Bonning, Shirley Luckhart, Otto Schmidt, David Schneider, and Uli Theopold for sharing their vision of priorities for the future of insect immunology as a discipline. Anita Gordillo, Cathy Cathers, and Dyan MacWilliam of the Beckage Laboratory deserve special accolades as they provided critiques of many portions of the text. The author heartily thanks Book Acquisitions/Life Sciences Editor Christine Minihane, Books Development/Life Sciences Editor Carrie Bolger, and Books Production Editor Karthikeyan Murthy of Elsevier/Academic Press for their enthusiastic support of this project, and their patience and dedication in overseeing publication of this volume. Lastly, I offer special thanks to my family members John, Ross, and Ian for their inspirations and encouragement shared with me during completion of this book.

1

INSECT AND VERTEBRATE IMMUNITY: KEY SIMILARITIES VERSUS DIFFERENCES

OTTO SCHMIDT*, ULRICH THEOPOLD** AND
NANCY E. BECKAGE†

**Insect Molecular Biology, School of Agriculture, Food and Wine, University of Adelaide,
Glen Osmond, SA 5064, Australia*
***Department of Molecular Biology and Functional Genomics, Stockholm University,
S10691 Stockholm, Sweden*
†*Departments of Entomology and Cell Biology and Neuroscience and Center for Disease
Vector Research, University of California-Riverside, Riverside, CA 92521, USA*

ABSTRACT: Historically, the mammalian adaptive immune system was the first to be analyzed in depth, providing strong paradigms on mechanisms of immune recognition and the distinction of self and non-self. However, the differentiation power of the innate immune system and the possible diversity of defense devices emerging in non-mammalian organisms may offer new perspectives on how multi-cellular organisms recognize potentially damaging objects or substances.

Abbreviations:

LM   = leverage mediated
LPS   = lipopolysaccharide
MHC   = major histocompatibility complex
PAMP = pathogen-associated molecular patterns
PAMS = pathogen-associated molecular structures
RNAi = RNA interference.

1.1   INTRODUCTION

Insects are a large and diverse group of animals that have adapted to extreme environments, including endoparasitic lifestyles, where one insect develops inside another insect. Compared to insects, with more than 32 orders, some containing hundreds of thousands of species, vertebrates are a relatively small group and rather homogeneous in morphology and physiology. Nevertheless, higher vertebrates have evolved a unique defense system in the form of an anticipatory immune response in addition to the innate immune system, which is common to all animals. The adaptive immune system is developed during early mammalian ontogeny, where gene rearrangements create thousands of gene variants encoding a repertoire of binding proteins that are clonally selected against self-recognition in immune-specific cells. After induction by foreign immunogens, a process which requires the involvement of the innate arm of the immune system, immune cells remain in readiness for future encounters with a potentially damaging object or dangerous substance.

While the adaptive immune system has impressive features involving the onto-genetic generation of antibody diversity and immunological memory, we have only started to analyze the differentiation power of the innate recognition system in vertebrates (Vivier and Malissen, 2005). Likewise, given the broad range of extreme and challenging environments in which some invertebrates live (Loker et al., 2004), we have not yet grasped the possible diversity of defense devices that may exist in non-mammalian organisms (Little et al., 2005). Comparisons of immune genes among related insect species (*Drosophila* and *Anopheles*) belong-ing to the same order of Diptera suggest very different defense strategies, using different protein families as recognition proteins and receptors (Christophides et al., 2004; Zdobnov et al., 2002). While immune strategies in most invertebrates remain to be analyzed, some already reveal highly specific and effective defense mechanisms that rival those of higher vertebrates.

Historically, the mammalian adaptive immune system was the first to be analyzed in depth using molecular biology tools. This has provided strong conceptual paradigms on how we perceive the mechanisms of immune recognition and the distinction of self and non-self. In fact, the notion that the specificity of immune recognition is determined exclusively by the nature of protein–epitope interactions has emerged from the formative power of antibody–antigen interactions that are highly specific and crucial for the immune response in specific immune cells. It is only recently that cell-free defense reactions uncovered multi-protein complexes upstream of cellular receptors (Schmidt and Theopold, 2004) that are relevant to recognition processes in insects. For example, cellular uptake reactions may be based on multi-protein complexes with enhanced detection capabilities due to combinatorial interactions between phagocytic receptors (Stuart and Ezekowitz, 2005) and upstream regulatory processes (Rahman et al., 2006). While these regulatory cascades were known to exist in many animal species (Krem and Cera, 2002), their relevance to insect immunity (Lemaitre et al., 1996) and development (Anderson, 2000) was uncovered first in *Drosophila*.

Another consequence of the conceptual preeminence of the adaptive immune system is the habit of some immunologists to use particular mammalian gene functions synonymous with general immune functions. Such is the strength of the mammalian paradigm that the presence of major histocompatibility complex (MHC) genes is sometimes correlated with the functionality of histocompatibility and the absence of MHC-like genes in other organisms is taken as evidence that these mechanisms do not exist in those organisms. But histocompatibility has been shown to be performed in many multi-cellular organisms, such as sponges (Fernandez-Busquets et al., 2002; Muller and Muller, 2003) or primitive chordates (De Tomaso et al., 2005) as part of a self-recognition mechanism, using different sets of proteins to achieve it (Litman, 2005).

In this chapter we use a comparison between insect and higher vertebrate immune reactions to highlight some of the progress that has been made in our understanding of how innate immune recognition has evolved to protect multi-cellular organisms against potentially damaging organisms. This also provides an opportunity to remind us of the large gaps that exist in our basic conceptual framework of how biological recognition processes work, particularly when it comes to the integration of immune functions with developmental and basic cellular functions, such as recognition of self.

1.2   SIMILARITIES

1.2.1   SENSING MECHANISMS

1.2.1.1   Recognition of Pathogen-Associated Molecular Structures

Biological recognition processes are generally perceived to be performed exclusively by the specific interactions of proteins with molecular structures that indicate non-self or altered-self. For example, the binding properties of antibodies, enabling the variable binding domains to attach to almost any molecular structure (antigens) and the ontogenic elimination process of self-binding antibodies by clonal selection, allow higher vertebrates with adaptive immune systems to signal non-self structures (Boehm, 2006; Mak and Saunders, 2006). Likewise, the evolution of specific binding proteins that interact with protein, lipid or sugar determinants (epitopes) that are unique to other organisms, identifies potentially damaging organisms in plants (Jones and Dangl, 2006) and animals (Janeway and Medzhitov, 2002). Since the innate immune system lacks clonal elimination mechanisms of self-binding proteins, the distinction between self and non-self relies on the selective accumulation of recognition protein repertoires that bind to target sites that are diagnostic for potentially damaging organisms (Janeway, 1989). These pathogen-associated molecular patterns (PAMPs) (Janeway, 1989) or pathogen-associated molecular structures (PAMSs) (Beutler, 2003) were selected in host organisms as target sites for potential recognition proteins. Thus, the distinction between self and non-self is based on the absence of microbe-specific structures in the host. A precondition for microbe-specific recognition by a host organism is that these structures are essential to the target organism and conserved enough to allow the host to evolve binding proteins before the pathogen has been able to eliminate or modify the target site. While the PAMS approach has been useful in providing a conceptual underpinning of innate self versus non-self recognition, it is not the complete picture, given that innate immune responses are more differentiated toward microbes than PAMSs predict, such as commensal microbes displaying PAMS without eliciting a response.

Another interaction with potentially damaging organisms is based on the so-called gene-for-gene interactions. Some pathogens require host-specific sites of interaction to gain access to host cells, which allows a host to become protected to pathogen attack by eliminating potential binding sites, or by adapting the interaction into a signal that identifies the presence of a damaging organism to the host. These gene-for-gene interactions have been identified in pathogenic or parasitic relationships with strong selection pressure, such as plant/pathogen (Jones and Dangl, 2006), insect parasitoid (Carton et al., 2005) or insect pathogen (Gottar et al., 2006) interactions.

One of the shortcomings of the PAMS-based concept of non-self recognition is that it lacks an explanation to the observed immune recognition of synthetic or altered-self structures. This highlights the question of what are the actual selection processes that produce specific recognition proteins. The evolutionary mechanisms associated with these processes have recently been brought into focus by the discovery of recognition molecules that produce diversity through alternative splicing (Dong et al., 2006; Watson, 2005). While this is a mechanism that can potentially produce enough diversity to match the genetic variability of short-lived microbes and parasites, the chances of self-recognition by one of more of the variant proteins are increased accordingly. At this stage we do not know whether alternative splicing is regulated by mechanisms that exclude self-binding proteins or whether the selection is at the level of individuals that produce novel splice products.

This raises the more general question of how organisms evolve specific binding proteins (Beutler and Hoffmann, 2004). How do organisms fight newly emerging pathogens and what happens when recognition proteins become obsolete because the pathogens have changed? Are hosts defenseless without specific recognition pathways?

1.2.1.2   Extracellular Sensor Particles

As indicated above, a major conundrum for microbe-specific recognition models is the documented recognition of objects or substances that have not been encountered by the organism or any of its ancestors. In fact, innate immune responses to artificial objects, such as plastic beads (Lavine and Strand, 2001) or some adjuvance substances (Matzinger, 1998), constitute a conceptual problem that is difficult to solve in the context of recognition systems that are based on the evolution of microbe-specific proteins.

Recent observations in insects suggest that lipid particles may serve a functional role not only as a lipid carrier (Canavoso et al., 2001; Rodenburg and Van der Horst, 2005), but also as sensor particles, being involved in the recognition and detoxification of lipopolysaccharide (LPS) (Kato et al., 1994) and other toxins (Vilcinskas et al., 1997) as well as in the production of reactive oxygen (Arakawa et al., 1996). Thus, lipid metabolism and lipid carrier proteins appear to play critical roles in the systemic immune response to parasites and pathogens (Cheon et al., 2006).

The role of the exchangeable apolipoprotein in insect lipid particles has long been implied in immune reactions, given that apolipophorin III was not only sensitive to particle–lipid composition (Sahoo et al., 2000; Soulages et al., 1996), but also to immune elicitors (Halwani et al., 2000; Niere et al., 1999) changing its structure (Niere et al., 2001) and affecting plasma-derived (Halwani and Dunphy, 1999; Wiesner et al., 1997) and cellular (Iimura et al., 1998; Whitten et al., 2004) immune activities.

Moreover, an association between lipid particles and typical immune proteins, such as phenoloxidase (Mullen and Goldsworthy, 2003), was further uncovered by the observation that lipid particles carry prophenoloxidase, its activating proteases and recognition proteins, such as LPS- and peptidoglycan-binding proteins (Rahman et al., 2006). It appears that in immune-induced insects lipid particles become associated with immune proteins and turn into adhesive particles with the potential to form cell-free aggregates (Ma et al., 2006; Rahman et al., 2006) or interact with cells (see below).

Genetic evidence for regulatory processes involved in immune recognition came from the observation that immune signals are often developed upstream of membrane-bound receptors and used in very different contexts. For example, the *Drosophila* Toll ligand Spaetzle is required for developmental and immune signaling (Lemaitre et al., 1996) and is activated to form a dimer as an outcome of regulatory processes that are reminiscent of proteolytic enzyme cascades activating pro-coagulants (Krem and Cera, 2002) and prophenoloxidase (Soderhall and Cerenius, 1998; see chapter 4 by Mike Kanost). In fact, the activated Spaetzle ligand resembles arthropod coagulogens at the structural level (Bergner et al., 1997; Delotto and Delotto, 1998).

What is the significance of these extracellular regulatory cascades to recognition processes? It appears that invertebrates with an open circulatory system are using coagulation reactions not only for wound healing, but also for the inactivation of pathogens (Theopold et al., 2004). While these extracellular defense reactions have long escaped our notice due to the difficulty of analyzing covalently linked coagulation products at the biochemical level, it has become apparent that the regulatory cascades controlling coagulation and melanization (Cerenius and Soderhall, 2004) are part of an ancestral defense reaction that has been adapted to multiple functions in different organisms (Krem and Cera, 2002). But while lipid-containing particles, such as lipophorin (Duvic and Brehelin, 1998; Li et al., 2002) and vitellogenin (Hall et al., 1999), are known to be the pro-coagulants in arthropods, we have only recently become aware that some plasma components, including immune proteins, are associated with lipid particles (Ma et al., 2006) changing their properties in the presence of elicitors (Schmidt et al., invited review). For example, lipophorin particles in insects interact with exchangeable lipoproteins and other plasma proteins, such as apolipoprotein III (Niere et al., 2001), prophenoloxidase and its activating proteases (Rahman et al., 2006), imaginal disk growth factors (Ma et al., 2006) and morphogens (Panakova et al., 2005). These modified particles may be involved in multiple processes, including lipid metabolism (Canavoso et al., 2001), immunity (Whitten et al., 2004), growth and development (Panakova et al., 2005).

Likewise, proteomic analysis of mammalian lipid particle composition revealed an association with immune proteins (Vaisar et al., 2007). Moreover, apolipoprotein E knock-out mice are highly susceptible to bacterial attack (de Bont et al., 1999) and protection against LPS by lipid particles (Kato et al., 1994; Vreugdenhil et al., 2003) may involve LPS transfer to and distribution among lipid particles (Kitchens et al., 2003; Levels et al., 2001). Finally, recent observations in mammals revealed upstream processes involving soluble CD14 and MD-2 proteins (Miyake, 2004; Visintin et al., 2001) that resemble some of the upstream processes observed in insects. Given the closed circulatory system with its threat of thrombosis, the most likely place to find aggregation-based defense reactions in vertebrates is probably outside the blood vessels, in the gut lumen, lung and other epithelia and in secretions, such as the milk.

The presence of all-pervading lipid particles involved in homeostatic functions inside and outside of cells is compatible with a 'sensor' function. In fact, some exchangeable proteins, required for lipid uptake and transport, such as apolipoprotein III in insects, register lipid composition (Van der Horst, 2003) and the presence of LPS (Whitten et al., 2004). Together with the observation that purified lipid particles respond to LPS by aggregation (Ma et al., 2006), it is conceivable that the first steps in the recognition process involve the incorporation of LPS into the lipid moiety of the particle (Rahman et al., 2006). Likewise, these particles can respond to other changes, such as lipid modification under oxidizing conditions, with changes in their properties. For example, some of the associated proteins can become activated, either directly (Mellroth et al., 2005; Yu and Kanost, 2002) or indirectly through regulatory cascades (Krem and Cera, 2002) to switch from non-adhesive to adhesive forms (Fig. 1.1).

Lipid particles are ideal biological sensors, being ubiquitous in and around all cells, containing lipids and glycolipids that are easily modified and oxidized, providing the ability to respond readily to many changes in the environment. Moreover, being similar to cellular membranes they are able to mimic damaging effects by responding to environmental clues and by changing their functional properties into immune effectors by becoming adhesive particles that can aggregate around toxins or pathogens (Rahman et al., 2006). Cell-free defense capacities combining afferent and efferent immune functions (Beutler, 2004) have long been implicated in invertebrate hemolymph, vertebrate blood plasma and milk, but whether these reactions are all based on coagulation reactions remains to be elucidated. The difficulty with identifying lipid particles involved in cell-free defense using genetic and RNAi screens is that abrogation of lipid carrier functions are deleterious to the organism hiding any possible immune phenotype. Linking lipid particles to immune functions may prove even more difficult in vertebrates, given that the closed circulatory system of vertebrates may have evolved lytic and phagocytic defense functions to inactivate potential pathogens, rather than coagulation reactions (see below).

Taken together these observations are compatible with the existence of cell-free defense reactions, involving coagulation and melanization reactions upstream of

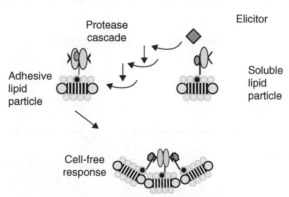

Cell-free immune recognition

FIGURE 1.1   Cell-free defense reactions involving recognition and aggregation around elicitors or damaging objects. Lipid particles, which can act as circulating sensor particles, are schematically depicted as a disk of lipid bilayer surrounded by ring-shaped apolipoproteins. Associated proteins respond to elicitors (LPS) or environmental cues by becoming adhesive either directly (Mellroth et al., 2005) or indirectly through regulatory cascades (Krem and Cera, 2002). Adhesive lipid particles aggregate by cross-linking lipid particles around damaging objects or substances.

cell-bound receptors. The importance of cell-free defense reactions is the notion that an immune signal is developed upstream of membrane-bound receptors by extracellular sensors, such as lipid-containing particles that change properties and become adhesive after exposure to damaging environments and elicitors. Adhesive lipid particles that self-assemble around damaging objects (Rahman et al., 2006) can potentially inactivate potential pathogens and toxins before they reach the cell surface. Conversely, the existence of particles with the capacity to inactivate damaging objects and substances raises the question of how adhesive particles are removed from circulation by cells.

1.2.1.3   Recognition of Self (Histocompatibility and Self-incompatibility)

Two cells from the same organism or cells from identical twins form shared flat membranes when they interact. As much as this constitutes the most visible manifestation of self-recognition that is common to all multi-cellular organisms (Burnet, 1971), we nevertheless have problems in explaining the process using instructive models of cell recognition (Fig. 1.2). Firstly, there is no apparent signaling involved in self-recognition, even though the exchange of information anticipated for the formation of a flat shared membrane between two cells could be formidable, involving the fine-tuning of mutual cell surface processes, such as adhesive forces and cell turgor. Secondly, while there are cytoplasmic scaffolds involved in receptor anchorage (Takeichi and Abe, 2005), the driving forces shaping self-based cell interactions appear to originate from processes on the cell surface following rules of thermodynamics and energy minimization as seen in air-filled soap bubbles (Janmey and Discher, 2004). Thirdly, self, altered-self and non-self recognition are performed by the same group of gene functions, such as those found in the vertebrate MHC locus or other similar gene functions in invertebrates. The fact that some of these genes are involved in 'self' as well as in 'non-self'

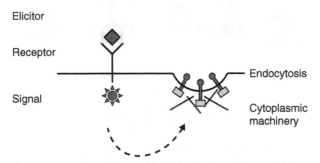

FIGURE 1.2   Instructive model of innate immune recognition and cellular response using the receptor-mediated endocytosis as an example. Elicitors bind to receptors using 'lock and key' interactions. Subsequent structural changes of the elicitor–receptor complex initiate signal pathways in the cytoplasm. In this model the signal is required to activate cytoplasmic machineries for the cell response.

processes is difficult to reconcile with unique receptor functions triggering instructive signaling pathways (Pradeu and Carosella, 2006). Finally, the detection of incompatibility between cells from closely related individuals of the same species is usually based on small allelic differences involving the absence of epitopes that are difficult to recognize even with the most sophisticated anticipatory receptor repertoire. The fact that incompatibility is seen in 'primitive' multi-cellular organisms, such as sponges (Fernandez-Busquets et al., 2002; Muller and Muller, 2003) or primitive chordates (De Tomaso et al., 2005), suggests that self-recognition is not based on anticipatory recognition processes identifying dominant epitopes, but deeply embedded in the biology and functionality of cell–cell interaction that establish multi-cellular organisms or tissues in the first place.

This basic conundrum forces us to step back and ask some fundamental questions. For example, do we need specific immune recognition proteins to achieve self-recognition? Or are self-recognition processes part of a basic cellular functionality that are reflected in homeostatic sub-routines that only require signaling pathways when external or internal circumstances cause a departure from the sub-routine. For example, the shared flat membrane may not be the result of instructive signaling pathways, but the outcome of a dynamic interactive process on the cell surface involving adhesive and cellular uptake reactions that form a balance of receptor stabilization and receptor uptake reactions (Schmidt and Schreiber, 2006). This paradigm shift engenders the simple notion that some processes, such as self-recognition, are based on 'hard-wired' cell surface-driven processes that only require signaling when imbalances indicate altered-self or non-self.

In this interactive model the involvement of multi-protein assemblies on the cell surface engaging adhesive receptors in both attachments (adhesion) as well as in uptake reactions that potentially lead to de-adhesion provides a novel approach to recognition, where cells actively engage with outside objects and other cells and in the process decode self–non-self properties. In line with this model, a stable arrangement between two cells sharing flat shared membranes is the outcome of a 'tug of war', where self is defined by the absence of instabilities, whereas changes in cell properties (altered-self) or inherent genetic differences in surface properties will indicate non-self, resulting in the formation of vesicles or internalization of cells or objects.

What are these multi-protein assemblies and how can one and the same receptor engage in adhesion and de-adhesion? The crucial assumption is that extracellular driving forces can shape the membrane and internalize objects and receptors (Schmidt and Theopold, 2004). Most interactions of cells with outside influences can be described as instructive processes, where signals are generated by receptors and transmitted to the cytoplasm or nucleus, causing activation of cytoplasmic machineries or differential gene expression (Fig. 1.2). Yet all available data are also compatible with mechanisms, where signal transmission can be the result of extracellular driving forces that mediate cellular uptake reactions and receptor internalization. A number of observations, where receptors without a signaling capacity, such as GPI-anchored receptors, or recombinant receptors with deleted cytoplasmic domains are still performing uptake reactions, suggest that signal transmission may be a consequence of uptake reactions not a prerequisite.

What are these extracellular driving forces? A conceptual clue comes from an almost forgotten mechanisms implying a zipper-mediated phagocytosis of objects (Swanson and Baer, 1995), where surface receptors are wrapped around the object forming a phagosome by a velcro-like mechanism (Fig. 1.3). While velcro-like mechanisms are clearly visible only in a few cases, such as heavily opsonized objects, it is obvious that surface properties and opsonins recruit receptors to the phagocytotic cup (Stuart et al., 2007) and given the type of opsonin, the uptake can be achieved by multi-functional receptors, including GPI-anchored receptors, such as scavenger receptors (Stuart and Ezekowitz, 2005). This suggests that any particle or object potentially interacts with receptors to sculpture the cell membrane, provided it has receptor binding sites or cell adhesion proteins attached to it.

If we accept that lipid particles are sensors that change properties after encountering environmental or immunological cues (Fig. 1.1), the most 'instructive' change in properties is that lipid particles become adhesive (Ma et al., 2006), causing particles to engage with cellular receptors, depending on the adhesive properties of the particle (Rahman et al., 2006). In this context, adhesive lipid particles can be visualized as small opsonized objects, which interact with opsonin-specific receptors. Although lipid uptake has been described in the literature as a classical case of receptor-mediated endocytosis reactions (Goldstein et al., 1985), the size

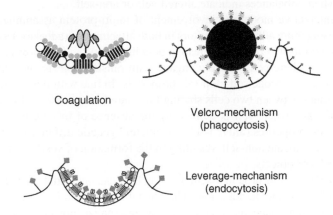

Coagulation

Velcro-mechanism
(phagocytosis)

Leverage-mechanism
(endocytosis)

FIGURE 1.3   Cellular uptake of non-self objects involving attachment and interactions leading to recognition of non-self. Adhesion and uptake of objects and microbes are driven by mechanisms that involve either velcro-mechanisms, where uptake is dependent on adhesive receptors wrapping the cell membrane around the object (black round object covered with lectins as opsonins), or cellular membrane invaginations dragging the object into the cell. Given the size differences of lipid particles and receptors adhesive lipid particles can be regarded as opsonized objects taken up by cells using velcro-like phagocytosis reactions. Since this involves a tilting of membrane-bound receptors around the particle, we called this an LM-uptake reaction (Schmidt and Theopold, 2004). Clustering of lipid particles on the cell surface may drive the uptake of solid and liquid cargo by a cellular clearance reaction based on dynamic adhesion processes on the cell surface (Schmidt and Schreiber, 2006).

and sensor properties of lipid particles are also compatible with a possible uptake by a velcro-like mechanism, where receptors are wrapped around the object creating the inverse curvature of the membrane required for the uptake reaction (Fig. 1.3). In an analogy with phagocytosis reactions, the interaction of adhesive receptors with lipid particles causes receptors to tilt, thereby producing a membrane curvature (Fig. 1.3). Clustering of particle–receptor complexes on the cell surface will drive uptake reactions in a so-called leverage-mediated (LM) process (Schmidt and Theopold, 2004), leading to membrane curvatures, receptor internalization, endocytosis and phagocytosis. Thus adhesive receptors that bind to external binding sites as well as to adhesive lipid particles can potentially become involved in two opposite reactions, the cell attachment to substrate or the detachment of spread cells due to receptor internalization (Fig. 1.4). It is this dynamic balance of forces between two cells that only form a shared flat membrane when the two cells are identical (Fig. 1.5). The unique properties of this 'tug of war' between two cells are that minute alterations in the composition of these multiprotein complexes may create an imbalance, causing vesicle formation or complete phagocytosis of altered-self or non-self cells and objects.

1.2.1.4   Recognition of Altered-Self (Apoptotic and Tumor Cells)

Histoincompatibility reactions to individuals from the same species are found in vertebrates and many invertebrates (Buss, 1987), including primitive chordates and sponges. This suggests that the mechanism involved in these processes is performed by the innate immune system, where self and altered-self recognition is paradoxically mediated by the same receptors (Stuart and Ezekowitz, 2005) that also act as

Dynamic adhesion and de-adhesion processes

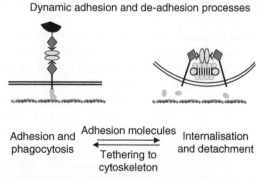

Adhesion and phagocytosis   Adhesion molecules   Internalisation and detachment

Tethering to cytoskeleton

FIGURE 1.4   Dynamic interaction of adhesion (attachment of receptors to external binding sites) and lateral cross-linking with adhesive particles on the cell surface, causing receptor internalization (Schmidt and Schreiber, 2006). Since lateral cross-linking of receptors (receptor movements in two-dimensional membrane) are thermodynamically favored over receptor binding to external binding sites (requiring movements in three dimensions), uptake by LM-mechanisms are favored over adhesion. To retain adhesive properties on the cell surface requires anchorage to cytoplasmic scaffolds, such as actin-cytoskeleton. Conversely, destabilization of cytoplasmic scaffolds may enhance macropinocytosis of existing clusters of LM-assemblies, but also prevent the formation of clusters.

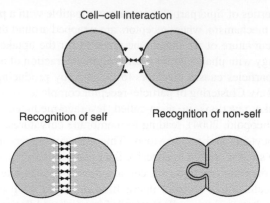

FIGURE 1.5   Cell adhesion and cell interaction leading to self and non-self recognition. After initial adhesive contact by adhesive receptors (black arrows) disruption of membrane attachments to cytoplasmic scaffolds allows lateral movement of receptors. This leads to LM-membrane invaginations pulling membranes inward (white arrows) bringing the cells closer together and producing more adhesive contacts. Eventually the two cells form a shared flat surface of two adjacent membranes if the combined adhesive and LM-mechanisms are equally balanced (below left). The implications of the model are that cells from two different genotypes produce forces at the membranes that are not equally balanced and generate vesicle formation and possible phagocytosis (below right).

effectors. For example, the detection of apoptotic cells and its uptake by macrophage-like immune cells or by neighboring epithelial cells is achieved by immune and scavenger receptors (Franc et al., 1996; Greenberg and Grinstein, 2002; Platt et al., 1999). This raises the question, what are the immunogens and how is phagocytosis of apoptotic cells accomplished? While a number of gene functions have been identified in *Drosophila* and *C. elegans* (Horvitz and Reddien, 2004; Mangahas and Zhou, 2005), the mechanism of recognition and apoptosis is still unclear (Seimon et al., 2006).

However, the unique properties of self-recognition as a 'tug of war' between two cells, where minute alterations in the composition of multi-protein complexes can create a difference in the balance of forces, provide a basis for the engulfment of apoptotic cells without a need for signaling. For example, epithelial cells that change membrane properties in the form of displaying phosphatidylserine on the outer membrane leaflet may generate instabilities at the site of cell–cell interactions, resulting in the formation of vesicles (Fig. 1.5) that indicate altered-self causing the uptake of apoptotic cells, by neighboring or macrophage-like cells.

A similar conundrum is the recognition of tumor cells (Houghton, 1994). In higher vertebrates tumor cells are inactivated when antibodies recognize dominant tumor markers from chromosome rearrangements (Argani et al., 2001), protein phosphorylation (Zarling et al., 2006), change of glycodeterminants (Kumar et al., 2005; Osinaga et al., 1996; Santos-Silva et al., 2005) or virus infection (Kurosaki et al., 2005). Thus the conceptual basis for tumor surveillance in higher vertebrates is the presence of new antigens in cancerous cells (Rosenberg, 2001). However, genetic analysis of insect tumors revealed mostly recessive tumor suppressor functions (Gateff, 1978; Sparrow, 1979) that can be used to target human genes involved in cancer (Bier, 2005; Brumby and Richardson, 2005). Moreover, investigations in the direct immunogenicity of mutated self suggest that the accumulation of small changes in proteins, mostly truncations, is responsible for detection of altered-self rather than single big changes (Engelhorn et al., 2006). While the loss of protein structures and functions is not easily detected by anticipatory recognition systems that are geared toward recognizing dominant 'gain of new antigens' (Pardoll, 2003), the problem seems to be impossible for PAMS-based innate recognition system. In this context it is conceivable that surveillance mechanisms based on 'tug of war'-like interactions can detect even minor changes in multi-protein complexes that form shared flat membranes.

1.2.2 EFFECTOR MECHANISMS

1.2.2.1 Antimicrobial Peptide Response

Antimicrobial proteins are important defense molecules against microbes with unique structural properties that allow the permeation and disruption of target membranes. Antimicrobial peptides are believed to kill microorganisms via non-receptor-mediated mechanisms, although peptides, such as nisin Z, exist that bind to bacterial cell wall components. According to Shai (2002) monomeric peptides that have random structure gain amphipathic structures and form oligomers in hydrophilic solutions, such that the hydrophobic regions are buried in the lumen of the oligomer and the hydrophilic regions are exposed to the solvent. Upon reaching the membrane the organization is reversed, i.e. the hydrophobic regions are exposed to the lipid constituents of the membrane and the hydrophilic regions are segregated in the lumen of the oligomer (if the peptide monomers insert into the membrane and oligomerize via the 'barrel' mechanism), or are exposed to the solution (if the peptide lies on the surface of the membrane and acts via the 'carpet' mechanism).

The main question is, why are these peptides inserted in bacterial but not in eukaryotic membranes? One suggestion is that the positive net charge of antibacterial peptides enables binding and permeation of the negatively charged phospholipid membranes of bacteria but not of zwitterionic membranes, which are the major constituents of the outer membrane leaflet of most eukaryotic membranes (Shai, 2002). However, there are important exceptions, which suggest that charge *per se* is not critical, but other factors contribute to target specificity, such as the secondary and tertiary structure of the peptide (Ahmad et al., 2006). Apart from the observation that alterations in a peptide's sequence are more likely to destroy its activity against eukaryotic than prokaryotic cells, there are no identifiable amino acid sequences that are responsible for the observed specificity (Hancock and Rozek, 2002). The fact that diastereomeric analogs (Papo et al., 2002) and hybrid antimicrobial peptides (Merrifield et al., 1995) are active suggests that target specificity is dependent on the sterical properties rather than specific protein binding activities. The main argument put forward to explain peptide activity against bacterial membranes is that cholesterol and ergosterol, which are found in eukaryotic membranes, protect against the insertion of antibacterial peptides (Boman, 2003). Although this has been confirmed in artificial membrane systems for most peptides, it raises the question why some antifungal peptides and antimicrobial peptides, such as mellitin, are toxic to some fungal species and many eukaryotic cells?

Some of the interactive mechanisms described above may be involved in peptide interactions with the cell membrane (Schmidt et al., 2005) and drive insertion into the membrane (Schmidt and Theopold, 2004). However, we are far from providing testable models for these processes.

1.2.2.2 Phagocytosis (Clearance of Damaging Objects)

Phagocytosis, the cellular uptake of particulate substrate, is a fundamental cellular process in all eukaryotic organisms and essential for the clearance of damaging objects in multi-cellular organisms (Stuart and Ezekowitz, 2005). While many organisms have specialized cells that engage in phagocytosis, such as the professional phagocytes in higher vertebrates and macrophage-like hemocytes in insects, it is important to remember that the capability to recognize and engulf objects and substances is inherent to all cells. This has been demonstrated in nematodes, where epithelial cells are able to recognize altered-self properties in neighboring cells undergoing programmed cell death and subsequently remove dead cells by phagocytosis (Mangahas and Zhou, 2005). Given the basic requirements of uptake reactions to cell function it is quite surprising how little is known about the molecular forces that drive this reaction. The perception is that phagocytosis, like endocytosis, is mediated by trigger mechanisms, where receptors interact with the object and subsequently instruct the cytoplasmic machinery to engulf the object (Scott et al., 2005).

Nevertheless, the most easily understood mechanism is the zipper (or velcro-) mechanisms of phagocytosis (Swanson and Baer, 1995), where receptors attach to the object and wrap the cellular membrane around it (Fig. 1.3). As discussed before, this represents an extracellular driving force, which is determined by receptor binding to surface determinants and opsonized surfaces. However, this velcro-mechanism is usually not detected in its pure form. Instead, most objects or microbes are connected to the membrane by few receptor attachments, with the cell membrane engaging in active curvature reactions (Aderem and Underhill, 1999), where vesicle (Sibley and Andrews, 2000) and other membrane addition processes from the endoplasmatic reticulum (Becker et al., 2005) support the engulfment of large objects.

In this model receptor clearance and uptake reactions are in a dynamic balance with adhesion reactions, where adhesive receptors engage in adhesion, phagocytosis and endocytosis, or clearance of receptors from the cell surface. For example, addition of soluble adhesion proteins, such as apolipoprotein III, can engage receptors in endocytosis-like uptake reactions causing de-adhesion of cells, while enhancing phagocytosis when immobilized on the surface of bacteria (Whitten et al., 2004). Similar observations where lectins that are immobilized on substrate cause cell spreading, whereas addition of soluble lectins cause detachment of spread cells (Glatz et al., 2004), suggest that receptor attachment to external binding sites and de-adhesion by receptor clearance from the cell surface are linked in a dynamic fashion (Schmidt and Schreiber, 2006).

Similar observations have been reported with vertebrate matricellular proteins, such as thrombospondin-1, mediating cell spreading when immobilized on substrate, but causing detachment when added to spread cells (Chandrasekaran et al., 2000). Opposite effects have also been reported in lung collectins that bind by their globular heads to cell receptors under normal conditions and suppress the production of proinflammatory mediators. However, when these same head domains interact with PAMPs on foreign organisms, apoptotic cells, or cell debris, presentation of the collagenous tails in an aggregated state to calreticulin/CD91 on the cells can initiate phagocytosis and/or proinflammatory and proimmunogenic responses (Gardai et al., 2003).

1.2.2.3 Endocytosis

Since attachments to external binding sites and to adhesive lipid particles can be performed by the same membrane-bound receptors, cell adhesion and de-adhesion can possibly be described in mechanistic terms, where the uptake of adhesive lipid particles by some receptors is viewed in the context of a velcro-like phagocytosis reactions (Fig. 1.3). Given that not much is known about the molecular details of lipid uptake and lipid shuttling (Van Hoof et al., 2005) in conjunction with the large size difference of lipid particles relative to membrane-bound receptors, it is conceivable to regard lipid particles as small objects that are taken up by phagocytosis reactions. In this context, receptors interact with lipid particles by attaching to associated adhesion proteins and assemble around the particle to form velcro-like attachments. The outcome of this interaction is the tilting of receptors around the particle (Fig. 1.3), causing an inverse curvature of the cell membrane, which is the first step toward receptor internalization and endocytosis. Lateral clustering of LM-assemblies forms vesicles in proportion to the size of the clusters on the cell surface. Again, like in phagocytosis, this process is affected by a number of extracellular and intracellular factors, such as the status of lipid particles and receptor anchorage to cytoplasmic scaffolds and actin fibers.

In principle, any large protein complex that can assemble receptors around objects with hinge-like properties can tilt receptors with the help of oligomeric adhesion proteins. To distinguish membrane sculpturing, involving receptors being tilted around hinge-like particles, from other mechanisms, we called this a leverage-mediated (LM) process, which has the capacity to sculpture membranes, internalize objects and substances, and dislocate receptors from cytoplasmic anchorage, using an extracellular driving force (Schmidt and Theopold, 2004).

The importance of extracellular driving forces is that the functionality of these assemblies is directly affected by outside factors and that cells are potentially able to decode these alterations while responding to it. Thus cell behavior can be the result of a dynamic balance between extracellular and cytoplasmic driving forces, where the proposed extracellular assemblies are able to produce configurational energy that sculpture membranes and dislocate membrane-anchored receptors tethered to cytoplasmic proteins, fibers and scaffolds. In this interactive model the intracellular rearrangements and activation of proteins by receptors are the result of mechanical dislocation of cytoplasmic receptor domains and their respective anchor proteins by LM-mechanisms. These LM-assemblies comprise adhesive proteins that can cross-link membrane-bound receptors (Fig. 1.3), which are bent around hinge-like lipid particles creating an inverse curvature of the membrane. In the process, receptors attached to cytoplasmic proteins or scaffolds may be dislocated from their cytoplasmic anchorage thereby triggering changes in cytoplasmic activities (Schmidt and Theopold, 2004). Whether or not these assemblies are able to perform membrane sculpturing depends on the overall configurational properties of the complex. Thus the specificity of LM-mechanisms is not only determined by the binding properties of the adhesion proteins as part of a 'lock and key' interaction, but also by the functionality of the assembly to produce configurational energy.

As a result of the dynamics of adhesive particle formation upstream of cell receptors (Fig. 1.1), extracellular recognition processes can integrate multiple cues, where the functional properties of the adhesive particle ultimately determine subsequent receptor interactions. Since modified adhesive lipid particles are able to self-assemble into cell-free coagulation products or with membrane-bound receptors into LM-complexes to perform cellular clearance reactions (Fig. 1.3), these sensors are also part of an extracellular machinery that has the potential to integrate 'self' and 'non-self' recognition processes by engaging the cell in dynamic activities that can be decoded while responding to it. This is best illustrated in extreme cases where external clues in the form of adhesive lipid particles constitute driving forces that engage cells in an extreme fashion, such as immunologically inert hemocytes in immune suppression (see chapter 11 by Schmidt, O.) or foam cells in atherosclerosis (Krieger and Herz, 1994).

1.3   DIFFERENCES

1.3.1   ACQUIRED (ADAPTIVE) IMMUNE SYSTEM IN HIGHER VERTEBRATES AND IMMUNOLOGICAL MEMORY INVOLVING CLONALLY SELECTED ANTIBODY-PRODUCING CELLS

While the innate immune recognition is probably more precise than we can account for by using the concept of pathogen-associated pattern recognition (Little et al., 2005) it is fundamentally different from the adaptive immune system (Klein, 1989) for a number of reasons. One is that the anticipatory nature of antibody repertoires is capable of binding epitopes never encountered by the organism or its predecessors using direct protein–epitope specific binding (Pancer and Cooper, 2006). Another is that self-recognizing antibody-producing cells are removed by clonal selection processes during ontogeny (Medzhitov and Janeway, 1998), and finally that the specific propagation of antibody-producing immune cells provides the basis for an immunological memory (Radbruch et al., 2006). None of these properties is realized in innate immune systems, whether in vertebrates or invertebrates (Nurnberger et al., 2004). Instead, recognition proteins are acquired through evolutionary adaptation processes resulting from repeated exposure to damaging organisms over many generations. Although highly diverse proteins can be the outcome of genetic recombination (Nair et al., 2005) and splice products (Watson, 2005), these mechanisms nevertheless represent innate adaptation processes involving selection at the population level to retain pathogen-binding proteins and remove self-recognizing proteins. Any process that generates diversity of recognition proteins during ontogeny requires clonal selection mechanisms to eliminate self-binding proteins. This is why Klein (1997) argued that there can be no anticipatory recognition system in invertebrates, because these organisms simply do not have the cellular capacity to create appropriate recognition repertoires via clonal selection of cells producing individual recognition proteins.

As discussed below, there are other mechanisms that provide immunological memory based on inducible immune mechanism that are retained at an elevated status during ontogeny, but can also be transmitted to subsequent generations by epigenetic mechanisms, such as immune-related DNA modifications or supply of immune components to offspring by a maternal effect (Lemke et al., 2004).

1.3.2 INDUCIBLE TOLERANCE AND MEMORY IN INVERTEBRATES

While insects lack anticipatory defense capabilities, they are able to induce immune activity after sub-lethal encounters with damaging objects or substances. For example, the ingestion of sub-lethal concentrations of Bt-formulations, containing *Bacillus thuringiensis* bacteria, spores and toxins, enables lepidopteran larvae to survive lethal doses at a later stage (Rahman et al., 2004), confirming earlier findings that an elevated immune status protects against pathogens (Reeson et al., 1998). Likewise, immune challenge in other organisms, such as bumble bees (Sadd and Schmid-Hempel, 2006) and Daphnia (Little et al., 2003), has been shown to provide specific protection at a later stage in development, suggestive of an imprinting (Little and Kraaijeveld, 2004; Schmid-Hempel, 2005).

One of the questions associated with immune induction and protection is the size of the fitness cost imposed on the insect (Kraaijeveld and Godfray, 1997; Moret and Schmid-Hempel, 2000), which is difficult to explain by the induction of immune proteins alone. However, the recent observation that lipid particles are pro-coagulants and involved in defense (Ma et al., 2006; Rahman et al., 2006) and detoxification reactions (Kato et al., 1994; Vilcinskas et al., 1997) could indicate immune-related post-translational modifications of lipid particles that affect lipid transport and metabolism. Moreover, the modification of lipid particles in immune-induced insects (Rahman et al., 2006) is likely to affect growth and development imposing immune-related fitness penalties (Kraaijeveld and Godfray, 1997; Moret and Schmid-Hempel, 2000; Rahman et al., 2004) as an outcome of a dynamic regulatory integration of immune, growth and developmental functions upstream of cellular receptors. The most common expression of fitness costs associated with an elevated immune status in insects is a delay in development (Rahman et al., 2006), which may leave induced individuals behind in numbers and eventually being out-grown by non-induced individuals of the same species. This not only provides an explanation for the transient nature of immune activation, but also suggests a strong interconnection between immunological and developmental functions in insects.

The most unexpected outcome of experimental studies involving the induction of the immune defense is that the elevated immune status can be transmitted to sub-sequent generations (Kurtz and Franz, 2003; Little and Kraaijeveld, 2004; Little et al., 2003; Rahman et al., 2004; Sadd and Schmid-Hempel, 2006). The fact that most systems report a maternal effect suggests that the transmission occurs by epigenetic mechanisms, involving the sex-specific modification of DNA or the incorporation of female-derived immune-inducible material into the oocyte with a possible immune induction during embryogenesis. Whatever the mechanism, the implications are that insects are able to acquire tolerance to damaging objects or substances by an incremental increase of an induced immune status in subsequent generations (Rahman et al., 2004).

REFERENCES

Aderem, A., and Underhill, D. M. (1999). Mechanism of phagocytosis in macrophages. *Annu. Rev. Immunol.* **17**, 593–623.

Ahmad, A., Yadav, S. P., Asthana, N., Mitra, K., Srivastava, S. P., and Ghosh, J. K. (2006). Utilization of an amphipathic leucine zipper sequence to design antibacterial peptides with simultaneous mod-ulation of toxic activity against human red blood cells. *J. Biol. Chem.* **281**, 22029–22038.

Anderson, K. V. (2000). Toll signaling pathways in the innate immune response. *Curr. Opin. Immunol.* **12**, 13–19.

Arakawa, T., Kato, Y., Hattori, M., and Yamakawa, M. (1996). Lipophorin: A carrier for lipids in insects participates in superoxide production in the haemolymph plasma. *Insect Biochem. Mol. Biol.* **26**, 403–409.

Argani, P., Rosty, C., Reiter, R. E., Wilentz, R. E., Murugesan, S. R., Leach, S. D., Ryu, B. W., Skinner, H. G., Goggins, M., Jaffee, E. M., et al. (2001). Discovery of new markers of cancer through serial analysis of gene expression: Prostate stem cell antigen is overexpressed in pancreatic adenocarcinoma. *Cancer Res.* **61**, 4320–4324.

Becker, T., Volchuk, A., and Rothman, J. E. (2005). Differential use of endoplasmic reticulum membrane for phagocytosis in J774 macrophages. *Proc. Natl. Acad. Sci. USA* **102**, 4022–4026.

Bergner, A., Muta, T., Iwanaga, S., Beisel, H. G., Delotto, R., and Bode, W. (1997). Horseshoe crab coagu-logen is an invertebrate protein with a nerve growth factor-like domain. *Biol. Chem.* **378**, 283–287.

Beutler, B. (2003). Not 'molecular patterns' but molecules. *Immunity* **19**, 155–156.

Beutler, B. (2004). Innate immunity: An overview. *Mol. Immunol.* **40**, 845–859.

Beutler, B., and Hoffmann, J. (2004). Innate immunity. *Curr. Opin. Immunol.* **16**, 1–3.

Bier, E. (2005). *Drosophila*, the golden bug, emerges as a tool for human genetics. *Nat. Rev. Genet.* **6**, 9–23.

Boehm, T. (2006). Quality control in self/nonself discrimination. *Cell* **125**, 845–858.

Boman, H. G. (2003). Antibacterial peptides: Basic facts and emerging concepts. *J. Intern. Med.* **254**, 197–215.

Brumby, A. M., and Richardson, H. E. (2005). Using *Drosophila melanogaster* to map human cancer pathways. *Nat. Rev. Cancer* **5**, 626–639.

Burnet, F. M. (1971). 'Self-recognition' in colonial marine forms and flowering plants in relation to the evolution of immunity. *Nature* **232**, 230–235.

Buss, L. W. (1987). *The Evolution of Individuality*. Princeton University Press, Princeton.

Canavoso, L. E., Jouni, Z. E., Karnas, K. J., Pennington, J. E., and Wells, M. A. (2001). Fat metabolism in insects. *Annu. Rev. Nutr.* **21**, 23–46.

Carton, Y., Nappi, A. J., and Poirie, M. (2005). Genetics of anti-parasite resistance in invertebrates. *Dev. Comp. Immunol.* **29**, 9–32.

Cerenius, L., and Soderhall, K. (2004). The prophenoloxidase-activating system in invertebrates. *Immunol. Rev.* **198**, 116–126.

Chandrasekaran, L., He, C.-Z., Al-Barazi, H., Krutzsch, H. C., Iruela-Arispe, M. L., and Roberts, D. D. (2000). Cell contact-dependent activation of alpha-3beta-1-integrin modulates endothelial cell responses to thrombospondin-1. *Mol. Biol. Cell* **11**, 2885–2900.

Cheon, H. M., Shin, S. W., Bian, G., Park, J. H., and Raikhel, A. S. (2006). Regulation of lipid metabolism genes, lipid carrier protein lipophorin, and its receptor during immune challenge in the mosquito *Aedes aegypti*. *J. Biol. Chem.* **281**, 8426–8435.

Christophides, G. K., Vlachou, D., and Kafatos, F. C. (2004). Comparative and functional genomics of the innate immune system in the malaria vector *Anopheles gambiae*. *Immunol. Rev.* **198**, 127–148.

de Bont, N., Netea, M. G., Demacker, P. N. M., Verschueren, I., Kullberg, B. J., van Dijk, K. W., van der Meer, J. W. M., and Stalenhoef, A. F. H. (1999). Apolipoprotein E knock-out mice are highly susceptible to endotoxemia and *Klebsiella pneumoniae* infection. *J. Lipid Res.* **40**, 680–685.

De Tomaso, A. W., Nyholm, S. V., Palmeri, K. J., Ishizuka, K. J., Ludington, W. B., Mitchel, K., and Weissman, I. L. (2005). Isolation and characterization of a protochordate histocompatibility locus. *Nature* **438**, 454–459.

Delotto, Y., and Delotto, R. (1998). Proteolytic processing of the *Drosophila Spaetzle* protein by *Easter* generates a dimeric NGF-like molecule with ventralising activity. *Mech. Develop.* **72**, 141–148.

Dong, Y., Taylor, H. E., and Dimopoulos, G. (2006). AgDscam, a hypervariable immunoglobulin domain-containing receptor of the *Anopheles gambiae* innate immune system. *PLoS Biol.* **4**, e229.

Duvic, B., and Brehelin, M. (1998). Two major proteins from locust plasma are involved in coagulation and are specifically precipitated by laminarin, a beta-1,3-glucan. *Insect Biochem. Mol. Biol.* **28**, 959–967.

Engelhorn, M. E., Guevara-Patino, J. A., Noffz, G., Hooper, A. T., Lou, O., Gold, J. S., Kappel, B. J., and Houghton, A. N. (2006). Autoimmunity and tumor immunity induced by immune responses to mutations in self. *Nat. Med.* **12**, 198–206.

Fernandez-Busquets, X., Kuhns, W. J., Simpson, T. L., Ho, M., Gerosa, D., Grob, M., and Burger, M. M. (2002). Cell adhesion-related proteins as specific markers of sponge cell types involved in allogeneic recognition. *Dev. Comp. Immunol.* **26**, 313–323.

Franc, N. C., Dimarcq, J. L., Lagueux, M., Hoffmann, J., and Ezekowitz, R. A. B. (1996). Croquemort, a novel *Drosophila* hemocyte/macrophage receptor that recognizes apoptotic cells. *Immunity* **4**, 431–443.

Gardai, S. J., Xiao, Y.-Q., Dickinson, M., Nick, J. A., Voelker, D. R., Greene, K. E., and Henson, P. M. (2003). By binding SIRP[alpha] or calreticulin/CD91, lung collectins act as dual function surveillance molecules to suppress or enhance inflammation. *Cell* **115**, 13–23.

Gateff, E. (1978). Malignant neoplasms of genetic origin in *Drosophila melanogaster*. *Science* **200**, 1448–1459.

Glatz, R., Roberts, H. L. S., Li, D., Sarjan, M., Theopold, U. H., Asgari, S., and Schmidt, O. (2004). Lectin-induced haemocyte inactivation in insects. *J. Insect Physiol.* **50**, 955–963.

Goldstein, J. L., Brown, M. S., Anderson, R. G. W., Russell, D. W., and Schneider, W. J. (1985). Receptor-mediated endocytosis: Concepts emerging from the LDL receptor system. *Annu. Rev. Cell Biol.* **1**, 1–39.

Gottar, M., Gobert, V., Matskevich, A. A., Reichhart, J.-M., Wang, C., Butt, T. M., Belvin, M., Hoffmann, J. A., and Ferrandon, D. (2006). Dual detection of fungal infections in *Drosophila* via recognition of glucans and sensing of virulence factors. *Cell* **127**, 1425–1437.

Greenberg, S., and Grinstein, S. (2002). Phagocytosis and innate immunity. *Curr. Opin. Immunol.* **14**, 136–145.

Hall, M., Wang, R., van Antwerpen, R., Sottrup-Jensen, L., and Soderhall, K. (1999). The crayfish plasma clotting protein: A vitellogenin-related protein responsible for clot formation in crustacean blood. *Proc. Natl. Acad. Sci. USA* **96**, 1965–1970.

Halwani, A. E., and Dunphy, G. B. (1999). Apolipophorin-III in *Galleria mellonella* potentiates hemolymph lytic activity. *Dev. Comp. Immunol.* **23**, 563–570.

Halwani, A. E., Niven, D. F., and Dunphy, G. B. (2000). Apolipophorin-III and the interactions of lipoteichoic acids with the immediate immune responses of *Galleria mellonella. J. Inver. Pathol.* **76**, 233–241.

Hancock, R. E. W., and Rozek, A. (2002). Role of membranes in the activities of antimicrobial cationic peptides. *FEMS Microbiol. Lett.* **206**, 143–149.

Horvitz, H. R., and Reddien, P. (2004). Apoptosis in *C. elegans. Annu. Rev. Cell Dev. Biol.* **20**, 193–221.

Houghton, A. N. (1994). Cancer antigens – Immune recognition of self and altered self. *J. Exp. Med.* **180**, 1–4.

Iimura, Y., Ishikawa, H., Yamamoto, K., and Sehnal, F. (1998). Hemagglutinating properties of apolipophorin III from the hemolymph of *Galleria mellonella* larvae. *Arch. Insect Biochem. Physiol.* **38**, 119–125.

Janeway, C. A., Ed. (1989). *Approaching the Asymptode? Evolution and Revolution in Immunology.* Cold Spring Harbor Laboratory Press, Plainview.

Janeway, C. A., and Medzhitov, R. (2002). Innate immune recognition. *Annu. Rev. Immunol.* **20**, 197–216.

Janmey, P. A., and Discher, D. E. (2004). Developmental biology: Holding it together in the eye. *Nature* **431**, 635–636.

Jones, J. D. G., and Dangl, J. L. (2006). The plant immune system. *Nature* **444**, 323–329.

Kato, Y., Motoi, Y., Taniai, K., Kadonookuda, K., Yamamoto, M., Higashino, Y., Shimabukuro, M., Chowdhury, S., Xu, J. H., Sugiyama, M., et al. (1994). Lipopolysaccharide–lipophorin complex formation in insect hemolymph – A common pathway of lipopolysaccharide detoxification both in insects and in mammals. *Insect Biochem. Mol. Biol.* **24**, 547–555.

Kitchens, R. L., Thompson, P. A., Munford, R. S., and O'Keefe, G. E. (2003). Acute inflammation and infection maintain circulating phospholipid levels and enhance lipopolysaccharide binding to plasma lipoproteins. *J. Lipid Res.* **44**, 2339–2348.

Klein, J. (1989). Are invertebrates capable of anticipatory immune responses? *Scand. J. Immunol.* **29**, 499–505.

Klein, J. (1997). Homology between immune responses in vertebrates and invertebrates: Does it exist? *Scand. J. Immunol.* **46**, 558–564.

Kraaijeveld, A. R., and Godfray, H. C. J. (1997). Trade-off between parasitoid resistance and larval competitive ability in *Drosophila melanogaster. Nature* **389**, 278–280.

Krem, M. M., and Cera, E. D. (2002). Evolution of enzyme cascades from embryonic development to blood coagulation. *Trends Biochem. Sci.* **27**, 67–74.

Krieger, M., and Herz, J. (1994). Structures and functions of multiligand lipoprotein receptors. *Annu. Rev. Biochem.* **63**, 601–637.

Kumar, S. R., Sauter, E. R., Quinn, T. P., and Deutscher, S. L. (2005). Thomsen–Friedenreich and Tn antigens in nipple fluid: Carbohydrate biomarkers for breast cancer detection. *Clin. Cancer Res.* **11**, 6868–6871.

Kurosaki, M., Izumi, N., Onuki, Y., Nishimura, Y., Ueda, K., Tsuchiya, K., Nakanishi, H., Kitamura, T., Asahina, Y., Uchihara, M., and Miyake, S. (2005). Serum KL-6 as a novel tumor marker for hepatocellular carcinoma in hepatitis C virus infected patients. *Hepat. Res.* **33**, 250–257.

Kurtz, J., and Franz, K. (2003). Evidence for memory in invertebrate immunity. *Nature* **425**, 37–38.

Lavine, M. D., and Strand, M. R. (2001). Surface characteristics of foreign targets that elicit an encapsulation response by the moth *Pseudoplusia includens. J. Insect Physiol.* **47**, 965–974.

Lemaitre, B., Nicolas, E., Michaut, L., Reichhart, J. M., and Hoffmann, J. A. (1996). The dorsoventral regulatory gene cassette spatzle/Toll/cactus controls the potent antifungal response in *Drosophila* adults. *Cell* **86**, 973–983.

Lemke, H., Coutinho, A., and Lange, H. (2004). Lamarckian inheritance by somatically acquired maternal IgG phenotypes. *Trends Immunol.* **25**, 180–186.

Levels, J. H. M., Abraham, P. R., van den Ende, A., and van Deventer, S. J. H. (2001). Distribution and kinetics of lipoprotein-bound endotoxin. *Infect. Immun.* **69**, 2821–2828.

Li, D., Scherfer, C., Korayem, A. M., Zhao, Z., Schmidt, O., and Theopold, U. (2002). Insect hemolymph clotting: Evidence for interaction between the coagulation system and the prophenoloxidase activating cascade. *Insect Biochem. Mol. Biol.* **32**, 919–928.

Litman, G. W. (2005). Histocompatibility: Colonial match and mismatch. *Nature* **438**, 437–439.

Little, T. J., and Kraaijeveld, A. R. (2004). The ecological and evolutionary implications of immunological priming in invertebrates. *Trends Ecol. Evol.* **19**, 58–60.

Little, T. J., O'Connor, B., Colegrave, N., Watt, K., and Read, A. F. (2003). Maternal transfer of strain-specific immunity in an invertebrate. *Curr. Biol.* **13**, 489–492.

Little, T. J., Hultmark, D., and Read, A. F. (2005). Invertebrate immunity and the limits of mechanistic immunology. *Nat. Immunol.* **6**, 651–654.

Loker, E. S., Adema, C. M., Zhang, S.-M., and Kepler, T. B. (2004). Invertebrate immune systems – not homogeneous, not simple, not well understood. *Immunol. Rev.* **198**, 10–24.

Ma, G., Hay, D., Li, D., Asgari, S., and Schmidt, O. (2006). Recognition and inactivation of LPS by lipophorin particles. *Dev. Comp. Immunol.* **30**, 619–626.

Mak, T.W., and Saunders, M.E. (2006). *The Immune Response: Basic and Clinical Principles.* 1194 pp. Academic Press/Elsevier, San Diego, CA.

Mangahas, P. M., and Zhou, Z. (2005). Clearance of apoptotic cells in *Caenorhabditis elegans. Semin. Cell Dev. Biol.* **16**, 295–306.

Matzinger, P. (1998). An innate sense of danger. *Semin. Immunol.* **10**, 399–415.

Medzhitov, R., and Janeway, C. A. (1998). Innate immune recognition and control of adaptive immune responses. *Semin. Immunol.* **10**, 351–353.

Mellroth, P., Karlsson, J., Hakansson, J., Schultz, N., Goldman, W. E., and Steiner, H. (2005). Ligand-induced dimerization of *Drosophila* peptidoglycan recognition proteins *in vitro. Proc. Natl. Acad. Sci.USA* **102**, 6455–6460.

Merrifield, R., Juvvadi, P., Andreu, D., Ubach, J., Boman, A., and Boman, H. (1995). Retro and retroenantio analogs of cecropin–melittin hybrids. *Proc. Natl. Acad. Sci.USA* **92**, 3449–3453.

Miyake, K. (2004). Innate recognition of lipopolysaccharide by Toll-like receptor 4-MD-2. *Trends Microbiol.* **12**, 186–192.

Moret, Y., and Schmid-Hempel, P. (2000). Survival for immunity: The price of immune system activation for bumblebee workers. *Science* **290**, 1166–1168.

Mullen, L., and Goldsworthy, G. (2003). Changes in lipophorins are related to the activation of phenoloxidase in the haemolymph of *Locusta migratoria* in response to injection of immunogens. *Insect Biochem. Mol. Biol.* **33**, 661–670.

Muller, W. E. G., and Muller, I. M. (2003). Origin of the metazoan immune system: Identification of the molecules and their functions in sponges. *Integr. Comp. Biol.* **43**, 281–292.

Nair, S. V., Ramsden, A., and Raftos, D. (2005). Ancient origins: Complement in invertebrates. *Invertebr. Survival J.* **2**, 114–123.

Niere, M., Meisslitzer, C., Dettloff, M., Weise, C., Ziegler, M., and Wiesner, A. (1999). Insect immune activation by recombinant *Galleria mellonella* apolipophorin III. *Biochim. Biophys. Acta Protein Struct. Mol. Enzymol.* **1433**, 16–26.

Niere, M., Dettloff, M., Maier, T., Ziegler, M., and Wiesner, A. (2001). Insect immune activation by apolipophorin III is correlated with the lipid-binding properties of this protein. *Biochemistry* **40**, 11502–11508.

Nurnberger, T., Brunner, F., Kemmerling, B., and Piater, L. (2004). Innate immunity in plants and animals: Striking similarities and obvious differences. *Immunol. Rev.* **198**, 249–266.

Osinaga, E., Babino, A., Grosclaude, J., Cairoli, E., Batthyany, C., Bianchi, S., Signorelli, S., Varangot, M., Muse, I., and Roseto, A. (1996). Development of an immuno-lectin-enzymatic assay for the detection of serum cancer-associated glycoproteins bearing Tn determinant. *Int. J. Oncol.* **8**, 401–406.

Panakova, D., Sprong, H., Marois, E., Thiele, C., and Eaton, S. (2005). Lipoprotein particles are required for Hedgehog and Wingless signalling. *Nature* **435**, 58–65.

Pancer, Z., and Cooper, M. D. (2006). The evolution of adaptive immunity. *Annu. Rev. Immunol.* **24**, 497–518.

Papo, N., Oren, Z., Pag, U., Sahl, H.-G., and Shai, Y. (2002). The consequence of sequence alteration of an amphipathic alpha-helical antimicrobial peptide and its diastereomers. *J. Biol. Chem.* **277**, 33913–33921.

Pardoll, D. (2003). Does the immune system see tumors as foreign or self? *Annu. Rev. Immunol.* **21**, 807–839.

Platt, N., da, S. R., and Gordon, S. (1999). Class A scavenger receptors and the phagocytosis of apoptotic cells. *Immunol. Lett.* **65**, 15–19.

Pradeu, T., and Carosella, E. D. (2006). On the definition of a criterion of immunogenicity. *Proc. Natl. Acad. Sci. USA* **103**, 17858–17861.

Radbruch, A., Muehlinghaus, G., Luger, E. O., Inamine, A., Smith, K. G. C., Dorner, T., and Hiepe, F. (2006). Competence and competition: The challenge of becoming a long-lived plasma cell. *Nat. Rev. Immunol.* **6**, 741–750.

Rahman, M. M., Roberts, H. L. S., Sarjan, M., Asgari, S., and Schmidt, O. (2004). Induction and transmission of *Bacillus thuringiensis* tolerance in the flour moth *Ephestia kuehniella*. *Proc. Natl. Acad. Sci. USA* **101**, 2696–2699.

Rahman, M. M., Ma, G., Roberts, H. L. S., and Schmidt, O. (2006). Cell-free immune reactions in insects. *J. Insect Physiol.* **52**, 754–762.

Reeson, A. F., Wilson, K., Gunn, A., Hails, R. S., and Goulson, D. (1998). Baculovirus resistance in the Noctuid *Spodoptera exempta* is phenotypically plastic and responds to population density. *Proc. Roy. Soc. London B Biol. Sci.* **265**, 1787–1791.

Rodenburg, K. W., and Van der Horst, D. J. (2005). Lipoprotein-mediated lipid transport in insects: Analogy to the mammalian lipid carrier system and novel concepts for the functioning of LDL receptor family members. *Biochim. Biophys. Acta Mol. Cell Biol. Lipids* **1736**, 10–29.

Rosenberg, S. A. (2001). Progress in human tumour immunology and immunotherapy. *Nature* **411**, 380–384.

Sadd, B. M., and Schmid-Hempel, P. (2006). Insect immunity shows specificity in protection upon secondary pathogen exposure. *Curr. Biol.* **16**, 1206–1210.

Sahoo, D., Narayanaswami, V., Kay, C. M., and Ryan, R. O. (2000). Pyrene excimer fluorescence: A spatially sensitive probe to monitor lipid-induced helical rearrangement of apolipophorin III. *Biochemistry* **39**, 6594–6601.

Santos-Silva, F., Fonseca, A., Caffrey, T., Carvalho, F., Mesquita, P., Reis, C., Almeida, R., David, L., and Hollingsworth, M. A. (2005). Thomsen-Friedenreich antigen expression in gastric carcinomas is associated with MUC1 mucin VNTR polymorphism. *Glycobiology* **15**, 511–517.

Schmid-Hempel, P. (2005). Natural insect host–parasite systems show immune priming and specificity: Puzzles to be solved. *BioEssays* **27**, 1026–1034.

Schmidt, O., and Schreiber, A. (2006). Integration of cell adhesion reactions – a balance of forces? *J. Theor. Biol.* **238**, 608–615.

Schmidt, O., and Theopold, U. (2004). An extracellular driving force of endocytosis and cell-shape changes (hypothesis). *BioEssays* **26**, 1344–1350.

Schmidt, O., Rahman, M. M., Ma, G., Theopold, U., Sun, Y., Sarjan, M., Fabbri, M., and Roberts, H. S. L. (2005). Mode of action of antimicrobial proteins, pore-forming toxins and biologically active peptides (hypothesis). *Invertebr. Survival J.* **2**, 82–90.

Schmidt, O., Söderhäll, K., Theopold, U., and Faye, I. (invited review). The role of adhesion in immune recognition. *Annu. Rev. Entomol.* **52**.

Scott, C. C., Dobson, W., Botelho, R. J., Coady-Osberg, N., Chavrier, P., Knecht, D. A., Heath, C., Stahl, P., and Grinstein, S. (2005). Phosphatidylinositol-4,5-bisphosphate hydrolysis directs actin remodeling during phagocytosis. *J. Cell Biol.* **169**, 139–149.

Seimon, T. A., Obstfeld, A., Moore, K. J., Golenbock, D. T., and Tabas, I. (2006). Combinatorial pattern recognition receptor signaling alters the balance of life and death in macrophages. *Proc. Natl. Acad. Sci. USA* **103**, 19794–19799.

Shai, Y. (2002). Mode of action of membrane active antimicrobial peptides. *Biopolymers* **66**, 236–248.

Sibley, L. D., and Andrews, N. W. (2000). Cell invasion by un-palatable parasites. *Traffic* 1, 100–106.

Soderhall, K., and Cerenius, L. (1998). Role of the prophenoloxidase-activating system in invertebrate immunity. *Curr. Opin. Immunol.* 10, 23–28.

Soulages, J., van Antwerpen, R., and Wells, M. (1996). Role of diacylglycerol and apolipophorin-III in regulation of physiochemical properties of the lipophorin surface: Metabolic implications. *Biochemistry* 35, 5191–5198.

Sparrow, J. (1979). Melanotic tumors. In *Genetics and Biology of Drosophila* (M. Ashburner, Ed.), pp. 277–313. Academic Press, New York.

Stuart, L. M., and Ezekowitz, R. A. B. (2005). Phagocytosis: Elegant complexity. *Immunity* 22, 539–550.

Stuart, L. M., Boulais, J., Charriere, G. M., Hennessy, E. J., Brunet, S., Jutras, I., Goyette, G., Rondeau, C., Letarte, S., Huang, H., et al. (2007). A systems biology analysis of the *Drosophila* phagosome. *Nature* 445, 95–101.

Swanson, J. A., and Baer, S. C. (1995). Phagocytosis by zippers and triggers. *Trends Cell Biol.* 5, 89–93.

Takeichi, M., and Abe, K. (2005). Synaptic contact dynamics controlled by cadherin and catenins. *Trends Cell Biol.* 15, 216–221.

Theopold, U., Schmidt, O., Soderhall, K., and Dushay, M. S. (2004). Coagulation in arthropods: Defence, wound closure and healing. *Trends Immunol.* 25, 289–294.

Vaisar, T., Pennathur, S., Green, P. S., Gharib, S. A., Hoofnagle, A. N., Cheung, M. C., Byun, J., Vuletic, S., Kassim, S., Singh, P., et al. (2007). Shotgun proteomics implicates protease inhibition and complement activation in the antiinflammatory properties of HDL. *J. Clin. Invest.* 117, 746–756.

Van der Horst, D. J. (2003). Insect adipokinetic hormones: Release and integration of flight energy metabolism. *Comp. Biochem. Physiol. B Biochem. Mol. Biol.* 136, 217–226.

Van Hoof, D., Rodenburg, K. W., and Van der Horst, D. J. (2005). Intracellular fate of LDL receptor family members depends on the cooperation between their ligand-binding and EGF domains. *J. Cell Sci.* 118, 1309–1320.

Vilcinskas, A., Kopacek, P., Jegorov, A., Vey, A., and Matha, V. (1997). Detection of lipophorin as the major cyclosporin-binding protein in the hemolymph of the greater wax moth *Galleria mellonella*. *Comp. Biochem. Physiol. C Comp. Pharmacol. Toxicol.* 117, 41–45.

Visintin, A., Mazzoni, A., Spitzer, J. A., and Segal, D. M. (2001). Secreted MD-2 is a large polymeric protein that efficiently confers lipopolysaccharide sensitivity to Toll-like receptor 4. *Proc. Natl. Acad. Sci.USA* 98, 12156–12161.

Vivier, E., and Malissen, B. (2005). Innate and adaptive immunity: Specificities and signaling hierarchies revisited. *Nat. Immunol.* 6, 17–21.

Vreugdenhil, A. C. E., Rousseau, C. H., Hartung, T., Greve, J. W. M., van't Veer, C., and Buurman, W. A. (2003). Lipopolysaccharide (LPS)-binding protein mediates LPS detoxification by chylomicrons. *J. Immunol.* 170, 1399–1405.

Watson, F. L. (2005). Extensive diversity of Ig-superfamily proteins in the immune system of insects. *Science* 309, 1874–1878.

Whitten, M. M. A., Tew, I. F., Lee, B. L., and Ratcliffe, N. A. (2004). A novel role for an insect apolipoprotein (apolipophorin III) in {beta}-1,3-glucan pattern recognition and cellular encapsulation reactions. *J. Immunol.* 172, 2177–2185.

Wiesner, A., Losen, S., Kopacek, P., Weise, C., and Gotz, P. (1997). Isolated apolipophorin III from *Galleria mellonella* stimulates the immune reactions of this insect. *J. Insect Physiol.* 43, 383–391.

Yu, X.-Q., and Kanost, M. R. (2002). Binding of hemolin to bacterial lipopolysaccharide and lipoteichoic acid: An immunoglobulin superfamily member from insects as a pattern-recognition receptor. *Eur. J. Biochem.* 269, 1827–1834.

Zarling, A. L., Polefrone, J. M., Evans, A. M., Mikesh, L. M., Shabanowitz, J., Lewis, S. T., Engelhard, V. H., and Hunt, D. F. (2006). Identification of class I MHC-associated phosphopeptides as targets for cancer immunotherapy. *Proc. Natl. Acad. Sci. USA* 103, 14889–14894.

Zdobnov, E. M., von Mering, C., Letunic, I., Torrents, D., Suyama, M., Copley, R. R., Christophides, G. K., Thomasova, D., Holt, R. A., Subramanian, G. M., et al. (2002). Comparative genome and proteome analysis of *Anopheles gambiae* and *Drosophila melanogaster*. *Science* 298, 149–159.

2

INSECT HEMOCYTES AND THEIR ROLE IN IMMUNITY

MICHAEL R. STRAND

Department of Entomology and Center for Emerging and Tropical Global Diseases, University of Georgia, Athens, GA 30602, USA

ABSTRACT: The cellular immune response of insects refers to defense responses mediated by hemocytes such as phagocytosis, encapsulation, and clotting. In this chapter I review the classification of hemocytes, their origins, and functions. Although hemocytes have similar functions across all insects, the naming of different hemocyte types varies among species. However, comparative studies across different taxa suggest hemocytes are produced during two stages of insect development: embryogenesis from head or dorsal mesoderm and during the larval or nymphal stages in mesodermally derived hematopoietic organs. Studies in model species like *Drosophila melanogaster* have identified several signaling pathways regulating hemocyte proliferation and differentiation. Studies in different taxa have also provided important insights into the regulation of hemocyte-mediated defense responses. These include the identification of multiple receptors involved in recognition or opsonization of foreign targets, cytokines that regulate hemocyte function, adhesion proteins, and effector molecules involved in killing foreign invaders. Studies in different insect models also reveal evidence of cross-talk between hemocytes and other components of the immune system. In summary, hemocytes perform diverse functions and comprise an essential arm of the insect immune system.

Abbreviations:

AMP	=	antimicrobial peptide
Dscam	=	Down syndrome cell adhesion molecule
GRP	=	glucan recognition proteins
GNBP	=	Gram-negative bacteria recognition protein
JAK/STAT	=	Janus kinase/signal transducers and activators of transcription
JNK	=	Jun kinase
LPS	=	lipopolysaccharide
LPSBP	=	LPS-binding protein
LRIM	=	leucine-rich repeat protein
NF-κB	=	nuclear factor κB
PAP	=	phenoloxidase activating proteinases
PGRP	=	peptidoglycan recognition proteins
PI3K	=	phosphoinositol 3 kinase
PO	=	phenoloxidase
PSP	=	plasmatocyte spreading peptide
RNAi	=	RNA interference
SPH	=	serine proteinase homolog
SR	=	scavenger receptor

2.1 INTRODUCTION

As discussed elsewhere in this volume, the innate immune system of insects consists of humoral and cellular defense responses. Humoral defenses refer to soluble effector molecules including antimicrobial peptides (AMPs), complement-like proteins, and products generated by complex protealytic cascades such as the phenoloxidase (PO) pathway (Blandin and Levashina, 2004; Cornelis and Soderhall, 2004; Imler and Bulet, 2005; Theopold et al., 2004). Cellular defenses in contrast refer to responses like phagocytosis, encapsulation, and clotting that are directly mediated by hemocytes (Gillespie et al., 1997; Irving et al., 2005; Lackie, 1988; Strand and Pech, 1995). Current understanding of hemocyte development and cellular defense responses derives primarily from the study of model insects like *Drosophila melanogaster* and selected Lepidoptera. However, important data have also recently accrued in vector arthropods like mosquitoes.

In this chapter I summarize this literature with emphasis on studies conducted during the last decade.

2.2 HEMOCYTE TYPES

Insect hemocytes are identified by a combination of morphological, antigenic, and functional characteristics (Brehelin and Zachary, 1986; Gardiner and Strand, 1999; Gupta, 1985; Jung et al., 2005; Lanot et al., 2001; Willott et al., 1994). Characterization of the molecular processes that regulate hemocyte function has also been augmented by the study of hemocyte-like cell lines, such as S2 and Mbn cells from *Drosophila melanogaster*, High Five cells from the lepidopteran *Trichoplusia ni*, and NISES-BoMo-Cam1 cells from the silkmoth *Bombyx mori* (Beck and Strand, 2003; Cherry and Silverman, 2006; Stuart and Ezekowitz, 2005; Taniai et al., 2006; Tauszig et al., 2000).

Although hemocytes have similar functions in immunity across all insects, the naming of different hemocyte types varies somewhat among species and taxa. *Drosophila* larvae contain three terminally differentiated types of hemocytes named plasmatocytes, crystal cells, and lamellocytes (Evans et al., 2003; Lanot et al., 2001; Wertheim et al., 2005) (Fig. 2.1A). Plasmatocytes represent 90–95% of all mature hemocytes, are strongly adhesive *in vitro*, and function as the professional phagocytes that engulf pathogens, dead cells, and other entities that gain entry into the hemocoel. Molecular markers for plasmatocytes include the extracellular matrix protein peroxidasin and an uncharacterized surface factor called the P1 antigen (Asha et al., 2003; Nelson et al., 1994). Crystal cells are non-adhesive cells that comprise approximately 5% of the hemocytes in circulation. Crystal cells are recognized by their rounded morphology and the expression of PO cascade components such as proPO 1 (PPO1). Lamellocytes are virtually absent in healthy *Drosophila* larvae but rapidly differentiate from prohemocytes following attack by parasitoid wasps and during metamorphosis (Lanot et al., 2001). Lamellocytes are identified as large, flat, adhesive cells that express reporters related to Jun kinase (JNK) signaling and the L1 antigen (Asha et al., 2003; Lanot et al., 2001). The main function of lamellocytes is encapsulation of parasitoids and other large foreign targets. Each of these hemocyte types differentiate from precursor prohemocytes that originate from pre-prohemocytes (Fig. 2.1A). Although most prohemocytes reside in hematopoietic organs, a small number of prohemocytes are also observed in circulation (Lanot et al., 2001).

In most other insects, the main differentiated hemocytes described in circulation are named granulocytes (=granular cells), plasmatocytes, spherule cells, and oenocytoids (Lavine and Strand, 2002; Ribeiro and Brehelin, 2006). In the older literature, where only morphological or histochemical methods were used for hemocyte identification, several other names for hemocytes have also been used (see Jones, 1962; Lackie, 1988). However, more recent studies indicate that these cells are sometimes not hemocytes at all but contaminants from other tissues

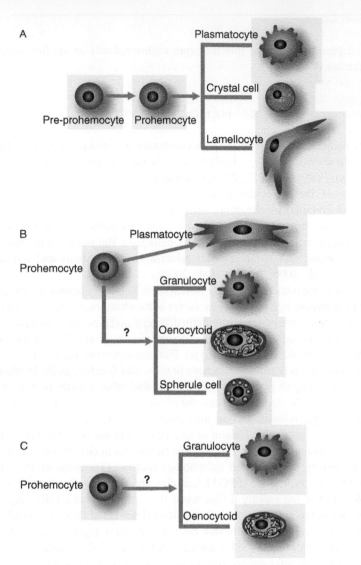

FIGURE 2.1 Hemocyte types and proposed lineage relationships in selected insects. (A) In *Drosophila* larvae, lymph glands contain pre-prohemocytes that transition to prohemocytes. Prohemocytes differentiate into phagocytic plasmatocytes, PO-containing crystal cells, and capsule-forming lamellocytes. Differentiated hemocytes in circulation also proliferate and contribute to maintenance of hemocyte populations. (B) Larval stage lepidopterans like *Pseudoplusia includens*, *Manduca sexta*, and *Bombyx mori* contain four differentiated hemocyte types in circulation: phagocytic granulocytes, PO-containing oenocytoids, spherule cells, and capsule-forming plasmatocytes. Hematopoietic organs contain prohemocytes which mature primarily into plasmatocytes. Granulocytes and other differentiated hemocyte types also proliferate in circulation. (C) Larval and adult stage mosquitoes like *Aedes aegypti* and *Anopheles gambiae* contain three hemocyte types in circulation: prohemocytes, phagocytic granulocytes, and PO-containing oenocytoids. Prohemocytes are putative progenitor cells but lineage relationships and sites of hemocyte proliferation are unclear. See text for further details.

(see Fig. 2.1), or are potentially one of the other major hemocyte types (i.e. granulocytes, plasmatocytes, spherule cells, or oenocytoids) that was given a different name by the investigator(s) conducting the study. In Lepidoptera, granulocytes are the most abundant hemocyte type in circulation (Fig. 2.1B). Granulocytes are morphologically distinguished by the granules in their cytoplasm, their ability to strongly adhere and spread on foreign surfaces in primary culture, and their tendency to spread symmetrically. Granulocytes also function as the professional phagocytes (Strand et al., 2006). Larger, granulocyte-like hemocytes called hyperphagocytic cells have also been described from the hawkmoth, *Manduca sexta* (Dean et al., 2004). Plasmatocytes are usually larger than granulocytes, spread asymmetrically on foreign surfaces, and are the main capsule-forming hemocytes (Strand et al., 2006). Non-adhesive hemocytes in larval stage Lepidoptera include oenocytoids that contain PO cascade components, and spherule cells that are potential sources of cuticular components (see Lavine and Strand, 2002). Prohemocytes have been described in hematopoietic organs from several Lepidoptera. A small proportion of hemocytes (usually <1%) in circulation of several lepidopterans have also been identified as prohemocytes. As with *Drosophila*, a number of molecular and antigenic markers have been developed that facilitate identification of different hemocyte types in model lepidopterans like *M. sexta* and *Pseudoplusia includens* (Beetz et al., 2004; Gardiner and Strand, 1999; Jiang et al., 1997; Lavine and Strand, 2003; Levin et al., 2005; Willott et al., 1994).

In mosquitoes, Hillyer and Christensen (2002) classified circulating hemocytes present in adult *Aedes aegypti* into granulocytes, oenocytoids, adipohemocytes, and thrombocytoids on the basis of morphology, binding of selected lectins, and enzymatic activity. In contrast, other investigators classified hemocytes from *Ae. aegypti* and *Culex quinquefasciatus* into only plasmatocytes and oenocytoids (Andreadis and Hall, 1976) or recognized multiple hemocyte types including prohemocytes (Foley, 1978; Kaaya and Ratcliffe, 1982; Drif and Brehelin, 1983). Castillo et al. (2006) used a combination of morphological, antigenic, and functional markers in comparative studies with *Anopheles gambiae* and *Ae. aegypti*, and concluded that both species contain three hemocyte types that they named granulocytes, oenocytoids, and prohemocytes (Fig. 2.1C). Granulocytes are strongly adhesive, phagocytic, and express several molecular markers that facilitate identification. They are also the most abundant cell type in both mosquito species while oenocytoids and prohemocytes together comprise less than 10% of the total hemocyte population. Oenocytoids are non-adhesive and are the only cell type that constitutively expresses PO activity. However, granulocytes inducibly express PO activity following immune challenge. The uniform size, rounded morphology, large nuclear to cytoplasmic ratio, and lack of labeling by differentiation markers are consistent with hemocytes named prohemocytes by Castillo et al. (2006) as being a type of progenitor cell although it is unknown whether they differentiate into granulocytes and/or oenocytoids (Fig. 2.1C). Castillo et al. (2006) also concluded that cells named adipohemocytes by Hillyer and Christensen (2002) are not hemocytes but fat body cells that can contaminate hemolymph samples collected from adult mosquitoes. In contrast to *Drosophila* or lepidopterans, mosquitoes seemingly do not produce any hemocyte type specialized for capsule formation (Castillo et al., 2006; Hillyer and Christensen, 2002).

Hemocytes have been described from species in many other taxa including Orthoptera, Blattaria, Coleoptera, Hymenoptera, Hemiptera, and Collembola (see Jones, 1962; Lackie, 1988 for summaries). However, most studies with insects in these groups are strictly morphological descriptions of hemocytes using light or electron microscopy with no accompanying functional data. Key challenges in functionally studying hemocytes in many species include the small size of many insects that often makes sample collection difficult. In addition, methods developed for maintaining hemocytes from *Drosophila*, Lepidoptera, or mosquitoes in primary culture are often unsuitable for other insects. The lack of genetic tools or molecular data in most insects is another obvious liability in conducting functional studies. There is also clearly a need to adopt more uniform terminology in naming insect hemocytes. For example, plasmatocytes, crystal cells, and lamellocytes in *Drosophila* are morphologically and functionally similar to granulocytes, oenocytoids, and plasmatocytes respectively in Lepidoptera (Lavine and Strand, 2002; Ribeiro and Brehelin, 2006). Yet, the different names being used for likely homologous cell types is confusing for individuals less familiar with the field or when attempting to make comparisons among species (Lavine and Strand, 2002; Ribeiro and Brehelin, 2006).

2.3   HEMATOPOIESIS

The hemocyte types described above arise during two stages of insect development (Akai and Sato, 1971; Feir, 1979; Holz et al., 2003; Jones, 1970; Ratcliffe et al., 1985; Traver and Zon, 2002). The first population of hemocytes is produced during embryogenesis from head or dorsal mesoderm while the second is produced during larval or nymphal stages in mesodermally derived hematopoietic organs. The hematopoietic organs of *Drosophila* are called lymph glands that form bilaterally along the anterior part of the dorsal vessel during embryogenesis (Jung et al., 2005). By the third instar, each lymph gland consists of an anterior primary lobe and several posterior secondary lobes that are separated by pericardial cells (Fig. 2.2A). The primary lobe consists of three zones: (1) a posterior signaling center (PSC) that contains a unique population of cells marked by expression of the transcription factor Collier and Notch ligand Serrate, (2) a medullary zone that contains quiescent prohemocytes, and (3) a cortical zone that contains plasmatocytes, crystal cells, and, following parasitoid attack, lamellocytes. Secondary lobes contain pre-prohemocytes, prohemocytes, and some plasmatocytes (Crozatier et al., 2004; Jung et al., 2005) (Fig. 2.2A).

The earliest lymph gland cells, called hemocyte precursor cells, are identified by expression of the GATA transcription factor homolog Serpent (Srp). As these

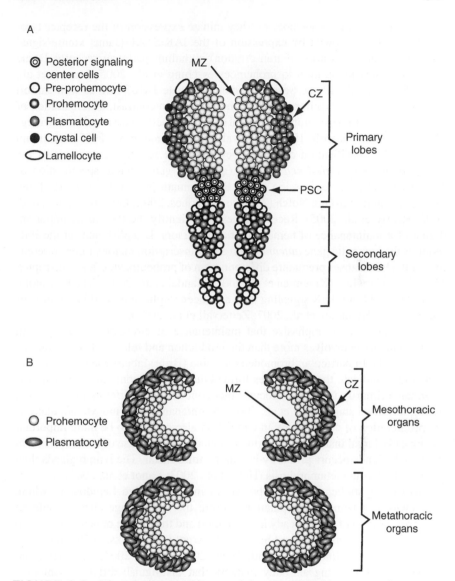

FIGURE 2.2 Schematic diagram of hematopoietic organs in third instar *Drosophila* (A) and sixth instar *Spodoptera frugiperda* (B). (A) Each lymph gland in *Drosophila* consists of a primary lobe and two or more secondary lobes. The primary lobe is further subdivided into a PSC that contains PSC cells, a medullary zone containing primarily prohemocytes, and a cortical zone containing primarily plasmatocytes. Secondary lobes contain a mixture of pre-prohemocytes, prohemocytes, and plasmatocytes (see Jung et al., 2005). (B) The hematopoietic organs in *S. frugiperda* larvae consist of two crescent-shaped meso- and metathoracic organs of similar size. Each organ is subdivided into an inner medullary zone containing prohemocytes and an outer cortical zone containing primarily plasmatocytes (see Gardiner and Strand, 2000).

cells transition to pre-prohemocytes, they initiate expression of the receptor tyrosine kinase Pvr followed by expression of the JAK/STAT (Janus kinase/signal transducers and activators of transcription) signaling pathway receptor Dome which identifies maturation to prohemocytes (Jung et al., 2005; Munier et al., 2002; Tepass et al., 1994). Several studies implicate JAK/STAT as well as Toll and Pvr signaling in proliferation of prohemocytes. In contrast, differentiation of prohemocytes into plasmatocytes, crystal cells, or lamellocytes requires downregulation of Dome (Dearolf, 1998; Jung et al., 2005; Munier et al., 2002; Sorrentino et al., 2004). Specification of plasmatocytes further requires expression of the transcription factors *glial cell missing* (*gcm*) and *gcm2*, while specification of crystal cells requires expression of the Runt-domain protein Lozenge (Lz) and Serrate signaling through Notch (Alfonso and Jones, 2002; Lebestky et al., 2000, 2003; Radtke et al., 2005). Recent studies also identify the PSC as an important domain for maintenance of hematopoietic precursors. Establishment of the PSC requires the homeotic gene *antennapedia* and transcription factor Collier, whereas loss of the PSC causes premature differentiation of prohemocytes due to disrupted JAK/STAT signaling (Krzemien et al., 2007; Mandal et al., 2007). The PSC along with JAK/STAT and JNK signaling have also been implicated in differentiation of lamellocytes (Krzemien et al., 2007; Zettervall et al., 2004).

It is important to emphasize that maintenance of circulating hemocytes in *Drosophila* larvae involves more than the production and release of cells from the lymph glands. In particular, bromodeoxyuridine (Brdu) labeling and transplantation studies indicate that hemocytes formed during embryonic development differentiate primarily into plasmatocytes that continue to proliferate in circulation during the larval stage (Holz et al., 2003). Maintenance of hemocyte populations via proliferation of cells already in circulation also appears to be more important during early larval instars, whereas the production and release of hemocytes from the lymph glands occurs primarily late in the third instar. The lymph glands then degenerate during metamorphosis (Holz et al., 2003; Lanot et al., 2001).

Most other studies on hematopoiesis in insects involve Lepidoptera where maintenance of hemocyte populations during the larval stage also depends on both proliferation of cells already in circulation and the release of hemocytes from hematopoietic organs (Akai and Sato, 1971; Arnold and Hinks, 1976; Hinks and Arnold, 1977; Gardiner and Strand, 1999; 2000; Nardi, 2004; Yamashita and Iwabuchi, 2001). Using antibody markers that distinguish different hemocyte types in combination with Brdu labeling, Gardiner and Strand (2000) found that all circulating hemocytes with the possible exception of oenocytoids proliferate in *Spodoptera frugiperda* and *P. includens* larvae. The hematopoietic organs in these and other lepidopterans are single-lobed structures that are located in the meso- and metathorax (four organs total) in close proximity to the imaginal wing disks (Fig. 2.2B). In *S. frugiperda*, the number of hemocytes in the hematopoietic organs increases greatly during larval development to a maximum of 300,000 cells per organ immediately prior to the onset of metamorphosis (Gardiner and Strand, 2000). These hemocytes consist primarily of prohemocytes that reside in an inner medullary region and plasmatocytes that occupy the outer cortex (Fig. 2.2B). Hematopoietic organs in *M. sexta* and the silkmoth *Bombyx mori* also predominantly contain prohemocytes and plasmatocytes (Ling et al., 2005; Nakahara et al., 2003, 2006; Nardi et al., 2003). Taken together, these results suggest that lepidopteran hematopoietic organs are important sources of plasmatocytes, particularly late in larval development, whereas the majority of granulocytes, spherule cells, and oenocytoids derive from proliferation of hemocytes already in circulation (Fig. 2.2B).

Studies with a number of insects indicate that the number of hemocytes in circulation can also rapidly increase in response to stress, wounding, or infection (Lackie, 1988; Ratcliffe et al., 1985). Some of these changes are due to hematopoietic events such as the rapid differentiation and release of lamellocytes from *Drosophila* lymph glands following parasitoid attack (Sorrentino et al., 2002; Wertheim et al., 2005). Other studies though indicate that already differentiated hemocytes are often sessile and weakly adhere to the surface of internal organs, yet can rapidly enter circulation and increase significantly the number of cells in the hemolymph following immune challenge (Castillo et al., 2006; Elrod-Erickson et al., 2000; Gardiner and Strand, 2000; Lanot et al., 2001; Moita et al., 2005). Little is known about the adhesion molecules and downstream signaling factors that regulate the entry of these sessile hemocytes into circulation. However, recent studies in *Drosophila* identify the GTPase Rac1, the JNK Basket, and actin stabilization as requirements for recruitment of plasmatocytes into circulation and encapsulation of parasitoids by lamellocytes (Williams et al., 2006). In Lepidoptera, the adhesive state of plasmatocytes is also affected by the titer of the cytokine plasmatocyte spreading peptide (PSP) (Clark et al., 1997, 1998, 2004). Local release and activation (see Fig. 2.2B) of PSP causes plasmatocytes to aggregate and form capsules, whereas systemic activation of PSP following a stress response causes plasmatocytes to drop out of circulation and become transiently sessile (Clark et al., 1997).

2.4   HEMOCYTE-MEDIATED DEFENSE RESPONSES

As noted in the introduction, the main immune defenses mediated by hemocytes are phagocytosis, encapsulation, and clotting. Phagocytosis is a widely conserved defense response in which individual cells internalize and destroy small targets. This depends first upon receptor-mediated recognition and binding of the target to a hemocyte followed by formation of a phagosome and engulfment of the target via actin polymerization-dependent mechanisms. The phagosome then matures to a phagolysosome by a series of fission and fusion events with endosomes and lysosomes (Stuart and Ezekowitz, 2005; Strochein-Stevenson et al., 2006). Insect hemocytes phagocytize a variety of microbial invaders including many types of bacteria, yeast, fungi, and protozoans. Hemocytes also recognize and internalize apoptotic bodies and inanimate objects like synthetic beads and India ink particles (Lanot et al., 2001; Lavine and Strand, 2002).

Encapsulation refers to the envelopment of larger invaders, like parasitoids and nematodes, by multiple hemocytes. The similar process of enveloping aggregated bacteria is called nodulation (Ratcliffe and Gagen, 1976, 1977). The capsules formed around parasitoids by *Drosophila* are comprised predominantly of lamellocytes but whether other cell types, like plasmatocytes, are involved in capsule formation is unclear. In contrast, while plasmatocytes are the main capsule-forming cell in Lepidoptera, recognition and encapsulation of some foreign targets also requires cooperation with granulocytes. This is best illustrated in studies with *P. includens* where *in vitro* experiments using purified populations of hemocytes indicate that plasmatocytes alone are capable of encapsulating some targets but encapsulation of other targets requires the presence of granulocytes (Lavine and Strand, 2001; Pech and Strand, 1996, 2000). Encapsulation of targets requiring granulocytes begins when a few granulocytes recognize and bind to the target forming a monolayer. These adherent granulocytes then recruit and activate plasmatocytes by releasing PSP and other unknown cytokines (Clark et al., 1998). Eventually an overlapping sheath of hemocytes consisting primarily of plasmatocytes develops which completely envelopes the target. Capsule formation ends when a monolayer of granulocytes attach to the periphery of the capsule and produce a basement membrane-like layer (Grimstone et al., 1967; Liu et al., 1998; Pech and Strand, 1996). This likely creates a self-surface that is refractory to plasmatocyte binding. Morphological studies indicate that capsules formed by other Lepidoptera have a similar architecture (see Lavine and Strand, 2002, for summary). In contrast, studies with *M. sexta* reveal an essentially random distribution of granulocytes and plasmatocytes in capsules making it unclear whether granulocytes play a role or not in recognition of foreign targets and recruitment of plasmatocytes (Wiegand et al., 2000).

Melanin is often deposited within and around the capsules produced by insects but whether hemocytes are the source of this material is uncertain (Schmidt et al., 2001; Strand and Pech, 1995; Wertheim et al., 2005). One possibility is that melanin forms in response to lysis of oenocytoids in Lepidoptera or crystal cells in *Drosophila* and release of PO cascade components at the site of capsule formation. Another is that melanin forms in response to activation of PO cascade components that are also present in plasma. Regardless of its precise source, characterization of the PO cascade in *M. sexta* indicates that proPO forms a complex with proPO activating proteinases (PAPs), serine proteinase homologs (SPHs), and pattern recognition immunolectins that function as pattern recognition receptors and bind to encapsulation targets like nematodes (Yu and Kanost, 2004). This finding provides a mechanism for how melanin can be selectively deposited around foreign targets and capsules rather than throughout the hemocoel during an immune response. It is also well known that melanin accumulates around many foreign targets in mosquitoes to form what are called melanotic capsules (Collins et al., 1986; Michel et al., 2005, 2006). Hemocytes do not appear to directly participate in the formation of melanotic capsules by binding to foreign targets. However, oenocytoids and granulocytes are potential sources of the PO cascade components required for melanotic encapsulation (Castillo et al., 2006).

Coagulation of insect hemolymph occurs at sites of external wounding (Bidla et al., 2005; Theopold et al., 2004). Microscopic studies in the lepidopteran *Galleria mellonella* and *Drosophila* reveal that soft clots initially consist of a fibrous matrix embedded with numerous hemocytes that are primarily granulocytes (*G. mellonella*) or plasmatocytes (*Drosophila*). This is followed by hardening of the clot due to cross-linking of proteins and melanization (Scherfer et al., 2006; Theopold et al., 2004). Experimental data are currently lacking but these observations suggest the possibility that hemocytes directly participate in clot formation by promoting matrix formation and/or releasing PO cascade components required for melanin formation.

2.5   RECEPTORS AND SIGNALING PATHWAYS MEDIATING HEMOCYTE FUNCTION

Phagocytosis and encapsulation depend upon recognition of the target as foreign followed by activation of downstream signaling and effector responses. Some foreign entities are recognized by humoral pattern recognition molecules that after binding to a target enhance its recognition by other receptors on the surface of hemocytes. This process is called opsonization. Other targets in contrast are recognized directly by hemocyte surface receptors.

2.5.1   OPSONIN-DEPENDENT AND -INDEPENDENT RECOGNITION

A number of humoral pattern recognition receptors potentially opsonize microorganisms by binding to lipopolysaccharides (LPSs), peptidoglycans, and glucans. These include hemolin, LPS-binding protein (LPSBP), Gram-negative bacteria recognition protein (GNBPs), soluble peptidoglycan recognition proteins (PGRP-SA and PGRP-SD), glucan recognition proteins (GRPs), soluble forms of Down's syndrome cell adhesion molecule (Dscam), and complement-like TEP proteins (Dong et al., 2006; Irving et al., 2005; Levashina et al., 2001; Moita et al., 2005; Terenius et al., 2007; Wang et al., 2006) (Fig. 2.3). Another subgroup of soluble PGRPs (PGRP-SB1, SC1a, 1b, and SC2) enzymatically degrade peptidoglycan. This activity kills some bacteria (see Bangham et al., 2006) as well as releases peptidoglycan fragments that potentially act as signaling molecules that mediate hemocyte effector responses (Garver et al., 2006). In the mosquito *An. gambiae*, the (leucine-rich repeat protein) LRIM1 is both an antagonist of *Plasmodium* and an opsonin of certain bacteria (Moita et al., 2005). Other humoral factors implicated in recognition and opsonization of foreign targets include a glutamine-rich protein purified from the beetle *Tenebrio molitor* (Cho et al., 1999), immunolectins from *M. sexta* that bind to diverse targets including bacteria and nematodes (Ling and Yu, 2006), and hemolin that is produced by several lepidopterans (Terrenius et al., 2007; Eleftherianos et al., 2007) (Fig. 2.3). Hemocytes themselves are a source for many of these humoral molecules although other immune tissues, like the fat body, are also potential sources.

Cell surface receptors with roles in opsonin-independent phagocytosis or encapsulation include the class B scavenger receptor (SR) (also called CD36 family members) Peste that binds certain intracellular bacteria (Fig. 2.3). The class C SR dSR-CI, and transmembrane protein Eater bind several Gram-positive and -negative bacteria (Kocks et al., 2005; Philips et al., 2005; Ramet et al., 2001), while membrane bound PGRPs (PGRP-LC and its co-receptor PGRP-LE) and transmembrane forms of Dscam mediate phagocytosis of bacteria in *Drosophila* and the mosquito *An. gambiae* (Dong et al., 2006; Moita et al., 2005; Ramet et al., 2002). The class B SR Croquemort in *Drosophila* and the low-density lipoprotein (LDL) receptor-related protein LRP1 in *An. gambiae* have been implicated in recognition and phagocytosis of apoptotic bodies (Franc et al., 1999; Moita et al., 2005) (Fig. 2.3). Scavenger receptors may also mediate the endocytic entry of dsRNA into insect cells (Saleh et al., 2006), whereas a long version of PGRP-LE was recently identified as an intracellular receptor with potential roles in recognition of intracellular bacteria (Kaneko et al., 2006). Lastly, integrins are dimeric transmembrane receptors that consist of a α and β subunit. Insects encode multiple α and β subunits with βPS and αPS4 in *Drosophila* and related integrin subunits in lepidopterans and mosquitoes participating in both encapsulation and phagocytosis (Irving et al., 2005; Lavine and Strand, 2003; Levin et al., 2005; Moita et al., 2005; Wertheim et al., 2005) (Fig. 2.3). Tetraspanin proteins have recently been shown to serve as ligands for integrins in *M. sexta* to mediate binding of hemocytes to one another as is required for capsule formation (Zhuang et al., 2007). Granulocyte-stimulated activation and binding of plasmatocytes as discussed above also appear to involve an immunoglobulin superfamily member called neuroglian (Nardi et al., 2006).

2.5.2 CYTOKINES AND SIGNALING PATHWAYS

Relatively few molecules have thus far been identified as extracellular signaling molecules (e.g. cytokines) that regulate hemocyte function. In *Drosophila*, infection by Gram-positive bacteria, fungi, and certain viruses induce cleavage of the cysteine-knot-like growth factor Spaetzle to its active form by a protealytic cascade. Processed Spaetzle then functions as a ligand for the Toll receptor that mediates via nuclear factor κB (NF-κB) transcription factors a number of immune genes including several AMPs (Imler and Bulet, 2005). Potential pattern recognition receptors involved in initiating the protealytic cascade that leads to Spaetzle processing include PGRP-SA and soluble GNBPs (Filipe et al., 2005; Imler and Zheng, 2005) (Fig. 2.3). In Lepidoptera, the cytokine PSP is also processed from a precursor protein by a protealytic cascade following immune challenge (Clark et al., 1998). After

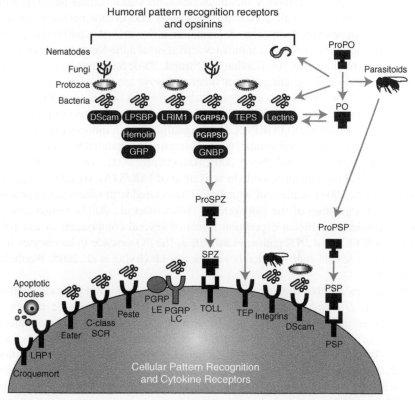

FIGURE 2.3 Humoral and cellular receptors involved in recognition of foreign invaders by insect hemocytes. In *Drosophila* and the mosquito *Anopheles gambiae*, soluble and cellular forms of Dscam and TEP proteins bind and opsonize several different microbes. Mosquito TEP and LRIM1 proteins are also involved in killing *Plasmodium* and possibly other protozoans. Other humoral receptors with recognition and opsonizing activities include LPSBPs, GNBPs, and PGRPs. Homologs of these proteins have been identified from several insect species and shown to bind different bacteria and fungi. In *Drosophila*, binding of microbial ligands by PGRP-SA and -SD also stimulates proteolytic processing of ProSpaetzle (ProSPZ) to form the cytokine Spaetzle (SPZ) which activates the Toll pathway. Another soluble receptor, hemolin, is known only from Lepidoptera. Immunolectins from Lepidoptera also bind bacteria and nematodes, and form complexes with PO that potentially facilitates localized deposition of melanin on the surface of microorganisms, nematodes, and parasitoids. In *Drosophila*, several cellular receptors including Eater, Peste, dSCRI, and PGRP-LC with its co-receptor PGRP-LE bind bacteria. The CD36 homolog Croquemort mediates uptake of apoptotic bodies in *Drosophila* while LRP1 has a similar function in *An. gambiae*. Dimeric integrins identified in *Drosophila*, mosquitoes, and Lepidoptera mediate phagocytosis of certain bacteria and encapsulation of large foreign targets like parasitoids. In Lepidoptera, parasitoids and other encapsulation targets also induce proteolytic processing of proPSP to the cytokine PSP which induces adhesion of plasmatocytes.

binding its receptor, PSP rapidly stimulates plasmatocytes to adhere and spread on foreign surfaces (Clark et al., 2004). In *P. includens*, cooperation between granulocytes and plasmatocytes during capsule formation is due at least in part to the release of PSP from granulocytes which stimulates activation of adhesion molecules including integrins by plasmatocytes (Lavine and Strand, 2003; Strand and Clark, 1999).

In addition to Toll signaling, several other pathways are also activated in hemocytes by cytokine stimulation and/or binding of foreign targets to surface receptors. These include the Imd pathway that is activated in response to binding of Gram-negative bacteria to PGRP-LC. Imd signaling in turn induces expression of novel effector genes as well as other genes that are also regulated by Toll signaling (Choe et al., 2002; Imler and Zheng, 2005; Ramet et al., 2002; Agaisse et al., 2005) (Fig. 2.2). TEP proteins play a role in activation of JAK/STAT signaling (Agaisse and Perrimon, 2004), while JNK signaling is associated with adhesion and phagocytosis via regulation of the cytoskeleton (Boutros et al., 2002). Genome-wide studies confirm significant expression levels of several components of the Toll, Imd, JAK/STAT, and JNK pathways as well as the PO cascade in hemocytes following immune challenge by microbes or parasitoids (Irving et al., 2005; Wertheim et al., 2005).

Recent studies also suggest that hemocytes orient to wound sites via chemotaxis in the *Drosophila* embryo. The identity of these chemotactic factors is unknown but the ability of hemocytes to respond to them involves phosphoinositol 3 kinase (PI3K) and Rac signaling activity (Stramer et al., 2005; Wood et al., 2006). The absence of signal peptides in insect proPOs has led to the suggestion that their release from oenocytoids (Lepidoptera) or crystal cells (*Drosophila*) depends upon cell lysis (see Cornelis and Soderhall, 2004; Nappi and Christensen, 2005). Precisely how recognition of a foreign target stimulates cell lysis and PO release is unclear but alterations in the function of the GTPase Rho A appear to block the response in crystal cells by interfering with rearrangements of the cytoskeleton (see Bangham et al., 2006).

2.6   HEMOCYTE-ASSOCIATED EFFECTOR RESPONSES

The signaling activities discussed above regulate the production and release of a number of extra- and intracellular effector molecules by hemocytes. These include several AMPs regulated by the Toll and Imd pathways (Imler and Bulet, 2005). Although the fat body is a primary site of AMP expression, studies in several insects reveal that hemocytes also produce many AMPs (Bartholomay et al., 2004; Boman, 2003; Dimopoulos et al., 2000; Hoffmann and Reichhart, 2002; Irving et al., 2005; Lackie, 1988; Lowenberger, 2001; Yamano et al., 1994). Studies in Lepidoptera further reveal that AMP expression varies among hemocyte types with granulocytes and plasmatocytes expressing multiple AMPs, and spherule cells and oenocytoids expressing only lysozyme (Lavine et al., 2005).

In the malaria vector *An. gambiae*, significant mortality of *Plasmodium* occurs when ookinetes transition to oocysts during passage through the midgut. The two most apparent events associated with parasite mortality are ookinete lysis and melanization (Whitten et al., 2006). Although incompletely understood, three molecules produced by hemocytes, TEP1, LRIM1, and APL1, are involved in parasite recognition and killing of parasites in the midgut (Blandin et al., 2004; Osta et al., 2004; Riehle et al., 2006). Oenocytoids and granulocytes are also primary sources of PO activity in *An. gambiae* (Castillo et al., 2006). As previously noted, POs are expressed as inactive zymogens (proPOs) in all insects and are converted to active PO by serine proteinases called proPO activating proteinases (PAPs). Knockdown of the serpin SRPN2, a PAP inhibitor, in *An. gambiae* increases melanization and reduces survival of *Plasmodium berghei* but has no apparent effect on survival of *P. falciparum* (Michel et al., 2006). This suggests the quinones and reactive intermediates produced following activation of the PO cascade are important in killing some parasites following melanotic encapsulation but not others.

Genome-wide analyses indicate that genes encoding several TEPs and PO cascade components are likewise expressed in hemocytes from *Drosophila* (Irving et al., 2005; Wertheim et al., 2005). Similar to the effects of SRPN2, loss of function mutants for the serpin Spn27A also results in increased melanization. Whether this effect increases resistance to infection is unclear but loss of Spn27A does increase susceptibility to wounding, suggesting that melanization may be more important for healing external wounds than direct killing of pathogens (DeGregorio et al., 2002; Ramet et al., 2001). Studies with Lepidoptera also suggest melanin or other factors generated by the PO cascade have anti-viral activity (Popham et al., 2004; Trudeau et al., 2001).

Intracellular killing of pathogens in mammalian macrophages usually occurs in the phagolysosome due to release of reactive oxygen intermediates, proteases, and other factors including PGRPs and AMPs (Reeves et al., 2002; Selsted and Ouellette, 1995; Parsons, 2003; Pearson et al., 2003; Tydell et al., 2006). Few functional studies have been conducted on phagolysosomes in insect hemocytes. However, proteomic analyses indicate that phagosomes formed in hemocyte-like S2 cells contain hundreds of proteins including several candidate effector molecules (Stuart and Ezekowitz, 2005; Stuart et al., 2007). Factors suggested in killing encapsulated invaders include asphyxiation, reactive oxygen and nitrogen species, and AMPs (Nappi and Christensen, 2005; Salt, 1970).

The mammalian immune system recognizes viral pathogens at the cell surface or in endosomal vesicles through pattern recognition receptors such as Toll-like receptors. Cytoplasmic helicase proteins also function as pattern recognition receptors that recognize dsRNA intermediates associated with replication by single-stranded RNA viruses and transcription by DNA viruses that induce the interferon pathway (see Cherry and Silverman, 2006; Li et al., 2004). dsRNA also induces the RNA interference (RNAi) pathway which directly degrades viral RNA. Cytoplasmic helicase proteins have not been identified in insects but have been demonstrated to function as a viral defense mechanism in mosquitoes and *Drosophila* (Galiana-Arnoux et al., 2006; Keene et al., 2004; Wang et al., 2006). Infection by some viruses also activate Toll, Imd, and JAK/STAT signaling (Dostert et al., 2005; Zambon et al., 2006). Recognition of viruses and intracellular anti-viral defense mechanisms could obviously involve a variety of cell types in addition to hemocytes. However, hemocytes are likely important in anti-viral defense given that most insect viruses disseminate through the hemocoel during the course of a systemic infection.

ACKNOWLEDGMENTS

I thank J. A. Johnson for assistance with the figures. Some work discussed here was funded by grants from the National Institutes of Health (AI38917 and AI063605) and US Department of Agriculture National Research Initiative (2005-05382) to MRS.

REFERENCES

Agaisse, H., and Perrimon, N. (2004). The roles of JAK/STAT signaling in *Drosophila* immune responses. *Immunol. Rev.* **198**, 72–82.

Agaisse, H., Burrack, L. S., Philips, J. A., Rubin, E. J., Perrimon, N., and Higgins, D. E. (2005). Genome-wide RNAi screen for host factors required for intracellular bacterial infection. *Science* **309**, 1248–1251.

Akai, H., and Sato, S. (1971). An ultrastructural study of the haemopoietic organs of the silkworm, *Bombyx mori. J. Insect Physiol.* **17**, 1665–1676.

Alfonso, T. B., and Jones, B. W. (2002). gcm2 promotes glial cell differentiation and is required with glial cells missing for macrophage development in *Drosophila. Dev. Biol.* **248**, 369–383.

Andreadis, T. G., and Hall, D. W. (1976). *Neoplectana carpocapsae*: Encapsulation in *Aedes aegypti* and changes in host hemocytes and hemolymph proteins. *Exp. Parasitol.* **39**, 252–261.

Arnold, J. W., and Hinks, C. F. (1976). Haemopoiesis in Lepidoptera. I. The multiplication of circulating haemocytes. *Can. J. Zool.* **54**, 1003–1012.

Asha, H., Nagy, I., Kovacs, G., Stetson, D., Ando, I., and Dearolf, C. R. (2003). Analysis of Ras-induced overproliferation in *Drosophila* hemocytes. *Genetics* **163**, 203–215.

Bangham, J., Jiggins, F., and Lemaitre, B. (2006). Insect immunity, the post-genomic era. *Immunity* **25**, 1–5.

Bartholomay, L. C., Cho, W. L., Rocheleau, T. A., Boyle, J. P., Beck, E. T., Fuchs, J. F., Liss, P., Rusch, M., Butler, K. M., Wu, R. C. C., Lin, S. P., Kuo, F. Y., Tsao, I. Y., Huang, C. Y., Liu, T. T., Hsiao, K. J., Tsai, S. F, Yang, U. C., Nappi, A. J., Perna, N. T., Chen, C. C., and Christensen, B. M. (2004). Description of the transcriptomes of immune response-activated hemocytes from the mosquito vectors *Aedes aegypti* and *Armigeres subalbatus. Infect. Immun.* **72**, 4114–4126.

Beck, M., and Strand, M. R. (2003). RNA interference silences *Microplitis demolitor* bracovirus genes and implicates *glc1.8* in blocking adhesion of infected host cells. *Virology* **314**, 521–535.

Beetz, S., Brinkmann, M., and Trenczek, T. (2004). Differences between larval and pupal hemocytes of the tobacco hornworm, *Manduca sexta*, determined by monoclonal antibodies and density centrifugation. *J. Insect Physiol.* **50**, 805–819.

Bidla, G., Lindren, M., Theopold, U., and Dushay, M. S. (2005). Hemolymph coagulation and phenoloxidase in *Drosophila* larvae. *Dev. Comp. Immunol.* **29**, 669–679.

Blandin, S., and Levashina, E. A. (2004). Thioester-containing proteins and insect immunity. *Mol. Immunol.* **40**, 903–908.

Blandin, S., Shiao, S., Moita, L., Janse, C., Waters, A., Kafatos, F., and Levashina, E. (2004). Complement-like protein TEP1 is a determinant of vectorial capacity in the malaria vector *Anopheles gambiae*. *Cell* **116**, 661–670.

Boman, H. G. (2003). Antibacterial peptides: Basic facts and emerging concepts. *J. Intern. Med.* **254**, 197–215.

Boutros, M., Agaisse, H., and Perrimon, N. (2002). Sequential activation of signaling pathways during innate immune responses in *Drosophila*. *Dev. Cell* **3**, 711–722.

Brehelin, M., and Zachary, D. (1986). Insect haemocytes, a new classification to rule out controversy. In *Immunity in Invertebrates* (M. Brehelin, Ed.), pp. 36–48. Springer Verlag, Berlin.

Castillo, J. C., Robertson, A. E., and Strand, M. R. (2006). Characterization of hemocytes from the mosquitoes *Anopheles gambiae* and *Aedes aegypti*. *Insect Biochem. Mol. Biol.* **36**, 891–903.

Chen, G.-C., Turano, B., Ruest, P. J., Hagel, M., Settleman, J., and Thomas, S. M. (2005). Regulation of rho and rac signaling to the actin cytoskeleton by paxillin during *Drosophila* development. *Mol. Cell Biol.* **25**, 979–987.

Cherry, S., and Silverman, N. (2006). Host–pathogen interactions in *Drosophila*, new tricks from an old friend. *Nat. Immunol.* **7**, 911–917.

Cho, M. Y., Lee, H. S., Lee, K. M., Homma, K., Natori, S., and Lee, B. L. (1999). Molecular cloning and functional properties of two early-stage encapsulation-relating proteins from the coleopteran insect, *Tenebrio molitor* larvae. *Eur. J. Biochem.* **262**, 737–744.

Choe, K.-M., Werner, T., Stoven, S., Hultmark, D., and Anderson, K. V. (2002). Requirement for a peptidoglycan recognition protein (PGRP) in Relish activation and antibacterial immune responses in *Drosophila*. *Science* **296**, 359–362.

Clark, K. D., Pech, L. L., and Strand, M. R. (1997). Isolation and identification of a plasmatocyte-spreading peptide from the hemolymph of the lepidopteran insect *Pseudoplusia includens*. *J. Biol. Chem.* **272**, 23440–23447.

Clark, K. D., Witherell, A., and Strand, M. R. (1998). Plasmatocyte spreading peptide is encoded by an mRNA differentially expressed in tissues of the moth *Pseudoplusia includens*. *Biochem. Biophys. Res. Comm.* **250**, 479–485.

Clark, K. D., Garczynski, S., Arora, A., Crim, J., and M. R. Strand. (2004). Specific residues in plasmatocyte spreading peptide are required for receptor binding and functional antagonism of insect immune cells. *J. Biol. Chem.* **279**, 33246–33252.

Collins, R., Sakai, R., Vernick, K., Paskewitz, S., Seeley, D., Miller, L., Collins, W., Campbell, C., and Gwadz, R. (1986). Genetic selection of a plasmodium-refractory strain of the malaria vector *Anopheles gambiae*. *Science* **234**, 607–610.

Cornelis, L., and Soderhall, K. (2004). The prophenoloxidse-activating system in invertebrates. *Immunol. Rev.* **198**, 116–126.

Crozatier, M., Ubeda, J. M., Vincent, A., and Meister, M. (2004). Cellular immune response to parasitization in *Drosophila* requires the EBF orthologue Collier. *PLoS Biol.* **2**, e196.

Dean, P., Potter, U., Richards, E. H., Edwards, J. P., Charnley, A. K., and Reynolds, S. E. (2004). Hyperphagocytic haemocytes in *Manduca sexta*. *J. Insect Physiol.* **50**(11), 1027–1036.

Dearolf, C. R. (1998). Fruit fly 'leukemia'. *Intl. J. Biochem. Biophys. Mol. Biol.* **1377**, M13–M23.

De Gregorio, E., Han, S., Lee, W., Baek, M., Osaki, T., Kawabata, S., Lee, B., Iwanaga, S., Lemaitre, B., and Brey, P. T. (2002). An immune-responsive serpin regulates the melanization cascade in *Drosophila*. *Dev. Cell* **3**, 581–592.

Dimopoulos, G., Casavant, T. L., Chang, S., Scheetz, T., Roberts, C., Donohue, M., Schultz, J., Benes, V., Bork, P., Ansorge, W., Bento Sores, M., and Kafatos, F. C. (2000). *Anopheles gambiae* pilot gene discovery project: Identification of mosquito innate immunity genes from expressed sequence tags generated from immune-competent cell lines. *Proc. Natl. Acad. Sci. USA* **97**, 6619–6624.

Dong, Y. M., Taylor, H. E., and Dimopoulos, G. (2006). AgDscam, a hypervariable immunoglobulin domain-containing receptor of the *Anopheles gambiae* immune system. *PLoS Biol.* **4**, 1137–1146.

Dostert, C., Jouanguy, E., Irving, P., Troxler, L., Galiana-Arnoux, D., Hetru, C., Hoffmann, J. A., and Imler, J.-L. (2005). The Jak-STAT signaling pathway is required but not sufficient for the antiviral response of *Drosophila*. *Nat. Immunol.* **6**, 946–953.

Drif, L., and Brehelin, M. (1983). The circulating hemocytes of *Culex pipiens* and *Aedes aegypti*: Cytology, histochemistry, hemograms, and functions. *Dev. Comp. Immunol.* **7**, 687–690.

Eleftherianos, I., Gokcen, F., Felfoldi, G., Millichap, P. J., Trenczek, T. E., French-Constant, R. H., and Reynolds, S. E. (2007). The immunoglobulin family protein Hemolin mediates cellular immune responses to bacteria in the insect *Manduca sexta*. *Cell. Microbiol.* **9**, 1137–1147.

Elrod-Erickson, M., Mishra, S., and Schneider, D. (2000). Interactions between the cellular and humoral immune responses in *Drosophila*. *Curr. Biol.* **10**, 781–784.

Evans, C. J., Hartenstein, V., and Banerjee, U. (2003). Thicker than blood: Conserved mechanisms in *Drosophila* and vertebrate hematopoiesis. *Dev. Cell* **5**, 673–690.

Feir, D. (1979). Multiplication of hemocytes. In *Insect Hemocytes* (A. P. Gupta, Ed.), pp. 67–82. Cambridge University Press, Cambridge, UK.

Filipe, S. R., Tomasz, A., and Ligoxygakis, P. (2005). Requirements of peptidoglycan structure that allow detection by the *Drosophila* Toll pathway. *EMBO Rep.* **6**, 327–333.

Foley, D. A. (1978). Innate cellular defense by mosquito hemocytes. In *Comparative Pathobiology* (L. A. Bulla Jr., and T. C. Cheng, Eds.), Vol. 4, pp. 114–143. Academic Press, New York.

Franc, N. C., Heitzler, P., Ezekowitz, R. A. B., and White, K. (1999). Requirement for croquemort in phagocytosis of apoptotic cells in *Drosophila*. *Science* **284**, 1991–1994.

Galiana-Arnoux, D., Dostert, C., Schneemann, A., Hoffmann, J. A., and Imler, J. L. (2006). Essential function *in vivo* for Dicer-2 in host defense against RNA viruses in *Drosophila*. *Nat. Immunol.* **7**, 590–597.

Gardiner, E. M. M., and Strand, M. R. (1999). Monoclonal antibodies bind distinct classes of hemocytes in the moth *Pseudoplusia includens*. *J. Insect Physiol.* **45**, 113–126.

Gardiner, E. M. M., and Strand, M. R. (2000). Hematopoiesis in larval *Pseudoplusia includens* and *Spodoptera frugiperda*. *Arch. Insect Bich. Physiol.* **43**, 147–164.

Garver, L. S., Wu, J., and Wu, L. P. (2006). The peptidoglycan recognition protein PGRP-SC1a is essential for Toll signaling and phagocytosis of *Staphylococcus aureus* in *Drosophila*. *Proc. Natl. Acad. Sci. USA* **103**, 660–665.

Gillespie, J. P, Kanost, M. R., Trenczek, T. (1997). Biological mediators of insect immunity. *Annu. Rev. Entomol.* **42**, 611–643.

Grimstone, A. V., Rotheram, S. and Salt, G. B. (1967). An electron-microscope study of capsule formation by insect blood cells. *J. Cell Sci.* **2**, 281–292.

Gupta, A. P. (1985). Cellular elements in hemolymph. In *Comprehensive Insect Physiology, Biochemistry, and Pharmacology* (G. A. Kerkut, L. I. Gilbert, Eds.), Vol. 3, pp. 401–451. Pergamon Press, Oxford.

Hillyer, J. F., and Christensen, B. M. (2002). Characterization of hemocytes from the yellow fever mosquito, *Aedes aegypti*. *Histochem. Cell Biol.* **117**, 431–440.

Hinks, C. F., and Arnold, J. W. (1977). Haemopoiesis in Leptidoptera II, the role of haemopoietic organs. *Can. J. Zool.* **55**, 1740–1755.

Hoffmann, J. A. and Reichhart, J.-M. (2002). *Drosophila* innate immunity, an evolutionary perspective. *Nat. Immunol.* **3**, 121–126.

Holz, A., Bossinger, B., Strasser, T., Janning, W., and Klapper, S. (2003). The two origins of hemocytes in *Drosophila*. *Development* **130**, 4955–4962.

Imler, J.-L., and Bulet, P. (2005). Antimicrobial peptides in *Drosophila*, structures, activities and gene regulation. In *Mechanisms of Epithelial Defense* (D. Kabelitz and J. M. Schroder, Eds.), Vol. 86, pp. 1–21. Karger, Basil.

Imler, J.-L., and Zheng, L. (2005). Biology of Toll receptors, lessons from insects and mammals. *J. Leuk. Biol.* **74**, 18–26.

Irving, P., Ubeda, J., Doucet, D., Troxler, L., Lagueux, M., Zachary, D., Hoffmann, J., Hetru, C., and Meister, M. (2005). New insights into *Drosophila* larval haemocyte functions through genome-wide analysis. *Cell. Microbiol.* **7**, 335–350.

Jiang, H., Wang, Y., Ma, C., and Kanost, M. R. (1997). Subunit composition of pro-phenoloxidase from *Manduca sexta*, molecular cloning of subunit proPO-P1. *Insect Biochem. Mol. Biol.* **27**, 835–850.

Jones, J. C. (1962). Current concepts concerning insect hemocytes. *Am. Zool.* **2**, 209–246.

Jones, J. C. (1970) Hemocytopoiesis in insects. In *Regulation of Hematopoiesis* (A. S. Gordon, Ed.), pp. 7–65. Appleton Press, New York.

Jung, S. H., Evans, C. J., Uemura, C., and Banerjee, U. (2005). The *Drosophila* lymph gland as a developmental model of hematopoiesis. *Development* **132**, 2521–2533.

Kaaya, G. P., and Ratcliffe, N. A. (1982). Comparative study of hemocytes and associated cells of some medically important dipterans. *J. Morph.* **173**, 351–365.

Kaneko, T., Yano, T., Aggarwall, K., Lim, J., Ueda, K., Oshima, Y., Peach, C., Erturk-Hasdemir, D., Goldman, W. E., Oh, B., Kurata, S., and Silverman, N. (2006). PGRP-LC and PGRP-LE have essential yet distinct functions in the *Drosophila* immune response to monomeric DAP-type peptidoglycan. *Nat. Immunol.* **7**, 715–723.

Keene, K. M., Foy, B. D., Sanchez-Vargas, I., Beaty, B. J., Blair, C. D., and Olson, K. E. (2004). RNA interference acts as a natural antiviral response to O'nong-nyong virus (Alphavirus; Togaviridae) infection of *Anopheles gambiae*. *Proc. Natl. Acad. Sci. USA* **101**, 17240–17245.

Kocks, C., Cho, J.-H., Nehme, N., Ulvila, J., Pearson, A. M., Meister, M., Strom, C., Conto, S. L., Hetru, C., Stuart, L. M., Stehle, T., Hoffmann, J. A., Reichhart, J.-M., Ferrandon, D., Ramet, M., and Ezekowitz, R. A. B. (2005). Eater, a transmembrane protein mediating phagocytosis of bacterial pathogens in *Drosophila*. *Cell* **123**, 335–346.

Krzemien, J., Dubois, L., Makki, R., Meister, M., Vincent, A., and Crozatier, M. (2007). Control of blood cell homeostasis in *Drosophila* larvae by the posterior signalling centre. *Nature* **446**, 325–328.

Lackie, A. M. (1988). Haemocyte behaviour. *Adv. Insect Physiol.* **21**, 85–178.

Lanot, R., Zachary, D., Holder, F., and Meister, M. (2001). Postembryonic hematopoiesis in *Drosophila*. *Dev. Biol.* **230**, 243–257.

Lavine, M. D., and Strand, M. R. (2001). Surface characteristics of foreign targets that elicit an encapsulation response by the moth *Pseudoplusia includens*. *J. Insect Physiol.* **47**, 965–974.

Lavine, M. D., and Strand, M. R. (2002). Insect hemocytes and their role in cellular immune responses. *Insect Biochem. Mol. Biol.* **32**, 1237–1242.

Lavine, M. D., and Strand, M. R. (2003). Hemocytes from *Pseudoplusia includens* express multiple alpha and beta integrin subunits. *Insect Mol. Biol.* **12**, 441–452.

Lavine, M. D., Chen, G., and Strand, M. R. (2005) Immune challenge differentially affects transcript abundance of three antimicrobial peptides in hemocytes from the moth *Pseudoplusia includens*. *Insect Biochem. Mol. Biol.* **35**, 1335–1346.

Lebestky, T., Chang, T., Hartenstein, V., Banerjee, U. (2000). Specification of *Drosophila* hematopoietic lineage by conserved transcription factors. *Science* **288**, 146–149.

Lebestky, T., Jung, S., and Banerjee, U. (2003). Serrate-expressing signaling center controls *Drosophila* hematopoiesis. *Genes Dev.* **17**, 348–353.

Levashina, E. A., Moita, L. F., Blandin, S., Vriend, G., Lagueux, M., and Kafatos, F. C. (2001). Conserved role of a complement-like protein in phagocytosis revealed by dsRNA knockout in cultured cells of the mosquito, *Anopheles gambiae*. *Cell* **104**, 709–718.

Levin, D. M., Breuer, L. N., Zhuang, S. F., Anderson, S. A., Nardi, J. B., and Kanost, M. R. (2005). A hemocyte-specific integrin required for hemocytic encapsulation in the tobacco hornworm, *Manduca sexta*. *Insect Biochem. Mol. Biol.* **35**, 369–380.

Li, W., Li, H., Lu, R., Li, F., Dus, M., Atkinson, P., Brydon, E. W. A., Johnson, K. L., García-Sastre, A., Ball, L. A., Palese, P., and Ding, S. (2004). Interferon antagonist proteins of influenza and vaccinia viruses are suppressors of RNA silencing. *Proc. Natl. Acad. Sci. USA* **101**, 1350–1355.

Ling, E., Shirai, K., Kanekatsu, R., and Kiguchi, K. (2005). Hemocyte differentiation in the hematopoietic organs of the silkworm, *Bombyx mori*: Prohemocytes have the function of phagocytosis. *Cell Tissue Res.* **320**, 535–543.

Ling, E. J., and Yu, X. Q. (2006). Cellular encapsulation and melanization are enhanced by immulectins, pattern recognition receptors from the tobacco hornworm *Manduca sexta*. *Dev. Comp. Immunol.* **30**, 289–299.

Liu, C. T., Hou, R. F., and Chen, C. C. (1998). Formation of basement membrane-like structure terminates the cellular encapsulation of microfilariae in the haemocoel of *Anopheles quadrimaculatus*. *Parasitology* **116**, 511–518.

Lowenberger, C. (2001) Innate immune response of *Aedes aegypti*. *Insect Biochem. Mol. Biol.* **31**, 219–229.

Mandal, L., Martinez-Agosto, J. A., Evans, C. J., Hartenstein, V., and Banerjee, U. (2007). A Hedgehog- and Antennapedia-dependent niche maintains *Drosophila* hematopoietic precursors. *Nature* **446**, 320–324.

Michel, K., Budd, A., Pinto, S., Gibson, T., and Kafatos, F. (2005). *Anopheles gambiae* SRPN2 facilitates midgut invasion by the malaria parasite *Plasmodium berghei*. *EMBO Rep.* **6**, 891–897.

Michel, K., Suwanchaichinda, C., Morlais, I., Lambrechts, L., Cohuet, A., Awono-Ambene, P. H., Simard, F., Fontenille, D., Kanost, M. R., and Kafatos, F. C. (2006). Increased melanizing activity in *Anopheles gambiae* does not affect development of *Plasmodium falciparum*. *Proc. Natl. Acad. Sci. USA* **103**, 16858–16863.

Moita, L. F., Wang-Sattler, R., Michel, K., Zimmermann, T., Blandin S., Levashina, E. A., and Kafatos, F. C. (2005). *In vivo* identification of novel regulators and conserved pathways of phagocytosis in *A. gambiae*. *Immunity* **23**, 65–73.

Munier, A. I., Doucet, D., Perrodou, E., Zchary, D., Meister, M., Hoffmann, J. A., Janeway, C. A. Jr., and Lagueux, M. (2002). PVF2, A PDGF/VEGF-like growth factor, induces hemocyte proliferation in *Drosophila* larvae. *EMBO Rep.* **3**, 195–1200.

Nakahara, Y., Kanamori, Y., Kiuchi, M., and Kamimura, M. (2003). *In vitro* studies of hematopoiesis in the silkworm, cell proliferation in and hemocyte discharge from the hematopoietic organ. *J. Insect Physiol.* **49**, 907–916.

Nakahara, Y., Matsumoto, H., Kanamori, Y., Kataoka, H., Mizoguchi, A., Kiuchi, M., and Kamimura, M. (2006). Insulin signaling is involved in hematopoietic regulation in an insect hematopoietic organ. *J. Insect Physiol.* **52**, 105–111.

Nappi, A. J., and Christensen, B. M. (2005). Melanogenisis and associated cytotoxic reactions, applications to insect innate immunity. *Insect Biochem. Mol. Biol.* **35**, 443–459.

Nardi, J. B. (2004). Embryonic origins of the two main classes of hemocytes – granular cells and plasmatocytes in *Manduca sexta*. *Dev. Genes Evol.* **214**, 19–28.

Nardi, J. B., Ujhelyi, E., Pilas, B., Garsha, K., and Kanost, M. R. (2003). Hematopoietic organs of *Manduca sexta* and hemocyte lineages. *Dev. Genes Evol.* **213**, 477–491.

Nardi, J. B., Pilas, B., Bee, C. M., Zhuang, S., Garsha, K., and Kanost, M. R. (2006). Neuroglian-positive plasmatocytes of *Manduca sexta* and the initiation of hemocyte attachment to foreign surfaces. *Dev. Comp. Immunol.* **30**, 447–462.

Nelson, R. E., Fessler, L. I., Takagi, Y., Blumberg, B., Keene, D. R., Olson, P. F., Parker, C. G., and Fessler, J. H. (1994). Peroxidasin: A novel enzyme-matrix protein of *Drosophila* development. *EMBO J.* **13**, 3438–3447.

Osta, M. A., Christophides, G. K., and Kafatos, F. C. (2004). Effects of mosquito genes on *Plasmodium* development. *Science* **303**, 2030–2032.

Parsons, J. T. (2003). Focal adhesion kinase, the first ten years. *J. Cell Sci.* **116**, 1409–1416.

Pearson, A. M., Baksa, K., Rämet, M., Protas, M., McKee, M., Brown, D., and Ezekowitz, R. A. B. (2003). Identification of cytoskeletal regulatory proteins required for efficient phagocytosis in *Drosophila*. *Microbes Infect.* **5**, 815–824.

Pech, L. L., and Strand, M. R. (1996). Granular cells are required for encapsulation of foreign targets by insect haemocytes. *J. Cell Sci.* **109**, 2053–2060.

Pech, L. L., and Strand, M. R. (2000). Plasmatocytes from the moth *Pseudoplusia includens* induce apoptosis of granular cells. *J. Insect Physiol.* **46**, 1565–1573.

Philips, J. A., Rubin, E. J., and Perrimon, N. (2005). *Drosophila* RNAi screen reveals C35 family member required for mycobacterial infection. *Science* **309**, 1248–1251.

Popham, H. J., Shelby, K. S., Brandt, S. L., and Coudron, T. A. (2004). Potent virucidal activity in larval *Heliothis virescens* plasma against *Helicoverpa zea* single capsid nucleopolyhedrovirus. *J. Gen. Virol.* **85**, 2255–2261.

Radtke, F., Wilson, A., and MacDonald, H. R. (2005). Notch signaling in hematopoiesis and lymphopoiesis: Lessons from *Drosophila. Bioessays* 27, 1117–1128.

Ramet, M., Pearson, A., Manfruelli, P., Li, X., Koziel, H., Gobel, V., Chung, E., Krieger, M., and Ezekowitz, R. A. B. (2001). *Drosophila* scavenger receptor CI is a pattern recognition receptor for bacteria. *Immunity* **15**, 1027–1038.

Ramet, M., Manfruelli, P., Pearson, A., Mathey-Prevot, B., and Ezekowitz, R. A. B. (2002). Functional genomic analysis and identification of a *Drosophila* receptor for *E. coli. Nature* **416**, 644–648.

Ratcliffe, N. A., and Gagen, S. J. (1976). Cellular defense reactions of insect hemocytes *in vivo*, nodule formation and development in *Galleria mellonella* and *Pieris brassicae* larvae. *J. Invert. Pathol.* **28**, 373–382.

Ratcliffe, N. A., and Gagen, S. J. (1977). Studies on the *in vivo* cellular reactions of insects, an ultrastructural analysis of nodule formation in *Galleria mellonella. Tissue Cell* **9**, 73–85.

Ratcliffe, N. A., Rowley, A. F., Fitzgerald, S. W., and Rhodes, C. P. (1985). Invertebrate immunity – basic concepts and recent advances. *Intl. Rev. Cytol.* **97**, 183–350.

Reeves, E. P., Lu, H., Jacobs, H. L., Messina, C. G., Bolsover, S., Gabella, G., Potma, E. O., Warley, A., Roes, J., and Segal, A. W. (2002). Killing activity of neutrophils is mediated through activation of proteases by K+ flux. *Nature* **416**, 291–297.

Ribeiro, C., and Brehelin, M. (2006). Insect haemocytes: What type of cell is what? *J. Insect Physiol.* **52**, 417–429.

Riehle, M. M., Markianos, K., Niare, O., Xu, J. N., Li, J., Toure, A. M, Podiougou, B., Oduol, F., Diawara, S., Diallo, M., Coulibaly, B., Ouatara, A., Kruglyak, L., Traore, S. F., and Vernick, K. D. (2006). Natural malaria infection in *Anopheles gambiae* is regulated by a single genomic control region. *Science* **312**, 577–579.

Saleh, M.-C., Rij, R. P. van, Hekele, A., Gillis, A., Foley, E., O'Farrell, P. H., Andino, R. (2006). The endocytic pathway mediates cell entry of dsRNA to induce RNAi silencing. *Nat. Cell Biol.* **8**, 793–802.

Salt, G. B. (1970). *The Cellular Defense Reactions in Insects.* Cambridge University Press, Cambridge.

Scherfer, C., Qazi, M. R., Takahashi, K., Ueda, R., Dushay, M. S., Theopold, U., and Lemaitre, B. (2006). The Toll immune-regulated *Drosophila* protein Fondue is involved in hemolymph clotting and puparium formation. *Dev. Biol.* **295**, 156–163.

Schmidt, O., Theopold, U., and Strand, M. R. (2001). Innate immunity and evasion by insect parasitoids. *Bioessays* **23**, 344–351.

Selsted, M. E., and Ouellette, A. J. (1995). Defensins in granules of phagocytic and non-phagocytic cells. *Trends Cell Biol.* **5**, 114–119.

Sorrentino, R. P., Carton, Y., and Govind, S., (2002). Cellular immune response to parasite infection in the *Drosophila* lymph gland is developmentally regulated. *Dev. Biol.* **243**, 65–80.

Sorrentino, R. P., Melk, J. P., and Govind, S. (2004). Genetic analysis of contributions of dorsal group and JAK-Stat92E pathway genes to larval hemocyte concentration and the egg encapsulation response in *Drosophila. Genetics* **166**, 1343–1356.

Stramer, B., Wood, W., Galko, M. J., Redd, M. J., Jacinto, A. Parkhurst, S. M., and Martin, P. (2005). Live imaging of wound inflammation in *Drosophila* embryos reveals key roles for small GTPases during *in vivo* cell migration. *J. Cell Biol.* **168**, 567–573.

Strand, M. R., and Clark, K. D. (1999). Plasmatocyte spreading peptide induces spreading of plasmatocytes but represses spreading of granulocytes. *Arch. Insect Bich. Physiol.* **42**, 213–223.

Strand, M. R., Beck, M. H., and Lavine, M. D. (2006). *Microplitis* demolitor bracovirus inhibits phagocytosis by hemocytes from *Pseudoplusia includens. Arch. Insect Bich. Physiol.* **61**, 134–145.

Strand, M. R., and Pech, L. L. (1995). Immunological basis for compatibility in parasitoid–host relationships. *Annu. Rev. Entomol.* **40**, 31–56.

Stroschein-Stevenson, S. L., Foley, E., O'Farrell, P. H., and Johnson, A. D. (2006). Identification of *Drosophila* gene products required for phagocytosis of *Candida albicans*. *PLoS Biol.* **4**, 87–99.

Stuart, L. M., and Ezekowitz, R. A. (2005). Phagocytosis: Elegant complexity. *Immunity* **22**, 539–550.

Stuart, L. M., Boulais, J., Charriere, G. M., Hennessy, E. J., Brunet, S., Jutras, I., Goyette, G., Rondeau, C., Letarte, S., Huang, H., Ye, P., Morales, F., Kocks, C., Bader, J. S., Desjardins, M., and Ezekowitz, R. A. (2007). A systems biology analysis of the *Drosophila* phagosome. *Nature* **445**, 95–101.

Taniai, K., Lee, J. H., and Lee, I. H. (2006). *Bombyx mori* cell line as a model of immune system organs. *Insect Mol. Biol.* **15**, 269–279.

Tauszig, S., Jouanguy, E., Hoffmann, J. A., and Imler, J.-L. (2000). Toll-related receptors and the control of antimicrobial peptide expression in *Drosophila*. *Proc. Natl. Acad. Sci. USA* **99**, 10520–10525.

Tepass, U., Fessler, L. I., Aziz, A., and Hartenstein, V. (1994). Embryonic origin of hemocytes and their relationship to cell death in *Drosophila*. *Development* **120**, 1829–1837.

Terenius, O., Bettencourt, R., Lee, S. Y., Li, W., Soderhall, K., and Faye, I. (2007). RNA interference of Hemolin causes depletion of phenoloxidase activity in *Hyalophora cecropia*. *Dev. Comp. Immunol.* **31**, 571–575.

Theopold, U., Schmidt, O., Soderhall, K., and Dushay, M. S. (2004). Coagulation in arthropods, defence, wound closure and healing. *Trends Immunol.* **25**, 289–294.

Traver, D., and Zon, L. I. (2002). Walking the walk: Migration and other common themes in blood and vascular development. *Cell* **108**, 731–734.

Trudeau, D., Washburn, J. O., and Volkman, L. E. (2001). Central role of hemocytes in *Autographa californica* M nucleopolyhedrovirus pathogenesis in *Heliothis virescens* and *Helicoverpa zea*. *J. Virol.* **75**, 996–1003.

Tydell, C. C., Yuan, J., Tran, P., and Selsted, M. E. (2006). Bovine peptidoglycan recognition protein-S, antimicrobial activity, localization, secretion, and binding properties. *J. Immunol.* **176**, 1154–1162.

Wang, X., Aliyari, R., Li, W., Li, H., Kim, K., Carthew, R., Atkinson, P., and Ding, S. (2006). RNA interference directs innate immunity against viruses in adult *Drosophila*. *Science* **312**, 45–454.

Wertheim, B., Kraaijeveld, A. R., Schuster, E., Blanc, E., Hopkins, M., Pletcher, S. D., Strand, M. R., Godfray, H. C. J., and Partridge, L. (2005). Genome wide expression in response to parasitoid attack in *Drosophila*. *Genome Biol.* **6**, R94, 1–20.

Whitten, M. M., Shiao, S. H., and Levashina, E. A. (2006). Mosquito midguts and malaria: Cell biology, compartmentalization and immunology. *Parasite Immunol.* **28**, 121–130.

Wiegand, C., Levin, D., Gillespie, J. P., Willott, E., Kanost, M. R., and Trenczek, T. (2000). Monoclonal antibody M13 identifies a plasmatocyte membrane protein and inhibits encapsulation and spreading reactions of *Manduca sexta* hemocytes. *Arch. Insect Bich. Physiol.* **45**, 95–108.

Williams, M. J., Wiklund, M. L., Wikman, S., and Hultmark, D. (2006). Rac1 signalling in the *Drosophila* larval cellular immune response. *J. Cell Sci.* **119**, 2015–2024.

Willott, E., Trenczek, T., Thrower, L. W., and Kanost, M. R. (1994). Immunochemical identification of insect hemocyte populations: Monoclonal antibodies distinguish four major hemocyte types in *Manduca sexta*. *Eur. J. Cell Biol.* **65**, 417–423.

Wood, W., Faria, C., and Jacinto, A. (2006). Distinct mechanisms regulate hemocyte chemotaxis during development and wound healing in *Drosophila melanogaster*. *J. Cell Biol.* **173**, 405–416.

Yamano, Y., Matsumoto, M., Inoue, K., Kawabata, T., and Morishima, I. (1994) Cloning of cDNAs for cecropins A and B, and expression of the genes in the silkworm, *Bombyx mori*. *Biosci. Biotech. Biochem.* **58**, 1476–1478.

Yamashita, M., and Iwabuchi, K. (2001). *Bombyx mori* prohemocyte division and differentiation in individual microcultures. *J. Insect Physiol.* **47**, 325–331.

Yu, X. Q., and Kanost, M. R. (2004). Immulectin-2, a pattern recognition receptor that stimulates hemocyte encapsulation and melanization in the tobacco hornworm, *Manduca sexta*. *Dev. Comp. Immunol.* **28**, 891–900.

Zambon, R. A., Vakharia, V. N., and Wu, L. P. (2006). RNAi is an antiviral immune response against a dsRNA virus in *Drosophila melanogaster*. *Cell. Microbiol.* **8**, 880–889.

Zettervall, C., Anderl, I., Williams, M. J., Palmer, R., Kurucz, E., Ando, I., and Hultmark, D. (2004). A directed screen for genes involved in *Drosophila* blood cell activation. *Proc. Natl. Acad. Sci. USA* **101**, 14192–14197.

Zhuang, S., Kelo, L., Nardi, J. B., and Kanost, M. R. (2007). An integrin–tetraspanin interaction required for cellular innate immune responses of an inset, *Manduca sexta*. *J. Biol. Chem.* (in press).

3

EICOSANOID ACTIONS IN INSECT IMMUNOLOGY

DAVID W. STANLEY* AND JON S. MILLER**

*USDA/Agricultural Research Service, Biological Control of Insects Research Laboratory, Columbia, MO 65203, USA
**Department of Biological Sciences, Northern Illinois University, DeKalb, IL 60115, USA

3.1 Introduction
3.2 Eicosanoids Act in Bacterial Clearance and Nodule Formation Reactions
3.3 The Biochemistry of Eicosanoids in Insect Immune Tissues
3.4 Eicosanoids Act in *In Vitro* Hemocyte Preparations
3.5 Eicosanoids in Hemocyte Spreading and Migration
3.6 A Frontier: Eicosanoids in Insect–Viral Interactions
3.7 Another Frontier: Prostaglandin Modes of Action

ABSTRACT: In this chapter we review eicosanoid actions in insect immunity. Eicosanoids are oxygenated metabolites of arachidonic acid and two other C20 polyunsaturated fatty acids. Groups of eicosanoids include prostaglandins, lipoxygenase products and epoxyeicosatrienoic acids. These compounds are most well studied in the context of biomedicine; however, we now know eicosanoids act in insect immune defense reactions. These include the cellular mechanisms responsible for clearing bacterial infection from hemolymph circulation and in microaggregation and nodulation reactions. Eicosanoids also act in plasmatocyte spreading and hemocyte migration toward a chemical source. Various eicosanoids act in insect defenses against bacteria, fungi, protozoan and parasitoid challenge. The most recent data indicate eicosanoids act in insect defenses against viral infection. With a view to a coming generation of insect scientists, we lift up insect–virus interactions and mechanisms of eicosanoid actions as two of the visible frontiers of insect immunology.

Abbreviations:

AA	=	arachidonic acid
BPB	=	*p*-bromophenacyl bromide
CM	=	conditioned medium
COX	=	cyclooxygenase
Dex	=	dexamethasone
DP, PGD	=	receptor
EBI	=	eicosanoid biosynthesis inhibitor
EGTA	=	glycol-bis(2-aminoethylether)-*N,N,N′,N′*-tetraacetic acid
EP, PGE	=	receptor
EtOH	=	ethanol
FP, PGF	=	receptor
GC/MS	=	gas chromatography/mass spectrometry
GPCR, G	=	protein coupled receptor
HETE	=	hydroxyeicosatetraenoic acid
HpETE	=	hydroperoxyeicosatetraenoic acid
HPLC	=	high performance liquid chromatography
IP, PGI	=	receptor
LdMNPV	=	*Lymantria dispar* nucleopolyhedrovirus
LOX	=	lipoxygenase
MAFP	=	methylarachidonyl fluorophosphates
PG	=	prostaglandin
PL	=	phospholipid
PLA_2	=	phospholipase A_2
TP	=	thromboxane receptor

3.1   INTRODUCTION

Our appreciation of eicosanoids dates back to the 1930s, when von Euler (1936) reported on a substance associated with prostate gland secretions stimulated contractions in uterine smooth muscle. He named the acidic, lipoidal substance 'prostaglandin' (PG). Ulf von Euler went on to study the chemistry of neurotransmitters, for which he shared the 1970 Nobel Prize in Physiology or Medicine, becoming one of the few 'Nobel families' as his father Hans von Euler-Chelpin was awarded the 1929 Nobel in Chemistry. von Euler was acquainted with Sune Bergström and he encouraged Bergström to investigate the chemistry of PGs. Bergström and his student, Bengt Samuelsson, worked out the chemistry of PGs. They reported on the chemical structures of PGs E and F (Bergström et al., 1962). It was immediately clear from their structures that PGs were oxygenated metabolites of arachidonic acid (AA) and this finding launched a lengthy research program into AA metabolism. We now recognize three groups of eicosanoids, so named from the Greek word 'eicos' for the 20 carbons in each compound. Cartoons of the biosynthetic pathways and actions of these compounds are presented in Figs 3.1 and 3.2, and eicosanoid structures are detailed elsewhere (Stanley, 2000, 2005). PGs are products of the cyclooxygenase (COX) pathways. The lipoxygenase (LOX) pathways yield a variety of products, including hydroxyeicosatetraenoic acids (HETEs) and leukotrienes. Epoxyeicosatrienoic acids are formed by epoxygenases, so far known in mammals, but not insects. Bergström and Samuelsson shared the 1982 Nobel Prize in Physiology or Medicine with Robert Vane for their work on eicosanoids.

The Nobel Prize indicates the significance of eicosanoids in mammalian biology and clinical medicine. Various eicosanoids are present in every mammalian tissue and body fluid, where they mediate an amazing array of events. Many of these events, such as ion transport functions, smooth muscle contraction and neurotransmission, are fundamental to biology. Eicosanoids also serve important roles in mammalian immunity, where they influence innate immune reactions including inflammation. Research over the last 25 years also revealed important

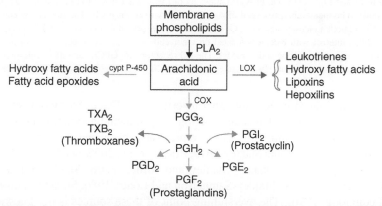

FIGURE 3.1   An outline of eicosanoid biosynthesis as understood from the biomedical background. Many cell stimulants (including other eicosanoids) initiate eicosanoid biosynthesis by activating PLA_2, which releases AA from membrane PLs. The free AA can enter any of three oxygenation pathways. (1) The COX pathway produces PLs and thromboxanes. (2) The LOX pathways yield leukotrienes, lipoxins and hydroperoxy- and hydroxyl fatty acids. The epoxy derivatives are products of cytochrome P-450 epoxygenases (from Stanley and Miller, 2006, with permission of Wiley-Blackwell).

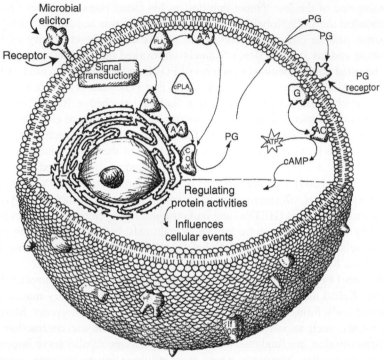

FIGURE 3.2 A model of PG action in cells, depicting a mechanism of PG action in biomedical systems. In this model a stimulating agent, perhaps a bacterial component such as lipopolysaccharide, influences intracellular events leading to up-regulation of eicosanoid biosynthesis. This would begin with translocation of a cytosolic PLA_2 (Stanley, 2006b) from the cytosol to intracellular and plasma membranes. The membrane-associated PLA_2 releases AA from PL pools. The free AA is converted into a PG or another eicosanoids. The PGs shown here are exported form the cell by PG transporters. The PGs can interact with receptors located on other cells (in a paracrine action mode) or on the exporting cell (in an autocrine action mode). The PG receptors are GPCRs and these entrain a series of intracellular regulator events (from Stanley and Miller, 2006, with permission of Wiley-Blackwell).

eicosanoid actions in insects and other invertebrates. These eicosanoid actions in invertebrates also are fundamental to biology, again including ion transport, reproductive biology and immunity. These topics have been reviewed from various points of view (Howard and Stanley, 1999; Rowley et al., 2005; Stanley and Howard, 1998; Stanley, 2000; 2006a, b; Stanley and Miller, 2006; Stanley-Samuelson, 1987; 1991; 1994a, b, c; Stanley-Samuelson and Loher, 1986; Stanley-Samuelson and Pedibhotla, 1996). The overarching point of these sources is our knowledge of eicosanoid actions in insect biology is steadily increasing. Drawing on the biomedical background, it is reasonable to expect a future of continued growth in this area.

The goal of this chapter is to review the development of our understanding of eicosanoid actions in insect immunology and then describe our most current work.

Finally, we want to highlight some of the important lacuna in our knowledge for those interested in advancing this area of insect science.

3.2 EICOSANOIDS ACT IN BACTERIAL CLEARANCE AND NODULE FORMATION REACTIONS

Stanley-Samuelson et al. (1991) first suggested that eicosanoids somehow act in the immune functions responsible for clearing injected bacteria from hemolymph circulation in tobacco hornworms, *Manduca sexta*. We followed a simple protocol. Large, 5th-instar hornworms were injected with a selected pharmaceutical eicosanoid biosynthesis inhibitor (EBI) and then injected with a red-pigmented strain of the insect pathogenic bacterium, *Serratia marcescens*. After brief incubation periods, hemolymph was withdrawn through a proleg, taken through a series of dilutions and plated on agar. We did not detect bacteria on plates prepared from ethanol (EtOH)-treated control hornworms. On the other hand, we recovered increasing numbers of bacterial colony-forming units (CFUs) throughout the 60 min experiments. The influence of one inhibitor, dexamethasone (Dex), was expressed in a dose-dependent manner and it was reversed by injecting hornworms with AA, the direct precursor for eicosanoid biosynthesis. We inferred from these results that eicosanoids mediate some or all of the cellular mechanisms responsible for clearing bacterial infection from hemolymph circulation (Stanley-Samuelson et al., 1991). We specified cellular mechanisms because most experiments lasted no more than 1 h, long before anti-bacterial proteins appear in the hemolymph of infected insects.

This ground-breaking work supported a very broad hypothesis that eicosanoids act in insect bacterial clearance, but it provided no insight into specific cellular clearance mechanisms mediated by eicosanoids. Based on the work of Dunn and Drake (1983), who reported that cells of the bacterium *Pseudomonas aeruginosa* P11-1 were cleared from hornworm circulation by nodules formation during the first 2 h following infection, we developed the hypothesis that eicosanoids mediate microaggregation and nodulation reactions to bacterial infection (Miller et al., 1994; Figs 3.3 and 3.4). To test this hypothesis we developed a simple protocol, which we later formalized (Miller and Stanley, 1998). Briefly, hornworms were injected with a selected EBI and then injected with *S. marcescens*. To assess the influence of the EBI on microaggregation reactions, we withdrew hemolymph and counted microaggregates on a hemacytometer under phase contrast optics. Nodulation reactions were assessed by dissecting anesthetized hornworms as described (Miller and Stanley, 1998). Dex-treated hornworms were severely crippled in their ability to form both microaggregates and nodules. For example, control hornworms produced about 120 nodules/insect by 6 h post-infection (PI), compared to about 20/insect in dex-treated hornworms. The Dex influence was recorded in a dose-related manner and, again, the Dex effect was reversed in

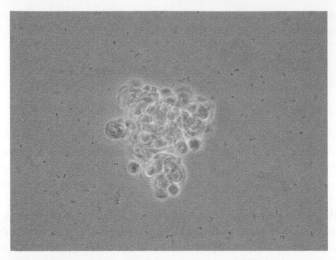

FIGURE 3.3　A photomicrograph (400 × magnification) of a hemocytic microaggregate, taken 1 h after injecting a tobacco hornworm, *M. sexta*, with the bacterium *Serratia marcescens*. The cells in this image are about 10–12 microns in diameter. (Prepared and photographed by Jon S. Miller.) (See Plate 3.3 for color version of this figure.)

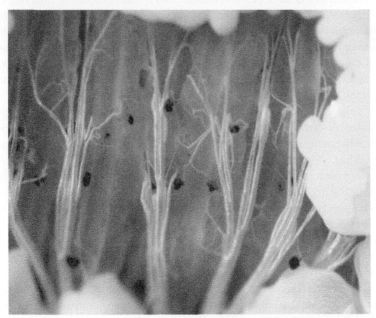

FIGURE 3.4　A photomicrograph (40 × magnification) of nodules formed in a tobacco hornworm, *M. sexta*, 4 h after injection with the bacterium *S. marcescens*. Nodules are typically darkened by a final melanization reaction and attached to body and organ surfaces. The nodules seen here are attached to the alimentary canal which is framed by malpighian tubules. The muscles surrounding the alimentary canal feature large tracheae, clearly visible in this image (from Stanley and Miller, 2006, with permission of Wiley-Blackwell). (See Plate 3.4 for color version of this figure.)

TABLE 3.1   Eicosanoids Act in Cellular Defense Reactions to Immune Challenge in Juveniles and Adults from At least Seven Major Insect Orders.

Order	Species	Life stage	Challenge	Reference
Lepidoptera				
	M. sexta	Larvae	*Serratia marcescens*	Miller et al. (1994)
			Beauveria bassiana	Lord et al. (2002)
			Metarhizium anisopliae	Dean et al. (2002)
	Agrotis ipsilon	Larvae	*S. marcescens*	Jurenka et al. (1997)
	Pseudaletia unipuncta	Larvae	*S. marcescens*	Jurenka et al. (1997)
	Galleria mellonella	Larvae	*Glass beads*	Mandato et al. (1997)
		Larvae	*BHSV-1*	Büyükgüzel et al. (2007)
	Bombyx mori	Larvae	*S. marcescens*	Stanley-Samuelson et al. (1997)
	Colias eurytheme	Larvae	*S. marcescens*	Stanley et al. (1999)
	Spodoptera exigua	Larvae	*Xenorhabdus nematophila*	Park and Kim (2000)
	Ostrinia nubialis	Larvae	*S. marcescens*	Tunaz et al. (2003a)
	Lymantria dispar	Larvae	*LdMNPV*	Stanley and Shapiro (2007)
Coleoptera				
	Zophobas attratus	Larvae	*S. marcescens*	Miller et al. (1996)
Hymenoptera				
	Apis mellifera	Adult	*S. marcescens*	Bedick et al. (2001)
	Pimpla turioinellae	Larvae	BHSV-1	Durmus et al. (2007)
Diptera				
	Drosophila melanogaster	Larvae	*Leptopilina boulardi egg*	Carton et al. (2002)
	Neobellieria bullata	Larvae	*Laminarin*	Franssens et al. (2005)
Orthoptera				
	Gryllus assimilis	Adult	*S. marcescens*	Miller et al. (1999)
	Gryllus firmus	Adult	*X. nematophila*	Park and Stanley (2006)
	Periplaneta americana	Adult	*S. marcescens*	Tunaz and Stanley (2000)
	Locusta migratoria	Adult	*Laminarin*	Goldsworthy et al. (2003)
Homoptera				
	Magicicada septendecim	Adult	*S. marcescens*	Tunaz et al. (1999)
	M. cassini	Adult	*S. marcescens*	Tunaz et al. (1999)
	Dactylopius coccus	Adult	*Laminarin*	de la Cruz Hernandez-Hernandez et al. (2003)
Hemiptera				
	Rhodnius prolixus	Nymph	*Trypanosoma rangeli*	Garcia et al. (2004)

hornworms treated with AA. We conducted experiments with EBIs that specifically probe the COX and LOX pathways and the results indicated that products of
both pathways act in microaggregation and nodulation reactions. We advanced
the idea that eicosanoids mediate these two related cellular defense actions (Miller
et al., 1994), formalized as the 'eicosanoid hypothesis'.

The nodulation discovery opened three lines of research. In one we considered
the hypothesis that eicosanoids mediate nodulation reactions in all insects that
express nodulation. In the second, we spent a few years on the biochemistry of
eicosanoids in insects. For the third, we searched for more specific cell steps that
depend on eicosanoid biosynthesis. We will complete this section with a brief
recapitulation of the first point.

Several research groups considered the eicosanoid hypothesis and experimental
results strongly support the hypothesis for at least 20 species representing seven
orders (Table 3.1). We emphasize the points that the hypothesis is supported for
juvenile and adult insects from hemi- and holometabolous orders. We infer that
eicosanoids mediate insect cellular immune reactions in all insects that express cellular immunity. Not all species do, as seen in our work with adult honeybees
(Bedick et al., 2001). We found that young adults, still in the in-house phases of their
behavioral ontogeny, express ordinary eicosanoid mediate nodulation reactions to
infection. Older, foraging adults, however, are virtually devoid of hemocytes and are
not able to mount nodulation reactions. We suppose that some very small insect
species may not be equipped for cellular immune reactions such as nodulation.

Table 3.1 also contains new information, recently published or now in press.
Eicosanoid-mediated cellular defenses against bacterial infections are well documented. Inspection of the table reveals that eicosanoids also act in cellular defenses
associated with challenge by fungal spores and conidia (Dean et al., 2002; Lord et al.,
2002), by protozoans (Garcia et al., 2004) and by parasitoid eggs (Carton et al.,
2002). Table 3.1 also shows that eicosanoids mediate melantoic nodulation reactions to viruses, as will be treated in a later section of this chapter.

3.3   THE BIOCHEMISTRY OF EICOSANOIDS
IN INSECT IMMUNE TISSUES

At the time of our work on eicosanoid actions in insect immune functions,
there was very little information on the presence of eicosanoids or its polyunsaturated
fatty acid precursor, AA, in insect tissues and no information in tobacco hornworms. The eicosanoid hypothesis rests on the foundation of the biochemistry of
eicosanoids. Clearly, if eicosanoids mediate cellular immune reactions, insect
immune tissues must be able to biosynthesize eicosanoids. In this brief section,
we present a sketch of the major points in the biochemistry of eicosanoids drawn
mostly from work on tobacco hornworms.

We began with analyses of the fatty acids associated with phospholipids (PLs) in
hornworm tissues. Earlier work had set the understanding that AA was present in

insect tissues, albeit in very low amounts, much lower than seen in mammals (Stanley-Samuelson and Dadd, 1983). Analysis of hemocyte and other tissues revealed that AA is present in hornworm tissues in the expected, low amounts (Ogg and Stanley-Samuelson, 1992; Ogg et al., 1991). The low proportions of AA are not due to nutritional inadequacies because we detected more AA in the hornworm culture medium than in the insect tissue. Later studies on fatty acid remodeling in hornworm hemocytes showed that hemocytes can quickly incorporate exogenous AA into cellular PLs, but within 2 h the AA is selectively removed from hemocyte PLs (Gadelhak and Stanley-Samuelson, 1994). We concluded that insect tissues selectively maintain fairly constant, low levels of AA.

We characterized eicosanoid biosynthesis in tobacco hornworm fat body (Stanley-Samuelson and Ogg, 1994) and hemocytes (Gadelhak et al., 1995). The fat body preparations produced mostly PGs, with PGA_2 the predominant product. In mammals PGA_2 is produced non-enzymatically as a rearrangement of PGE_2. The fat body preparations produce PGA_2 via another mechanism as incubating fat body preparations with radiolabeled PGE_2 did not yield labeled PGA_2. The hemocyte preparations differed from fat body because LOX products, possibly one of the HETEs, were the predominate yield from hemocyte preparations. Hence, both fat body and hemocytes are competent to biosynthesize eicosanoids, although these two tissues differ from one another in their patterns of biosynthesis.

We have characterized eicosanoid biosynthesis in other insect tissues, including midgut epithelia from tobacco hornworms (Büyükgüzel et al., 2002) and fat body from the beetle *Zophobas atratus* (Tunaz et al., 2002). These studies include rigorous chemical identification of PGs by GC/MS analysis. It is safe to say that insects are generally competent to produce PGs and other eicosanoids.

PLs are asymmetrical in animal cells, with a saturated fatty acid esterified to the *sn*-1 position and a polyunsaturated fatty acid, often AA, to the *sn*-2 position. This arrangement drew attention to phospholipase A_2 (PLA_2) as a critical step in eicosanoid biosynthesis. In the contemporary biomedical models, eicosanoid biosynthesis begins with hydrolysis of AA from biomembrane PLs, a single-step hydrolysis catalyzed by PLA_2. Using a routine radiometric protocol, we characterized PLA_2 in hornworm fat body (Uscian and Stanley-Samuelson, 1993) and hemocytes (Schleusener and Stanley-Samuelson, 1996). The fat body PLA_2 is interesting because it is located in the cytosol and does not depend on calcium for catalysis. The hemocyte PLA_2 is also cytosolic and calcium seems to somehow inhibit enzyme activity as adding glycol-bis(2-aminoethylether)-N,N,N',N'-tetraacetic acid (EGTA) to reactions increased PLA_2 activity. The hemocyte PLA_2 showed a marked preference for arachidonyl-containing PL substrate, as seen in some mammalian PLA_2s (Schleusener and Stanley-Samuelson, 1996).

In later work on PLA_2, we discovered two separate PLA_2s in tobacco hornworm hemocytes (Park et al., 2005). One is sensitive to the cytosolic PLA_2 inhibitor methylarachidonyl fluorophosphates (MAFP) and the other is sensitive to the secretory PLA_2 inhibitor *p*-bromophenacyl bromide (BPB). Both enzymes act in cellular immunity as separate inhibition of either enzyme impaired nodulation formation in experimental hornworms (Park et al., 2005). PLA_2s are regarded as a very large superfamily of enzymes (Schaloske and Dennis, 2006). Our work on PLA_2s associated with insect midgut contents, with oral secretions and in immune tissues indicates these enzymes will prove to be of great importance in insect biology (Stanley, 2006b). We regard this as a virtually untouched frontier for emerging insect scientists.

Overall, we have documented the major elements of eicosanoid biosynthesis in insect tissues, including the presence of AA, the direct substrate for eicosanoid biosynthesis, in cellular PLs, and the presence of PLA_2, and the biosynthesis of PGs and other eicosanoids in fat body and hemocytes, the immune-conferring tissues.

Aside from documenting the presence of eicosanoid systems in insect tissues, two studies suggest a direct link between eicosanoid biosynthesis and immune signaling. Jurenka et al. (1999) investigated the influence of bacterial infection on PG biosynthesis in true armyworms. They developed a fluorescence HPLC (high performance liquid chromatography) procedure to determine PG quantities and applied the procedure to record increased PG quantities in armyworm hemolymph 30 min after bacterial challenge. They also showed that treating armyworms with a COX inhibitor blocked the increase in hemolymph PG titers. The authors concluded that bacterial challenge stimulates PG production. In a similar vein, Tunaz et al. (2003b) showed that bacterial challenge stimulated increased hemocyte PLA_2 activity in tobacco hornworms. Together, these findings indicate that bacterial infection stimulates two components of the eicosanoid system, increased PLA_2 activity and increased PG production.

3.4   EICOSANOIDS ACT IN *IN VITRO* HEMOCYTE PREPARATIONS

The work described so far indicates that eicosanoids somehow mediate microaggregation and nodulation reactions to immune challenge. We explored the question of which tissues are responsible for producing immune-mediating eicosanoids. One possibility is that hemocytes produce all eicosanoids necessary to signal microaggregation and nodulation reactions, which would include the specific cell actions that amount to these visible reactions. Alternatively, it cannot be said with authority that other tissues do not contribute to eicosanoid-mediated immune signaling in whole insects. We set the hypothesis that challenge-stimulate hemocytic eicosanoid production was both necessary and sufficient to signal microaggregation reactions (Miller and Stanley, 2001).

For these experiments, tobacco hornworm hemolymph was collected by pericardial puncture and primary hemocyte cultures were prepared by diluting hemolymph with Grace's medium. Hemocytes were dispensed into microtiter plate wells, treated with a selected EBI and challenged with bacteria. At selected times post-challenge microaggregation was assessed by direct counting on a hemacytometer. We found that microaggregation reactions reached a maximum of about 3×10^4 microaggregates/ml at 30 min post-challenge and there was no increase after 30 min. The microaggregation reactions were potently inhibited in hemocyte preparations treated with Dex and several COX inhibitors but not in hemocyte preparations treated with esculetin, a specific LOX inhibitor. The inhibitory influence of Dex was ameliorated by AA, but not by palmitic acid, a saturated fatty acid that cannot be converted into eicosanoids. While intact bacteria were used in the experiments just described, we carried out a parallel line of work with lipopolysaccharide (LPS) prepared from *S. marcescens*. The results of these experiments show that specific bacterial components can evoke microaggregation reactions *in vitro* (Miller and Stanley, 2004). Overall, we inferred from these findings that isolated hemocyte preparations are competent to produce and secrete the eicosanoids required to mediate microaggregation reactions.

We conducted an additional experiment to determine whether hemocytes were able to produce and secrete the eicosanoids necessary to signal the microaggregation process. Reasoning that bacterial-challenged hemocytes would secrete eicosanoids, we prepared primary hemocyte cultures in 1 ml volumes. These were challenged with bacteria and after 1 h incubations, the hemocyte preparations were passed through 0.2 μm filters. The resulting filtrate was taken to be 'conditioned medium' (CM). We found that CM was sufficient to stimulate microaggregation in untreated hemocyte preparations and it also reversed the influence of Dex on microaggregation. The CM experiments indicate that the eicosanoids responsible for mediating microaggregation reactions can be recovered from the hemocyte primary culture medium.

Our findings with selective inhibitors of the COX and LOX pathways indicated that PGs, but not LOX products, mediated *in vitro* microaggregation processes. We probed this a little more thoroughly (Phelps et al., 2003). Again, we found that the COX inhibitors indomethacin and naproxen inhibited microaggregation reactions, but two LOX inhibitors, caffeic acid and esculetin, did not. We also found that the PG intermediate PGH_2 reversed the inhibitory influence of Dex, but the LOX product 5-hydroperoxyeicosatetraenoic acid (5-HpETE) did not. We infer that PGs are the predominant eicosanoids acting in the microaggregation phase of nodule formation. This would indicate that the roles of LOX products in nodulation are expressed sometime after formation of microaggregates.

3.5   EICOSANOIDS IN HEMOCYTE SPREADING AND MIGRATION

Microaggregation and nodulation reactions are large-scale, visible cellular defense reactions to microbial challenge. We take these large-scale reactions to be the culmination of an unknown number of small-scale, relatively invisible reactions. Plasmatocyte spreading on surfaces may be regarded as one of the component actions that comprise the overall nodulation process. Because of the important role in nodulation, Miller (2005) investigated the hypothesis that eicosanoids mediate cell spreading in primary hemocyte cultures prepared from tobacco hornworms.

Primary hemocyte cultures were prepared by pericardial puncture from hornworms that had been injected with EtOH (the drug vehicle for control larvae) or with a selected EBI diluted in EtOH. Hemocyte preparations were applied to glass cover slips and allowed to settle for 7 min. After washing and allowing hemocytes to settle for selected times, the hemocytes were fixed in formaldehyde. Digital images (Fig. 3.5) were analyzed using ImageJ software. Because plasmatocyte length, but not width, changed with incubation times the main focus was directed to length of the cells.

Plasmatocytes from control hornworms elongated to about 41 μm after 60 min incubations. The most rapid cell elongation took place during the first 30 min and length did not significantly increase in longer incubation periods. The elongation reactions were severely truncated in primary hemocyte cultures prepared from hornworms that had been treated with Dex and the Dex effect was expressed in a dose-related manner. The inhibitory influence of Dex was reversed by injecting AA into Dex-treated hornworms. Miller (2005) also found that inhibitors of COX and LOX pathways resulted in truncated elongation. Hence, the outcomes of these

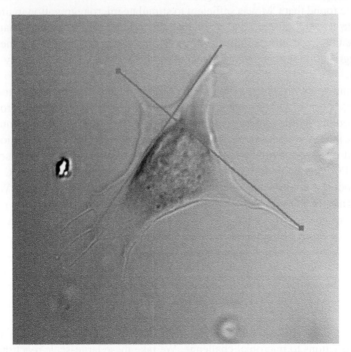

FIGURE 3.5 A photomicrograph of a plasmatocyte from an untreated tobacco hornworm, *Manduca sexta*, after spreading on a glass cover slip for 1 h. The red lines represent digital measurements of the cell dimensions. This photograph was taken through confocal optics at 400×. (Prepared and photographed by Jon S. Miller.) (See Plate 3.5 for color version of this figure.)

experiments allow the conclusion that eicosanoids mediate one of the smaller-scale steps in the overall nodulation process, namely plasmatocyte elongation.

Continuing in this line, Kwon et al. (2007) considered the influence of bacterial infection on plasmatocyte elongation. We found that hemocytes prepared from tobacco hornworms 15 and 60 min after infection were altered in size. Specifically, all hemocytes were smaller than 15 μm and none of the hemocytes exhibited cell spreading. On the idea that this change could result from an adventitious influence of *S. marcescens*, our standard bacterial challenge species, we performed these experiments with three additional bacterial species, *Escherichia coli*, *Bacillus subtilis* and *Micrococcus luteus*. The results were similar for all four species. In another set of experiments, we found that the influence of bacterial infection on cell spreading declined with time after infection. As seen before, the retarding influence of Dex was reversed by AA, PGH_2 and CM. For these experiments, we considered the possibility that CM prepared in the presence of bacteria could be contaminated with bacterial molecules that passed through the filter step. The appropriate control experiment was to prepare CM using only bacterial cells in the absence of hemocytes. We showed that CM prepared in the absence of hemocytes did not influence cell behavior. Overall, Kwon et al. (2007) demonstrated that bacterial infection exerts a strong effect on eicosanoid mediate cell spreading reactions to challenge.

As a final word in this section, we mention on emerging work on hemocyte chemotaxis. Chemotaxis is the ability to register and move toward or away from a chemical source, following a chemical gradient. This is a fundamental property of many cells, including bacteria and other single-celled organisms. In metazoans, chemotaxis is important in development and immunity, among other important areas. We used a modern form of a Boyden apparatus (Boyden, 1962) in which a lower chamber is charged with a chemotactic molecule and an upper chamber, separated from the lower by a porous membrane, is charged with cells. In our work, primary hemocyte cultures prepared from tobacco hornworms were placed in the upper chamber. We found that hemocytes are able to migrate toward the source of a bacterial peptide and to demonstrate that the chemotactic response depended on eicosanoids. A full report on this work is forthcoming.

3.6 A FRONTIER: EICOSANOIDS IN INSECT–VIRAL INTERACTIONS

Understanding insect–viral interactions is one of the frontiers of insect immunology. Insects raise a growing number of defense reactions to viral infection, although they are probably not expressed to the same extent in all species. Among the pioneering work in this area, Clem (2005) recently reviewed the roles of apoptosis in insect defense reactions to viral challenge. As most viral infections occur *per os*, infected midgut cells undergo apoptosis and are sloughed from the alimentary canal. Aside from sloughing the midgut cells, many viruses are not able to replicate in an apoptotic environment, which can slow or halt the progress of viral infection. Washburn et al. (1996) suggested insects also protect themselves from viral infection by hemocytic encapsulation of virus-infected cells. More recently, Ponnuvel et al. (2003) identified a lipase gene expressed only in silkworm, *Bombyx mori*, midgut. The purified lipase expressed anti-viral activity, from which the authors speculated the enzyme may act as a deterrent to viral infection. In a different line of work, Popham et al. (2004) suggested a novel anti-viral mechanism for larval lepidopterans. In their view, constitutive plasma phenoloxidase (PO) may act as an anti-viral defense, possibly killing viruses via reactive oxygen intermediates formed during the enzyme activity. Beyond these, the idea that viruses can be killed by anti-viral RNA silencing is drawing a great deal of attention (Saumet and Lecellier, 2006). Our understanding of insect anti-viral immune mechanisms is still growing.

Büyükgüzel et al. (2007) considered the possibility that eicosanoids act in insect cell–virus interactions. In one line of work they posed the hypothesis that eicosanoids mediate insect nodulation reactions to the vertebrate virus bovine herpes simplex virus-1 (BHSV-1). The authors began with larvae of the wax moth, *Galleria mellonella* (Büyükgüzel et al., 2007), first recording nodulation reactions to viral injections. Nodules reached a high of nearly 200 nodules/larva at 4 h after injecting 2×10^5 PFU/larva. The nodulation reactions were expressed in a challenge-dose manner. Treating experimental larvae by injection of the COX inhibitor, indomethacin, before injecting the viral challenge severely impaired nodulation reactions. The authors inferred from these data that products of the COX pathway, PGs, mediate nodulation reactions to BHSV-1.

We also designed an experiment to determine whether orally administered EBIs would influence the nodulation response to viruses injected into experimental larvae. Larvae were reared on media amended with indomethacin at 0, 0.01, 0.1 and 1.0 g/100 g diet and 7th instars were challenged by injection with a standard BHSV-1 challenge. We recorded about 225 nodules/larva from insects reared in the absence of dietary indomethacin, which declined to 115/larva, 88/larva and 21/larva with increasing dietary indomethacin concentrations. We determined PO activity levels in hemolymph taken from experimental larvae and, again, PO activity levels declined with increasing dietary indomethacin concentrations (Büyükgüzel et al., 2007). These data support the idea that orally administered EBIs can probe eicosanoid-mediated events.

Virtually all research on insect parasitoids has focused on host–parasitoid interactions (this volume). To broaden appreciation of parasitoid biology, we investigated the possibility that BHSV-1 infections also stimulate nodulation reactions in larvae of the parasitoid, *Pimpla turioinellae* (Durmus et al., 2007). In this work we found that parasitoid larvae produced nodules followed injections of BHSV-1, and the numbers of nodules increased with increasing challenge dosages to a high of about 50–55 nodules/larva. The highest numbers of nodules obtained at about 1 h post-injection. The nodulation reactions were inhibited in larvae that had been treated with inhibitors of PLA_2 as well as the COX and LOX pathways. The influence of the EBI treatments was expressed in dose-related ways and the effect of the PLA_2 inhibitor, Dex, was reversed in larvae injected with AA. These findings support the idea that eicosanoids act in insect–virus interactions and draw attention to the immune capacity of parasitoids.

Aside from the findings with nodulation, eicosanoids may act in other aspects of insect anti-viral immunity, as well, because recent work indicates that oral administration of EBIs increased susceptibility of gypsy moth larvae to *Lymantria dispar* nucleopolyhedrovirus (LdMNPV) (Stanley and Shapiro, 2007). For these experiments, gypsy moth larvae were reared on standard culture medium that had been surface treated with a mixture of LdMNPV and a selected EBI. For most of these test chemicals, inhibitors of the COX and LOX pathways and of PLA_2 substantially increased larval susceptibility to LdMNPV. Our on-going work indicates the EBI effect holds for other lepidopteran larvae–NPV combinations and that the effect is expressed in a dose-dependent manner.

Overall, these recent data strongly bolster the idea that, as seen with other microbial challengers, eicosanoids also act in insect defense reactions to viral infection.

3.7 ANOTHER FRONTIER: PROSTAGLANDIN MODES OF ACTION

The idea that eicosanoids mediate cellular (and some humoral) defense reactions to immune challenge raises the important issue of how PGs and other eicosanoids exert their influence on insect hemocytes. Drawing on the biomedical model, PGs and at least some LOX products interact with specific cell surface receptors, particularly G protein coupled receptors of the rhodopsin family (GPCRs; Breyer et al., 2001; Sugimoto and Narumiya, 2007). Figure 3.2 presents a model of how PGs can influence cell actions. The genes encoding GPCRs amount to relatively large proportions of animal genomes and these receptors are sorted into a variety of families (Vanden Broeck, 2001). These receptors selectively couple with various intracellular G proteins, each of which regulates the activities of specific cellular effector proteins, including adenylyl cyclase and others (Blenau and Baumann, 2001). Some mammalian PG receptors influence events related to homeostasis, such as ion transport physiology and others influence gene expression in target cells.

The detailed situation is still more complex. Whereas most PG receptors are thought to be located on cell surfaces, nuclear PG receptors also have been reported (Bhattacharya et al., 1999). Cell surface PG receptors also exist is various subtypes. PGs D, E, F, I and thromboxane A have counterpart receptors, abbreviated as DP, EP, FP, IP and TP receptors. The EP receptors occur in subtypes, EP1, EP2, EP3 and EP4 and each of these interacts with different G proteins. EP3 receptors are further subdivided into variants formed by splice variations (Hatae et al., 2002; Sugimoto and Narumiya, 2007). All these variations in PG receptors allow various PGs to exert multiple and even contradictory influences on cells. For example, EP3 variant EP3A can stimulate increased intracellular cAMP concentrations, while EP3 variant EP3B lowers intracellular cAMP concentrations. Although there is a great deal of information on PG receptors in the biomedical literature, there is only scant knowledge on invertebrates.

In the most detailed work so far, Qian et al. (1997) described a PGE_2 receptor from salivary glands of the tick, *Amblyomma americanum*. PGE_2 seems to mediate two physiological processes in tick salivary glands. PGE_2 acts in salivary secretions and acts in receptor-mediated protein exocytosis. It is appropriate to speculate that PGs exert their actions in insect and other invertebrates via receptors as seen in tick salivary glands.

PGs can influence gene expression in many mammalian systems (Sugimoto and Narumiya, 2007) and our most recent work indicates that PGE influences gene expression in established insect cell lines. In a paper now in preparation, we will report that treating a cell line established from *Helicoverpa zea* pupal ovaries (HzAM1 cells) with PGE and PGA altered expression of over 20 proteins, particularly heat shock proteins (Stanley, unpublished observations).

CONCLUSIONS

Eicosanoids play key roles in mediating and coordinating insect cellular immune reactions to infection and invasion. Pharmaceutical inhibitors of eicosanoid biosynthesis severely impair bacterial clearance, microaggregation and nodulation reactions to challenge. Within these large-scale reactions, eicosanoids act in cell spreading and chemotaxis, as well as other, now unknown cell actions. New information on possible roles of eicosanoids in insect–virus interactions and research into the mechanisms of eicosanoid action are among the visible frontiers of our knowledge of insect immunology.

ACKNOWLEDGMENTS

Many thanks to Nancy Beckage for an invitation to contribute to this volume. This chapter reports the results of research only and mention of a proprietary product does not constitute an endorsement or recommendation for its use by the USDA.

REFERENCES

Bedick, J. C., Tunaz, H., Nor Aliza, A. R., Putnam, S. M., Ellis, M. D., and Stanley, D. W. (2001). Eicosanoids act in nodulation reactions to bacterial infections in newly emerged adult honey bees, *Apis mellifera*, but not in older foragers. *Comp. Biochem. Physiol. Part C* **130**, 107–117.

Bergstrom, S., Ryhage, R., Samuelsson, B., and Sjovall, J. (1962). The structure of prostaglandin E, F_1 and F_2. *Acta Chemica Scandinavica* **16**, 501–502.

Bhattacharya, M., Varma, D. R., and Chemtob, S. (1999). Nuclear prostaglandin receptors. *Gene Ther. Mol. Biol.* **4**, 323–338.

Blenau, W., and Baumann, A. (2001). Molecular and pharmacological properties of insect biogenic amine receptors: Lessons from *Drosophila melanogaster* and *Apis mellifera*. *Arch. Insect Biochem. Physiol.* **48**, 13–38.

Boyden, S. (1962). The chemotactic effect of mixtures of antibody and antigen on polymorphonuclear leucocytes. *J. Exp. Med.* **115**, 453–466.

Breyer, R. M., Bagdassarian, C. K., Myers, S. A., and Breyer, M. D. (2001). Prostanoid receptors: Subtypes and signaling. *Annu. Rev. Pharmacol. Toxicol.* **41**, 661–690.

Büyükgüzel, K., Tunaz, H., Putnam, S. M., and Stanley, D. W. (2002). Prostaglandin biosynthesis by midgut tissue isolated from the tobacco hornworm, *Manduca sexta*. *Insect Biochem. Mol. Biol.* **32**, 435–443.

Büyükgüzel, E, Tunaz, H., Stanley, D., and Büyükgüzel, K. (2007). Eicosanoids mediate *Galleria mellonella* cellular immune response to viral infection. *J. Insect Physiol.* **53**, 99–105.

Carton, Y., Frey, F., Stanley, D. W., Vass, E., and Nappi, A. J. (2002). Dexamethasone inhibition of the cellular immune response of *Drosophila melanogaster* against a parasitoid. *J. Parasitol.* **288**, 405–407.

Clem, R. J. (2005). The role of apoptosis in defense against baculovirus infection in insects. *Curr. Top. Microbiol.* **289**, 113–130.

Dean, P., Gadsden, J. C., Richards, E. H., Edwards, J. P., Charnley, A. K., and Reynolds, S. E. (2002). Modulation by eicosanoids of immune responses by the insect *Manduca sexta* to the pathogenic fungus *Mearhizium anisopliae*. *J. Invertebr. Pathol.* **79**, 93–101.

de la Cruz Hernandez-Hernandez, F., Garcia-Gil de Munoz, F., Rojas-Martinez, A., Hernandez-Martinez, S., and Lanz-Mandoza, H. (2003). Carminic acid dye from the homopteran *Dactylopius coccus* hemolymph is consumed during treatment with different microbial elicitors. *Arch. Insect Biochem. Physiol.* **4**, 37–45.

Dunn, P. E., and Drake, D. R. (1983). Fate of bacteria injected in naive and immunized larvae of the tobacco hornworm *Manduca sexta*. *J. Invertebr. Pathol.* **41**, 77–85.

Durmus, Y., Büyükgüzel, E., Terzi, B., Tunaz, H., Stanley, D., and Büyükgüzel, K. (2007). Eicosanoids mediate melantoic nodulation reactions to viral infection in larvae of the parasitic wasp, *Pimpla turioinellae*. *J. Insect Physiol.* (in press) (March 2007).

Franssens, V., Simonet, G., Bronckaers, A., Claeys, I., De Loof, A., and Vanden Broeck, J. (2005). Eicosanoids mediate the laminarin induced nodulation response in larvae of the flesh fly, *Neobellieria bullata*. *Arch. Insect Biochem. Physiol.* **59**, 32–41.

Gadelhak, G. G., and Stanley-Samuelson, D. W. (1994). Incorporation of polyunsaturated fatty acids into phospholipids of hemocytes from the tobacco hornworm, *Manduca sexta*. *Insect Biochem. Mol. Biol.* **24**, 775–785.

Gadelhak, G. G., Pedibhotla, V. K., and Stanley-Samuelson, D. W. (1995). Eicosanoid biosynthesis by hemocytes from the tobacco hornworm, *Manduca sexta*. *Insect Biochem. Mol. Biol.* **25**, 743–749.

Garcia, E. S., Machado, E. M. M., and Azambuja, P. (2004). Effects of eicosanoid biosynthesis inhibitors on the prophenoloxidase-activating system and microaggregation reactions in the hemolymph of *Rhodnius prolixus* infected with *Trypanosoma rangeli*. *J. Insect Physiol.* **50**, 157–165.

Goldsworthy, G., Mullen, L., Opoku-Ware, K., and Chandrakant, S. (2003). Interactions between the endocrine and immune systems in locusts. *Physiol. Entomol.* **28**, 54–61.

Hatae, N., Sugimoto, Y., and Ichikawa, A. (2002). Prostaglandin receptors: Advances in the study of EP3 receptor signaling. *J. Biochem. (Tokyo)* **131**, 781–784.

Howard, R. W., and Stanley, D. W. (1999). The tie that binds: Eicosanoids in invertebrate biology. *Ann. Entomol. Soc. Am. (Special Issue Honoring Carl Schaefer)* **92**, 880–890.

Jurenka, R. A., Miller, J. S., Pedibhotla, V. K., Rana, R. L., and Stanley-Samuelson, D. W. (1997). Eicosanoids mediate microaggregation and nodulation responses to bacterial infections in black cutworms, *Agrotis ipsilon*, and true armyworms, *Pseudaletia unipuncta*. *J. Insect Physiol.* **43**, 125–133.

Jurenka, R. A., Pedibhotla, V. K., and Stanley, D. W. (1999). Prostaglandin production in response to bacterial infection in true armyworm larvae. *Arch. Insect Biochem. Physiol.* **41**, 225–232.

Kwon, H. S., Stanley, D. W., and Miller, J. S. (2007). Bacterial challenge and eicosanoids act in plasmatocyte spreading. *Entomol. Exp. Appl.* (in press) (March 2007).

Lord, J. C., Anderson, S., and Stanley, D. W. (2002). Eicosanoids mediate *Manduca sexta* cellular response to the fungal pathogen *Beauveria bassiana*: A role for the lipoxygenase pathway. *Arch. Insect Biochem. Physiol.* **51**, 46–54.

Mandato, C. A., Diehl-Jones, W. L., Moore, S. J., and Downer, R. G. H. (1997). The effects of eicosanoid biosynthesis inhibitors on prophenoloxidase activation, phagocytosis and cell spreading in *Galleria mellonella. J. Insect Physiol.* **43**, 1–8.

Miller, J. S. (2005). Eicosanoids influence *in vitro* elongation of plasmatocytes from the tobacco horn-worm, *Manduca sexta. Arch. Insect Biochem. Physiol.* **59**, 42–51.

Miller, J. S., and Stanley, D. W. (1998). The nodule formation reaction to bacterial infection: Assessing the role of eicosanoids. In *Techniques in Insect Immunology* (A. Wiesner, Ed.), pp. 265–270. SOS Publications, Fair Haven.

Miller, J. S., and Stanley, D. W. (2001). Eicosanoids mediate microaggregation reactions to bacterial challenge in isolated insect hemocyte preparations. *J. Insect Physiol.* **47**, 61–69.

Miller, J. S., and Stanley, D. W. (2004). Lipopolysaccharide evokes microaggregation reactions in hemocytes isolated from tobacco hornworms, *Manduca sexta. Comp. Biochem. Physiol. Part A* **137**, 285–295.

Miller, J. S., Nguyen, T., and Stanley-Samuelson, D. W. (1994). Eicosanoids mediate insect nodulation responses to bacterial infections. *Proc. Natl. Acad. Sci. USA* **91**, 12418–12422.

Miller, J. S., Howard, R. W., Nguyen, T., Nguyen, A., Rosario, R. M. T., and Stanley-Samuelson, D. W. (1996). Eicosanoids mediate nodulation responses to bacterial infections in larvae of the tenebrionid beetle *Zophobas atratus. J. Insect Physiol.* **42**, 3–12.

Miller, J. S., Howard, R. W., Rana, R. L., Tunaz, H., and Stanley, D. W . (1999). Eicosanoids mediate nodulation reactions to bacterial infections in adults of the cricket, *Gryllus assimulis. J. Insect Physiol.* **45**, 75–83.

Ogg, C. L., and Stanley-Samuelson, D. W. (1992). Phospholipid and triacylglycerol fatty acid compositions of the major life stages and selected tissues of the tobacco hornworm *Manduca sexta. Comp. Biochem. Physiol.* **101B**, 345–351.

Ogg, C. L., Howard, R. W., and Stanley-Samuelson, D. W. (1991). Fatty acid composition and incorporation of arachidonic acid into phospholipids of hemocytes from the tobacco hornworm *Manduca sexta. Insect Biochem.* **21**, 809–814.

Park, Y., and Kim, Y. (2000). Eicosanoids rescue *Spodoptera exigua* infected with *Xenorhabdus nematophilus*, the symbiotic bacteria to the entomopathogenic nematode *Steinernema carpocapsae. J. Insect Physiol.* **46**, 1469–1476.

Park, Y., and Stanley, D. (2006). The entomopathogenic bacterium, *Xenorhabdus nematophila*, impairs insect immunity by inhibition of eicosanoid biosynthesis in adult crickets, *Gryllus firmus. Biol. Control* **38**, 247–253.

Park, Y., Nor Aliza, A. R., and Stanley, D. W. (2005). A secretory PLA_2 associated with tobacco horn-worm hemocytes membrane preparations acts in cellular immune reactions. *Arch. Insect Biochem. Physiol.* **60**, 105–115.

Phelps, P. K., Miller, J. S., and Stanley, D. W. (2003). Prostaglandins, not lipoxygenase products, mediate insect microaggregation reactions to bacterial challenge in isolated hemocyte preparations. *Comp. Biochem. Physiol.* **136**, 409–416.

Ponnuvel, K. M., Nakazawa, H., Furukawa, S., Asaoka, A., Ishibashi, J., Tanaka, H., and Yamakawa, M. (2003). A lipase isolated from the silkworm *Bombyx mori* shows antiviral activity against nucle-opolyhedrovirus. *J. Virol.* **77**, 10725–10729.

Popham, H. J. R., Shelby, K. S., Brandt, S. L., and Coudron, T. A. (2004). Potent virucidal activity in larval *Heliothis virescens* plasma against *Helicoverpa zea* single capsid nucleopolyhedrovirus. *Virol.* **85**, 2255–2261.

Qian, Y., Essenberg, R. C., Dillwith, J. W., Bowman, A. S., and Sauer, J. R. (1997). A specific prostaglandin E_2 receptor and its role in modulating salivary secretion in the female tick, *Amblyomma americanum* (L.). *Insect Biochem. and Molec. Biol.* **27**, 387–395.

Rowley, A. F., Vogan, C. L., Taylor, G. W., and Clare, A. S. (2005). Prostaglandins in non-insectan invertebrates: Recent insights and unsolved problems. *J. Exp. Biol.* **208**, 3–14.

Saumet, A., and Lecellier, C.-H. (2006). Anti-viral RNA silencing: Do we look like plants? *Retrovirol.* **3**, 3. Available at http://www.retrovirology.com/content/3/1/3

Schaloske, R. H., and Dennis, E. A. (2006). The phospholiopase A_2, superfamily and its group numbering system. *Biochem Biophys Acta* **1761**, 1246–1259.

Schleusener, D. R., and Stanley-Samuelson, D. W. (1996). Phospholipase A_2 in hemocytes of the tobacco hornworm, *Manduca sexta. Arch. Insect Biochem. Physiol.* **33**, 63–74.

Stanley, D. W. (2000). *Eicosanoids in Invertebrate Signal Transduction Systems.* Princeton University Press, Princeton.

Stanley, D. W. (2005). Eicosanoids. In *Comprehensive Molecular Insect Science* (L. I. Gilbert, K. Iatrou, and S. Gill, Eds.), Vol. 4, pp. 307–339. Elsevier, Oxford.

Stanley, D. W. (2006a). Prostaglandins and other eicosanoids in insects: Biological significance. *Annu. Rev. Entomol.* **51**, 25–44.

Stanley, D. W. (2006b). The non-venom insect phospholipases A_2. *Biochem. Biophys. Acta* **1761**, 1383–1390.

Stanley, D. W., and Howard, R. W. (1998). The biology of prostaglandins and related eicosanoids in invertebrates: Cellular, organismal and ecological actions. *Am. Zool.* **38**, 369–381.

Stanley, D. W., and Miller, J. S. (2006). Eicosanoid actions in insect cellular immune functions. *Entomol. Exp. Appl.* **119**, 1–13.

Stanley, D., and Shapiro, M. (2007). Eicosanoid biosynthesis inhibitors increase the susceptibility of *Lymantria dispar* to nucleopolyhedrovirus LdMNPV. *J. Invertebr. Pathol.* **95**, 119–124.

Stanley, D. W., Hoback, W. W., Bedick, J. C., Tunaz, H., Rana, R. L., Nor Aliza, A. R., and Miller, J. S. (1999). Eicosanoids mediate nodulation reactions to bacterial infections in larvae of the butterfly, *Colias eurytheme. Comp. Biochem. Physiol.* **41**, 225–232.

Stanley-Samuelson, D. W. (1991). Comparative eicosanoid physiology in invertebrate animals. *Am. J. Physiol.* **260** (*Reg. Integ. Comp. Physiol.* **29**), R849–R853.

Stanley-Samuelson, D. W. (1994a). Assessing the physiological significance of prostaglandins and other eicosanoids in insect physiology. *J. Insect Physiol.* **40**, 3–11.

Stanley-Samuelson, D. W. (1994b). Prostaglandins and related eicosanoids in insects. *Adv. Insect Physiol.* **24**, 115–212.

Stanley-Samuelson, D. W. (1994c). The biological significance of prostaglandins and other eicosanoids in invertebrates. *Am. Zool.* **34**, 589–598.

Stanley-Samuelson, D. W., and Dadd, R. H. (1983). Long-chain polyunsaturated fatty acids: Patterns of occurrence in insects. *Insect Biochem.* **13**, 549–558.

Stanley-Samuelson, D. W., and Loher, W. (1986). Prostaglandins in insect reproduction. *Ann. Entomol. Soc. Am.* **79**, 841–853.

Stanley-Samuelson, D. W., and Pedibhotla, V. K. (1996). What can we learn from prostaglandins and other eicosanoids in insects? *Insect Biochem. Mol. Biol.* **26**, 223–234.

Stanley-Samuelson, D. W., Jensen, E., Nickerson, K. W., Tiebel, K., Ogg, C. L., and Howard, R. W. (1991). Insect immune response to bacterial infection is mediated by eicosanoids. *Proc. Natl. Acad. Sci. USA* **88**, 1064–1068.

Stanley-Samuelson, D. W., Jurenka, R. A., Blomquist, G. J., and Loher, W. (1987). Sexual transfer of prostaglandin precursor in the field cricket, *Teleogryllus commodus. Physiol. Entomol.* **12**, 347–354.

Stanley-Samuelson, D. W., and Ogg, C. L. (1994). Prostaglandin biosynthesis by fat body from the tobacco hornworm, *Manduca sexta. Insect Biochem. Mol. Biol.* **24**, 481–491.

Stanley-Samuelson, D. W., Pedibhotla, V. K., Rana, R. L., Nor Aliza, A. R., Hoback, W. W., and Miller, J. S. (1997). Eicosanoids mediate nodulation responses to bacterial infections in larvae of the silkmoth, *Bombyx mori. Comp. Biochem. Physiol.* **118A**, 93–100.

Sugimoto, Y., and Narumiya, S. (2007). Prostaglandin E receptors. *J. Biol. Chem.* **282**, 11613–11617.

Tunaz, H., and Stanley, D. W. (2000). Eicosanoids mediate nodulation reactions to bacterial infections in adults of the American cockroach, *Periplaneta americana* (L.). *Proc. Entomol. Soc. Ont.* **130**, 97–108.

Tunaz, H., Bedick, J. C., Miller, J. S., Hoback, W. W., Rana, R. L., and Stanley, D. W. (1999). Eicosanoids mediate nodulation reactions to bacterial infections in adults of two 17-year periodical cicadas, *Magicicada septendecim* and *M. cassini. J. Insect Physiol.* **45**, 923–931.

Tunaz, T., Putnam, S. M., and Stanley, D. W. (2002). Prostaglandin biosynthesis by fat body from larvae of the beetle *Zophobas atratus*. *Arch. Insect Biochem. Physiol.* **49**, 80–93.

Tunaz, H., Isikber, A. A., and Er, M. K. (2003a). The role of eicosanoids on nodulation reactions to bacterium *Serratia marcescens* in larvae of *Ostrinia nubilalis*. *Turk. J. Agric. Forest.* **27**, 269–275.

Tunaz, H., Park, Y., Büyükgüzel, K., Bedick, J. C., Nor Aliza, A. R., and Stanley, D. W. (2003b). Eicosanoids in insect immunity: Bacterial infection stimulates hemocytic phospholipase A_2 activity in tobacco hornworms. *Arch. Insect Biochem. Physiol.* **52**, 1–6.

Uscian, J. M., and Stanley-Samuelson, D. W. (1993). Phospholipase A_2 activity in the fat body of the tobacco hornworm *Manduca sexta*. *Arch. Insect Biochem. Physiol.* **24**, 187–201.

Vanden Broeck, J. (2001). Insect G protein-coupled receptors and signal transduction. *Arch. Insect Biochem. Physiol.* **48**, 1–12.

von Euler, U. S. (1936). On the specific vasodilating and plain muscle stimulating substances from accessory gential glands in men and certain animals (prostaglandin and vestiglandin). *J. Physiol.* (London) **88**, 213–234.

Washburn, J. O., Kirkpatrick, B. A., and Volkman, L. E. (1996). Insect protection against viruses. *Nature* **383**, 767–768.

4

PHENOLOXIDASES IN INSECT IMMUNITY

MICHAEL R. KANOST AND MAUREEN J. GORMAN

Department of Biochemistry, Kansas State University, Manhattan, KS 66506, USA

ABSTRACT: Phenoloxidases are present as zymogens in insect hemolymph, and they become activated upon wounding or infection as part of the innate immune response. These enzymes are similar to mammalian tyrosinases in their ability to use reactive sites containing copper atoms to catalyze two types of reactions that require molecular oxygen as a substrate. They can hydroxylate tyrosine to form dihydroxyphenylalanine, and they oxidize *o*-diphenols to form quinones. The quinones undergo additional reactions leading to synthesis of melanin, which is deposited on the surface of encapsulated parasites, hemocyte nodules, and wound sites. The melanin itself and reactive chemical species produced during melanin synthesis appear to help kill invading pathogens and parasites. Rather than having sequence similarity to mammalian tyrosinases, insect prophenoloxidases (proPOs) are homologous with arthropod hemocyanins and insect hexamerin storage proteins. ProPOs are synthesized primarily by hemocytes and released into plasma by cell lysis. They are activated by proteolytic cleavage at a specific site near their amino-terminus through the action of hemolymph serine proteases containing amino-terminal clip domains. These proPO activating proteases are themselves activated as part of a protease cascade stimulated by recognition of microbial infection. At least in some cases, activation of proPO also requires the participation of serine protease homolog (SPH) cofactors, which lack proteolytic activity. The proPO activation cascade is regulated by plasma serine protease inhibitors, including members of the serpin superfamily, and active phenoloxidase (PO) may be directly inhibited by proteinaceous factors. Such regulation is essential because the products of PO activity are potentially toxic to the host. Further research is required to gain a more detailed understanding at a molecular level of the assembly of plasma protein complexes leading to proPO activation and melanin deposition on foreign surfaces and the regulation of these processes.

Abbreviations:

βGRP	=	beta-1,3-glucan recognition protein
CPC	=	cetyl pyridinium chloride
DOPA	=	3,4-dihydroxyphenylalanine
DOPAC	=	3,4-dihydroxyphenylacetic acid
HP	=	hemolymph proteinase
NADA	=	*N*-acetyldopamine
NBAD	=	*N*-β-alanyldopamine
NBANE	=	*N*-acetylnonepinephrine
PAP	=	prophenoloxidase activating proteinase
PGRP	=	peptidoglycan recognition protein
PO	=	phenoloxidase
ppA	=	prophenoloxidase activating enzyme
PPAE	=	prophenoloxidase activating enzyme
PPAF	=	prophenoloxidase activating factor
SPH	=	serine protease homolog
SRPN2	=	serpin-2.

4.1   INTRODUCTION

Individuals studying insects have long known that hemolymph exposed to air gradually darkens and eventually becomes quite black. This phenomenon was suspected as long ago as 1898 to be due to an oxidative enzyme with tyrosinase-like phenoloxidase (PO) activity (Biedermann, 1898). Early efforts at biochemical characterization of the enzyme responsible for this activity are described in reviews by Sussman (1949) and Wyatt (1961). We know now that insects contain two types of phenoloxidases. Laccase-type enzymes (EC 1.10.3.2) oxidize *o*- or *p*-diphenols to quinones, which function in sclerotization and tanning of cuticle (Arakane et al., 2005; Dittmer et al., 2004). Enzymes in hemolymph that have tyrosinase-like activity can hydroxylate tyrosine (EC 1.14.18.1) and also oxidize o-diphenols to quinones (EC 1.10.3.1) (Gorman et al., 2007a). The latter type of enzyme, which we refer to as PO, and its regulation will be the topic of this chapter. The quinones produced by PO undergo a series of additional enzymatic and non-enzymatic reactions leading to polymerization and melanin synthesis.

Arthropod POs are synthesized as zymogens (prophenoloxidase, proPO), which are activated by proteolytic cleavage at a specific site as a response to infection or wounding. Active PO catalyzes the formation of quinones, which undergo further reactions to form melanin (Cerenius and Söderhäll, 2004; Nappi and Christensen, 2005). The earliest studies of insect phenoloxidases were focused on their presumed role in cuticle tanning. Later, the observation of melanin-like capsules formed during many insect immune responses led Taylor (1969) to propose that phenoloxidases may function in invertebrate immunity, an idea soon supported by experimental studies of insect host immune responses to parasitoids (Nappi, 1974) and demonstration that proPO is activated upon exposure of hemolymph to microorganisms (Pye, 1974). In the intervening years, the function of proPO in arthropod immunity and regulation of its activating system have been discussed in numerous reviews, including the following in the last 10 years (Ashida and Brey, 1997; Cerenius and Söderhäll, 2004; Christensen et al., 2005; Kanost et al., 2004; Lemaitre and Hoffmann, 2007; Nappi and Christensen, 2005; Nappi and Vass, 2001; Söderhäll and Cerenius, 1998; Sugumaran, 2002).

Synthesis of melanin occurs in response to many types of infections by parasites and pathogens, and in most cases, this defense appears to help protect the host insect. Melanin is deposited within nodules, composed of aggregated hemocytes and microorganisms, that form in the hemocoel of heavily infected insects (Fig. 4.1A) (Koizumi et al., 1999; Ratcliffe and Gagen, 1977; Ratcliffe et al., 1991; Stanley et al., 1998). Deposition of a melanin coat is also frequently observed in the encapsulation response of insects to eukaryotic parasites or in experimental injections of beads or other foreign objects that provoke encapsulation (Fig. 4.1B and C). Such encapsulation usually involves adherence of multiple layers of hemocytes on the surface of the invader (see chapter 2 by M. R. Strand), but in some cases, the capsule is formed of only acellular material including plasma proteins, molecules released from hemocytes, and melanin (Gillespie et al., 1997; Lavine and Strand, 2002).

The melanization of encapsulated parasites is thought to be an important defensive response in insects, including insect vectors of human diseases (Christensen et al., 2005; Gillespie et al., 1997; Michel and Kafatos, 2005; Nappi and Christensen, 2005). The melanin capsule can block absorption of nutrients by parasites and thus may contribute to their killing by starvation (Chen and Chen, 1995). In addition, cytotoxic reactive oxygen and nitrogen intermediates formed during melanin synthesis may help to kill invading organisms (Nappi and Christensen, 2005). Treatments that block melanin synthesis can result in increased susceptibility of insects to parasites (Liu et al., 1997; Shiao et al., 2001;

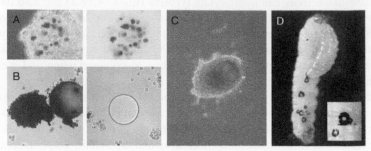

FIGURE 4.1   Melanin formation in response to infection or wounding: (A) A hemocyte nodule dissected from a *M. sexta* larva 1 day after injection of *Micrococcus luteus*. The left image was obtained with phase contrast optics and highlights the presence of multiple layers of hemocytes. The right image was obtained with bright field conditions and shows the presence of spots of melanin within the nodule. (B) Encapsulation and melanization of two nickel–agarose chromatography beads coated with a 6-His tagged domain from *M. sexta* immulectin-2 and incubated with hemolymph *in vitro* (left panel). Control beads coated with a different recombinant protein (*M. sexta* cuticular protein CP36) were not encapsulated or melanized (right panel). (C) Melanization and encapsulation *in vivo* of a *Cotesia congregata* egg in an unsuitable host, the sphingid lepidopteran *Pachysphinx occidentalis*. (D) Melanized wounds on a *Plodia interpunctella* larva injured by the feeding of an ectoparasitoid larva, *Habrobracon hebetor* (Baker and Fabrick, 2002). The insert shows such melanized regions at higher magnification. (B is from Xiao-Qiang Yu, University of Missouri, Kansas City, MO; C is from Nancy Beckage, University of California, Riverside, CA; D is from Jeffrey Fabrick, USDA-ARS Arid-Land Agricultural Research Center, Maricopa, AZ.) (See Plate 4.1 for color version of this figure.)

Tamang et al., 2004), and successful parasites and parasitoids disrupt the host melanization response (Beckage, 1998; Brivio et al., 2002; Shelby et al., 2000; Stoltz and Cook, 1983). Insect strains deficient in the melanization response typically have impaired ability to defend against infection (Nappi et al., 1991, 1992, 2005; Paskewitz and Riehle, 1994; Paskewitz et al., 1989), although a *Drosophila melanogaster* mutant deficient in proPO activation was able to survive microbial infections (Leclerc et al., 2006). ProPO activation and melanization also appear to be a component of a defensive response of lepidopteran larvae to viral infection (Ourth and Renis, 1993; Shelby and Popham, 2006; Trudeau et al., 2001; see chapter 9 by W. O. Sparks, L. C. Bartholomay, and B. C. Bonning).

Melanized regions often occur in the integument at the sites of wounds, even in apparent absence of pathogens (Fig. 4.1D). This is likely due to activation of proPO that has moved from hemolymph into the cuticle (Asano and Ashida, 2001). The action of PO at such sites may help to plug and heal wounds (Lai et al., 2002). PO and putative components of its activating system have been identified within hemolymph clots (Karlsson et al., 2004; Li et al., 2002), suggesting that there may be some interaction between PO activation and the coagulation system, although the latter is not yet well understood.

In recent years, understanding at a molecular level of the PO activating system and its regulation has developed rapidly, and many of the components have now been characterized through molecular cloning. The field owes much to the pioneering biochemical studies of Masaaki Ashida, whose laboratory first characterized many of the proteins involved in proPO activation, through purification and careful analysis of these molecules from the silkworm, *Bombyx mori* (e.g. Ashida and Dohke, 1980; Ashida et al., 1974, 1983; Ochiai and Ashida, 1988; Ohnishi et al., 1970; Yoshida et al., 1986, 1996). Many of the ideas in this chapter depend heavily on a fundamental base of knowledge first developed through work with the silkworm proPO system. We will describe the current state of knowledge about the activation and function of insect PO, a complex component of the innate immune system.

4.2 PROPO SEQUENCES, TISSUE SOURCE, AND LOCATION

ProPOs are polypeptides of approximately 80 kDa and contain 0.14–0.22% copper, indicating two atoms of copper per protein molecule (Ashida and Brey, 1997; Aspán et al., 1995). Amino acid sequences for proPO (deduced from cloned cDNAs) were first reported in 1995 from four arthropod species, the crayfish *Pacifastacus leniusculus*, and three insect species, *Manduca sexta*, *B. mori*, and *Drosophila melanogaster* (Aspán et al., 1995; Fujimoto et al., 1995; Hall et al., 1995; Kawabata et al., 1995). Since that time, cDNA or gene sequences for proPO have been determined for at least 26 insects and 7 crustacean species. The arthropod proPO sequences are homologous with arthropod hemocyanins, oxygen-transport proteins that use two copper atoms to form an oxygen-binding site, and with insect hexamerin storage proteins, which lack copper (Burmester, 2001, 2002; Decker and Terwilliger, 2000). The copper-binding motifs are particularly well conserved in the hemocyanin and PO sequences, and it is reasonable to conclude that a similar oxygen-binding site exists for transport by hemocyanins and for catalysis by phenoloxidases. Burmester suggests that the phenoloxidases existed first and that hemocyanins evolved from PO-like ancestral genes, with storage hexamerins diverging later from hemocyanins (Burmester, 2002). Thus, these three groups form a large protein family with different physiological functions but a common structural framework. In fact, PO activity has been detected in some arthropod hemocyanins (Adachi et al., 2003; Decker and Jaenicke, 2004; Decker and Rimke, 1998; Decker and Tuczek, 2000; Decker et al., 2001; Lee, S. Y. et al., 2004; Nagai et al., 2001; Pless et al., 2003; Zlateva et al., 1996).

Arthropod hemocyanins exist as hexamers (or multimers of hexamers), and the name hexamerin for the insect storage proteins is derived from their quarternary structure. Insect proPO may exist predominantly as dimers, although monomers as well as higher multimers can form, depending on ionic strength (Ashida and Brey, 1997; Jiang et al., 1997; Sezaki et al., 2001; Yasuhara et al., 1995). A crustacean proPO was reported to form native hexamers, similar to hemocyanin (Jaenicke and Decker, 2003).

Two or more proPO genes have been identified in most insect species that have been investigated, although the honeybee genome contains a single proPO gene

(Evans et al., 2006). For lepidopteran species, two proPO genes that fall into two distinct clades can be identified. *D. melanogaster* (Waterhouse et al., 2007) and the beetle *Tribolium castaneum* (Haobo Jiang, personal communication) each have three proPO genes. There has been an expansion of the proPO gene family in mosquitoes, with 9 proPO genes in *Anopheles gambiae* and 10 in *Aedes aegypti* (Waterhouse et al., 2007). It is not known how the different proPOs in a species differ enzymatically. Products of different proPO genes may associate to form heterodimers (Jiang et al., 1997), although it is unknown what effect that may have on function.

Most reports indicate that proPO is synthesized predominantly, if not exclusively, by hemocytes of insects and crustaceans (Cerenius and Söderhäll, 2004). In lepidopterans, oenocytoids are identified as the cell-type producing proPO (Iwama and Ashida, 1986; Jiang et al., 1997), whereas the crystal cells of *D. melanogaster* produce proPO (Rizki et al., 1985; Williams, 2007), suggesting that oenocytoids and crystal cells may be homologous cell types (see chapter 2 by M. R. Strand). In mosquitoes, proPO is synthesized in oenocytoids and granular hemocytes (Castillo et al., 2006; Hernandez et al., 1999; Hillyer and Christensen, 2002; Hillyer et al., 2003). Arthropod proPOs lack a secretion signal peptide, yet the protein occurs in plasma, presumably due to hemocyte lysis. A similar mechanism is thought to be true for release of arthropod hemocyanins into plasma (Voit and Feldmaier-Fuchs, 1990). Oenocytoids from different species vary in stability, some rupturing extremely easily and spontaneously upon injury (Ashida and Brey, 1997) and perhaps at a low rate even in the absence of injury or infection. In lepidopterans, the bulk of proPO appears in plasma (Ashida and Brey, 1997; Saul et al., 1987), but in hemimetabolous insects such as locusts and cockroaches, proPO appears to be mostly retained in hemocytes until microbial exposure stimulates its release (Brehélin et al., 1989; Durrant et al., 1993; Hoffmann et al., 1970; Leonard et al., 1985). Perhaps oenocytoid lysis occurs *in vivo* in the absence of injury or infection to maintain a steady state level of proPO in plasma, although this idea is difficult to test in insects that have fragile oenocytoids. Lysis of proPO-producing hemocytes at a site of wounding or infection could result in a local high concentration of proPO. A signal transduction pathway stimulating *D. melanogaster* crystal cell lysis has recently been identified (Bidla et al., 2007).

4.3　PROPO ZYMOGEN ACTIVATION

ProPO zymogens can be activated by specific proteolysis or by interaction with amphipathic molecules. Both types of activation must involve a conformational change in the protein, making the active site accessible to substrate or rearranging the active site to a structure required for catalyzing oxidation reactions. Activation of proPO by treating hemolymph with detergents, fatty acids, or alcohols has been observed (Sugumaran and Kanost, 1993). Purified proPO from *D. melanogaster* can be reversibly activated by 2-propanol (Asada, 1998; Asada et al., 1993). The cationic detergent cetyl pyridinium chloride (CPC) at low concentration is an efficient activator of purified proPO and a useful reagent for investigating PO activity (Chase et al., 2000; Hall et al., 1995). The binding and activation kinetics of CPC with *Sarcophaga bullata* proPO have recently been investigated (Xie et al., 2007). ProPO activation by naturally occurring amphipathic lipids also occurs. Lysolecithin was found to be a potent activator of a lobster proPO (Sugumaran and Nellaiappan, 1991), which suggests the interesting possibilities that phospholipids released from damaged cells might be a factor in activating proPO at wound sites and that phospholipase A2 may be involved in stimulating proPO activation (see chapter 3 by D. W. Stanley and J. S. Miller).

Cleavage of a conserved Arg–Phe bond about 50 residues from the amino-terminus results in activation of insect proPOs, whereas crayfish proPO is activated by cleavage at a more distant site, after Arg 176 (Cerenius and Söderhäll, 2004). ProPO from a beetle, *Holotrichia diomphalia*, undergoes a second cleavage after Arg 152 during its activation (Lee et al., 1998b). Early work on this subject was done by Ashida and coworkers who demonstrated clearly the activation of purified silkworm proPO by specific proteolysis (Ashida and Dohke, 1980; Ashida et al., 1974; Dohke, 1973). The first proPO activating proteases to be purified were extracted from cuticle (Aso et al., 1985; Dohke, 1973; Satoh et al., 1999), and perhaps these have a role in proPO activation upon wounding. Serine proteases that directly activate proPO have been purified from hemolymph in addition to cuticle, and corresponding cDNAs of proteases from both groups have been cloned, including those from *M. sexta* (Jiang et al., 1998, 2003), *B. mori* (Satoh et al., 1999), *H. diomphalia* (Lee et al., 1998a), and a crayfish (Wang et al., 2001). All of these enzymes contain a carboxyl-terminal serine protease catalytic domain linked by a 20–100 residue region to one or two amino-terminal clip domains. Clip domains are 35–55 amino acid residue sequences that contain three conserved disulfide bonds, and these domains may function to mediate interactions between members of protease cascade pathways (Jiang and Kanost, 2000). *M. sexta* proPO activating proteinase-1 (PAP-1) and the *H. diomphalia* and *P. leniusculus* proteases contain a single clip domain, while *M. sexta* PAP-2 and PAP-3 and *B. mori* proPO activating enzyme (PPAE) each have two clip domains. They are synthesized as zymogens and must be activated by another protease as part of a serine protease cascade.

B. mori PPAE is expressed in integument, hemocytes, and salivary glands but not in fat body (Satoh et al., 1999). Its putative *M. sexta* ortholog, PAP-1, is expressed in larval fat body, tracheae, and nerve tissue, and is upregulated in fat body and hemocytes after injection of bacteria (Zou et al., 2005). PAP-2 expression was detected in fat body and hemocytes only after larvae were injected with bacteria (Jiang et al., 2003a). PAP-3 mRNA is present at a low level in fat body and hemocytes of naïve larvae, and its expression is significantly upregulated in those tissues after injection of bacteria (Jiang et al., 2003b). At the prepupal stage it is highly expressed in integument, fat body, and hemocytes (Zou and Jiang, 2005).

The sequenced insect genomes contain large numbers of genes that encode a carboxyl-terminal protease domain and an amino-terminal clip domain (46 in *D. melanogaster*, 55 in *A. gambiae*, 71 in *A. aegypti*, 18 in *Apis mellifera*) (Evans et al., 2006; Ross et al., 2003; Waterhouse et al., 2007). Complete genomic DNA sequences for *D. melanogaster* and *A. gambiae* have facilitated the completion of genetic screens that identified about a dozen clip domain proteases that are somehow involved in proPO activation; however, their substrates and positions in the pathway are not yet known (Castillejo-Lopez and Hacker, 2005; Leclerc et al., 2006; Paskewitz et al., 2006; Tang et al., 2006; Volz et al., 2005, 2006). We have cloned cDNAs for more than 20 clip domain proteases expressed in the fat body or hemocytes of *M. sexta* (Jiang et al., 1999, 2005; Kanost et al., 2001) and have used these cDNAs as part of a candidate gene approach to identify members of the proPO activation pathways.

When we initially characterized a proPO-activating protease from *M. sexta*, the highly purified protease could not efficiently activate proPO, but required participation of a non-proteolytic protein fraction (Jiang et al., 1998). This protein cofactor was identified in *H. diomphalia* (Kwon et al., 2000) and *M. sexta* (Yu et al., 2003) as a protein with a clip domain and a serine protease domain, in which the active site serine residue is changed to glycine. These serine protease homologs (SPHs) lack protease activity due to the incomplete catalytic triad. There are 13 such clip domain SPH genes in the *D. melanogaster* genome, 12 in *A. aegypti*, 15 in *A. gambiae*, and 6 in *A. mellifera* (Ross et al., 2003; Waterhouse et al., 2007; Zou et al., 2006). The active form of the SPHs that function as cofactors for proPO activation is themselves activated through specific cleavage by a serine protease in hemolymph (Kim et al., 2002; Lee et al., 2002; Yu et al., 2003). *B. mori* PPAE and *P. leniusculus* ppA do not require SPH cofactors for activating proPO (Satoh et al., 1999; Wang et al., 2001).

The SPHs from *M. sexta* that stimulate proPO activation bind to a hemolymph lectin that is a recognition protein for bacterial lipopolysaccharide and to proPO and proPO-activating proteinase (Yu et al., 2003). The interaction between the lectin and a proPO activation complex may serve to localize melanin synthesis to the surface of invading bacteria. Activated *M. sexta* SPHs bind to proPO, and active PAPs, and promote proPO cleavage by the PAPs by mechanisms that are not yet clear (Gupta et al., 2005a; Wang and Jiang, 2004b). A three-dimensional structure of an SPH (prophenoloxidase activating factor-II, PPAF-II) from *H. diomphalia* has been determined by X-ray crystallography (Piao et al., 2005). This is the first available structure of a clip domain, and provides a suggestion that a cleft in the clip domain may bind to PO. The proteolytic activation of 45 kDa PPAF-II results in the formation of a large oligomer (~600 kDa) of the protein, as demonstrated by size exclusion chromatography. This oligomeric PPAF-II was visualized by electron microscopy as a putative structure made of two stacked hexameric rings (Piao et al., 2005). This is consistent with the observation that the active form of *M. sexta* SPH also forms large oligomers (~790 kDa) (Wang and Jiang, 2004b). It appears that these structures may function to bring the components of protease cascades close together and to localize protease complexes on the surface of microorganisms. The parasitoid wasp *Cotesia rubecula* produces an SPH as a venom protein, which disrupts proPO activation in its lepidopteran host, perhaps by interfering with the correct association of the proPO activation complex (Asgari, 2006; Zhang et al., 2004).

The initiation of a proPO activation cascade is thought to depend on self-activation of a protease zymogen, stimulated by an infection or wound. Proteins that bind to microbial surfaces (pattern recognition proteins) are expected to be involved in this step, through interaction with microbial polysaccharides and an initiating protease (Ashida et al., 1983; Yu et al., 2002). Plasma proteins that bind to bacterial lipopolysaccharide (C-type lectins), peptidoglycan (peptidoglycan recognition proteins, PGRP), or fungal glucans (glucan recognition proteins) can stimulate activation of the proPO cascade (Ashida and Brey, 1997; Chen et al., 1995, 1999; Fabrick et al., 2003; Kurata, 2004; Lee et al., 2004; Ma and Kanost, 2000; Ochiai and Ashida, 1988, 1999, 2000; Wang et al., 2004, 2005; Yoshida et al., 1986, 1996; Yu and Kanost, 2004; Yu et al., 2002; Zhang et al., 2003). In *M. sexta* and perhaps in many insect species, peptidoglycan and β-1,3-glucans are more potent stimulators of proPO activation than is lipopolysaccharide.

An initiating protease, hemolymph protease-14 (HP14), which triggers proPO activation in response to Gram-positive bacteria and β-1,3-glucans has been identified in *M. sexta* (Ji et al., 2004). In addition to a carboxy-terminal protease domain, HP14 contains five low density lipoprotein receptor class A repeats, a Sushi domain, and a unique Cys-rich region. It autoactivates in the presence of β-1,3-glucan (curdlan) and *M. sexta* glucan recognition protein (Wang and Jiang, 2006). Orthologs of HP14 in the *D. melanogaster* and *A. gambiae* genomes (Ji et al., 2004) are speculated to also function as cascade-initiating proteases. *M. sexta* HP14 activates a clip domain protease, proHP21, which cleaves and activates proPAP-2 and proPAP-3 (Gorman et al., 2007b; Wang and Jiang, 2007). However, HP21 does not activate proPAP-1 (Wang and Jiang, 2007). The *M. sexta* pathway including βGRP, proHP14, proHP21, proPAP-2 or proPAP-3, SPH-1 and SPH-2, and proPO constitutes nearly the complete cascade for one proPO activation pathway (the activation of the SPHs in *M. sexta* is not yet characterized) (Fig. 4.2). It is apparent that additional intersecting or perhaps even distinct pathways for proPO activation exist in *M. sexta*, suggesting an intriguing degree of complexity in this innate immune response.

4.4 ENZYMATIC ACTIVITY OF INSECT HEMOLYMPH PHENOLOXIDASES

For many years it was supposed that insect POs were orthologous to mammalian tyrosinases. Thus, many models of PO activity are based on our understanding of tyrosinase activity. Later it was discovered that insect POs lack similarity in amino acid sequence to mammalian tyrosinases (reviewed in van Holde et al., 2001);

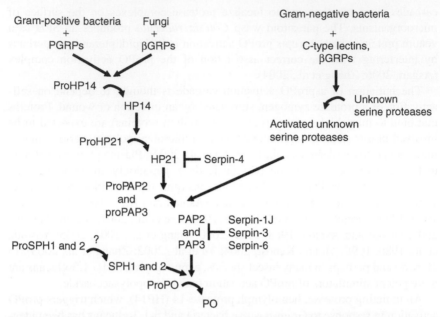

FIGURE 4.2 A model for a proPO activation pathway in *M. sexta*. Hemolymph plasma proteins known as pattern recognition proteins (βGRP, PGRP, C-type lectins) bind to polysaccharides on the surface of microorganisms. This interaction leads to activation of initiator protease(s) such as *M. sexta* HP14 by a mechanism not yet understood, which triggers a protease cascade. Activation of proHP21 by HP14 leads to activation of prophenoloxidase activating proteinase (PAP). SPHs (themselves activated by as yet unknown proteases) function together with PAPs to form a functional proPO activator, which cleaves proPO to form active PO. PO catalyzes the oxidation of hemolymph catecholic phenols to corresponding quinones, which can undergo further reactions to form melanin. Proteases in the pathway are regulated by serine protease inhibitors known as serpins. Abbreviations: βGRP, beta-1,3-glucan recognition protein; HP, hemolymph proteinase; PAP, prophenoloxidase activating proteinase; PGRP, peptidoglycan recognition protein; PO, phenoloxidase; SPH, serine protease homolog.

however, their enzymatic activities are similar. Mammalian tyrosinases have been studied in great depth and are known to catalyze three types of reactions during melanin synthesis: hydroxylation of a monophenol, dehydrogenation of a catechol (often described as oxidation of an *o*-diphenol), and dehydrogenation of a dihydroxyindole (Korner and Pawelek, 1982). During melanin synthesis, mammalian tyrosinase hydroxylates tyrosine to generate 3,4-dihydroxyphenylalanine (DOPA), which is then oxidized to DOPA-quinone (Korner and Pawelek, 1982). The kinetics of the first reaction are complex, and a plot of activity over time is not linear but begins with a variable lag phase (Molina et al., 2007; Pomerantz, 1966). There is some disagreement as to whether DOPA is released from the active site after the hydroxylation step or whether DOPA is simply an intermediate in the generation of DOPA-quinone from tyrosine (Garcia-Borron and Solano, 2002). Oxidation of DOPA by tyrosinase proceeds without a lag phase and is a much faster reaction (Rodriguez-Lopez et al., 1992).

Many of the first investigations of insect POs were hindered by the difficulty of purifying active POs, which have an unusually sticky character (as discussed in Ashida and Dohke, 1980). To circumvent this problem, some studies were done by purifying proPO and then activating the enzyme by proteolytic cleavage (e.g. by incubating with chymotrypsin or a protease partially purified from insect tissues) or by inducing a conformational change in the enzyme with a detergent, phospholipid, or alcohol (e.g. Asada and Sezaki, 1999; Aso et al., 1985; Hall et al., 1995; Ohnishi et al., 1970; Pau and Kelly, 1975). An additional complication was the fact that insects synthesize multiple enzymes with similar functions. This group of enzymes includes tyrosine hydroxylase, which hydroxylates monophenols but does not typically oxidize diphenols; laccase, which directly oxidizes both o- and p-diphenols but not monophenols (tyrosinase does not oxidase p-diphenols); and peroxidase, which can oxidize both monophenols and diphenols (Nappi and Vass, 1993; Okun, 1996; Thomas et al., 1989; Vie et al., 1999). Problems with the classification of purified enzymes were partly solved by substrate specificity experiments. In addition, various chemical inhibitors were used to categorize purified enzymes, e.g. phenylthiourea is a better inhibitor of tyrosinase than of laccase (Barrett, 1991) (although truly specific inhibitors of diphenoloxidase activity were not available).

All of the POs purified from insect tissues oxidized diphenols, but only a subset was found to hydroxylate monophenols; thus, despite the long-held view that insect POs have both activities, the data have been somewhat equivocal. One likely reason for the confusion is that some of the enzymes being tested were actually a mix of more than one type of enzyme. A second explanation is the difficulty of detecting activity toward monophenols compared with diphenols because tyrosinases are less active toward monophenols than diphenols and because tyrosinases exhibit a lag phase if the substrate is a monophenol. Chase et al. (2000) presented convincing evidence for the tyrosinase-like hydroxylation activity of a well-characterized PO. In their study, highly purified proPO from the hemolymph of a flesh fly, *Sarcophaga bullata*, was activated with a detergent, incubated with tyrosine methyl ester or various diphenols, and oxygen consumption in the reaction was assayed. The *S. bullata* PO oxidized diphenols such as dopamine more rapidly than it hydroxylated tyrosine methyl ester, and a lag phase occurred only with the monophenol substrate. A monophenol-specific lag phase was observed for two other POs: one from the hemolymph of a blowfly, *Calliphora erythrocephala*, with tyrosine as the substrate, and one from the hemolymph of *M. sexta*, when the substrate was tyrosine or tyramine (Aso et al., 1985; Pau and Kelly, 1975). In these studies, proPO was activated with a partially purified protease fraction that was not tested for the ability to hydroxylate monophenols; however, it is likely that PO was the source of the observed activity. The third function of tyrosinases, oxidation of dihydroxyindole, has not been a focus of insect PO studies. *M. sexta* PO can oxidize dihydroxyindole (Aso et al., 1985). In addition, the incubation of dihydroxyindole with PO purified from the hemolymph of a grasshopper, *Locusta migratoria*, led to the production of black pigments, suggesting that PO may have this type of activity (Cherqui et al., 1998).

Several studies of the substrate specificity of insect POs have generated a list of substrates, only some of which are likely to be biologically relevant. The monophenols that have been found to be substrates of at least some insect POs are tyrosine (and its methyl ester) and tyramine (Asada and Sezaki, 1999; Aso et al., 1985; Chase et al., 2000). Catechol substrates include 1,2-dihydroxybenzene (catechol), 4-methyl catechol, DOPA, dopamine, N-acetyldopamine (NADA), N-β-alanyldopamine (NBAD), N-acetylnonepinephrine (NBANE), 3,4-dihydroxy-phenylacetic acid (DOPAC), norepinephrine, 3,4-dihydroxymandelic acid, and 3,4-dihydroxybenzoic acid (Asada and Sezaki, 1999; Aso et al., 1985; Chase et al., 2000; Hall et al., 1995). Although DOPA frequently has been used to assay for PO activity, the enzyme oxidizes dopamine, NADA, and NBAD more efficiently (indicated by lower Km values). This discovery combined with the fact that DOPA is much less soluble than dopamine and N-acetyldopamines has led to the suggestion that these catecholamines are the natural substrates for PO (Sugumaran, 2002), although we are unaware of studies that rule out DOPA as a natural substrate. Tyrosine and most of the diphenols mentioned above (catechol, DOPA, dopamine, NADA, NBAD, and DOPAC) have been detected in hemolymph of one or more insect species (Czapla et al., 1990; Hopkins et al., 1984; Munkirs et al., 1990; Zhao et al., 1995). They may be present in a non-substrate form, conjugated to sulfate, glucose, or phosphate (Kramer and Hopkins, 1987). Such conjugates are a storage form of tyrosine and diphenols that can accumulate in the hemolymph until they are needed during molting, when newly synthesized cuticle undergoes rapid tanning (Kramer and Hopkins, 1987). It is unknown if the enyzmes required to release tyrosine and diphenols from their conjugated forms are activated during an immune response. In at least some insects, tyrosine is stored in vacuoles or inclusions in fat body cells (Kramer and Hopkins, 1987; McDermid and Locke, 1983), but again, whether this pool of tyrosine is used for immune reactions in addition to cuticle tanning is unknown.

The available data strongly support the view that the melanin-like materials that are formed in hemolymph are composed of cross-linked polyphenols and proteins, but which diphenols are the actual substrates oxidized by PO during an immune response is not entirely clear, in part due to the difficulty of chemical analysis of such small amounts of cross-linked, insoluble material. The melanotic capsules that have been tested were insoluble in organic solvents and strong acids and resistant to various enzymes that degrade proteins, lipoproteins, or polysac-charides, but the capsules could be disrupted with strong bases, and their color could be bleached with strong oxidizing agents (Götz, 1986). Melanin-like capsules made by *Chironomus* (midge) larvae tested positive in histochemical tests for melanin and protein (Götz and Vey, 1974). The color of melanotic capsules varies, depending in part on the insect species; yellowish brown, reddish brown, brown, and black capsules have been observed (Götz, 1986), likely due to differences in the mixture of catechols available to PO. Proof that PO is involved in melanotic capsule formation comes from experiments in which decreased PO activity (by chemical inhibitors or RNA-mediated gene silencing) resulted in diminished melanization (Liu et al., 1997; Nappi, 1974; Shiao et al., 2001).

The most commonly described biochemical path to melanin is the PO-catalyzed production of DOPA-quinone from tyrosine, followed by additional enzymatic and non-enzymatic reactions leading to the formation of eumelanin (dark brown or black) or, in the presence of thiol compounds, pheomelanins (yellow to reddish brown) (reviewed in Nappi and Christensen, 2005). This pathway may be used during melanization of parasitoid eggs in *D. melanogaster* (reviewed in Nappi and Sugumaran, 1993). Good evidence for the participation of dopamine in immune type melanization comes from studies demonstrating a requirement for dopa decarboxylase during melanization (reviewed in Christensen et al., 2005). Upregulation of dopa decarboxylase gene expression in response to infection has been observed in several insect species (Kim et al., 2000; Zhu et al., 2003). Since dopa decarboxylase converts DOPA to dopamine, it seems likely that during an immune response, dopamine is formed and then PO oxidizes dopamine (and possibly its derivatives such as NADA and NBAD) to generate quinones. Involvement of NBAD is suggested by a study demonstrating the synthesis of NBAD in response to immune challenge in the larval hemolymph of a species of beetle, *Tenebrio molitor* (Kim et al., 2000). It is possible that other diphenols found in the hemolymph also participate in immune type melanization reactions.

4.5   REGULATION OF PO ACTIVITY

The PO activation system produces several types of molecules that could harm the host insect if produced in excess. These include proteases that could degrade host proteins, cytotoxic quinones, and reactive oxygen and nitrogen species. The system is regulated under most conditions to produce a local melanization response at a specific site and for limited duration. *D. melanogaster* mutants that produce spontaneous melanotic masses indicate the pathology that can occur if PO is unleashed abnormally (Minakhina and Steward, 2006). It is possible to induce systemic melanization, with massive production of melanin in hemolymph and melanin deposition on internal tissues, by injecting large amounts of dead bacteria or peptidoglycan into the hemocoel (Fig. 4.3). Such excessive PO activation may be analogous to disseminated intravascular coagulation in humans, systemic pathology due in part to dysregulation of thrombin activation (Levi et al., 2003). Systemic melanization can also be elicited in *M. sexta* by injecting large amounts (>1 mg) of serine proteases such as trypsin or chymotrypsin (Fig. 4.3) or by injecting small amounts (micrograms) of metalloproteases such as thermolysin (Kanost, unpublished results). Activation of proPO by small amounts of thermolysin in *Galleria mellonella* was recently reported (Altincicek et al., 2007). Melanization stimulated by exposing hemolymph to proteases is probably due to non-specific activation of protease zymogens in the proPO cascade or of proPO

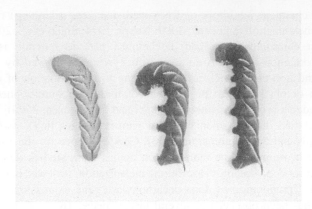

FIGURE 4.3 Systemic melanization resulting from uncontrolled phenoxidase activation. *Manduca sexta* larvae were injected with water (left), 200 μg of *Micrococcus luteus* (middle), or with 2 mg of bovine chymotrypsin (right) and photographed 24 h later. (See Plate 4.3 for color version of this figure.)

itself. The high concentration of serine protease inhibitors in insect hemolymph (Polanowski and Wilusz, 1996) is the likely reason that orders of magnitude more serine protease than metalloprotease is required to provoke proPO activation.

Serine protease inhibitors from several gene families have been identified in hemolymph as regulators of proPO activation (Kanost, 1999; Kanost and Jiang, 1996). A protease inhibitor named pacifastin from crayfish blocks proPO activation by inhibiting the proPO activating protease (Hergenhahn et al., 1987; Liang et al., 1997). Pacifastin is composed of a heavy chain similar in sequence to transferrin and a light chain with nine related Cys-rich domains (pacifastin domains) that have the protease inhibitory activity. Members of a family of 4 kDa proteins composed of a single pacifastin domain present in hemolymph of locusts are able to block proPO activation, but the proteases they inhibit have not yet been identified (Boigegrain et al., 1992; Brehelin et al., 1991; Kellenberger and Roussel, 2005; Simonet et al., 2002b, 2005). Genes encoding pacifastin domains exist in holometabolous insect species (Simonet et al., 2002a, 2003), but such proteins have not yet been investigated. Small proteins from the Kunitz family of protease inhibitors from *M. sexta*, *S. bullata*, and *B. mori* can interfere with proPO activation (Aso et al., 1994; Ramesh et al., 1988; Saul and Sugumaran, 1986; Sugumaran et al., 1985) and may have a regulatory role *in vivo*, but their protease targets in hemolymph are not known.

Serpins are a family of ∼50 kDa proteins functioning primarily as serine protease inhibitors, which regulate vertebrate and invertebrate plasma proteases (Gettins, 2002; Silverman et al., 2001). Serpins from several insect species have been demonstrated to regulate proPO activation. We have been investigating the functions of serpins from *M. sexta* (Kanost, 2007). Three serpins from *M. sexta* hemolymph (serpin-1J, serpin-3, and serpin-6), which disrupt proPO activation when added as recombinant proteins to plasma, directly inhibit proPO activating proteases (Jiang et al., 2003; Wang and Jiang, 2004; Zhu et al., 2003b) (Fig. 4.2). The reactive site loop of serpin-3, the bait that interacts with the active site of target proteases, contains a sequence very similar to the conserved activation site in proPOs (Zhu et al., 2003b). This sequence is likely an excellent fit in the active site of PAPs, and it is probable that serpin-3 is a physiologically important regulator of PAP activity. Serpin-3 and its orthologs in other species have an amino-terminal sequence extension not present in other serpins, and perhaps this has a role in its function. An ortholog of serpin-3 from another lepidopteran, *Hyphantria cunea*, has been shown to disrupt proPO activation (Park et al., 2000). A variety of experiments have shown that *D. melanogaster* serpin-27A (an apparent ortholog of *M. sexta* serpin-3) regulates melanization, and it can inhibit the proPO activating protease from *H. diomphalia* (De Gregorio et al., 2002; Ligoxygakis et al., 2002; Nappi et al., 2005), suggesting that it is likely to inhibit a *D. melanogaster* PAP. In *A. gambiae*, decreased expression of a serpin-3 ortholog (SRPN2) resulted in formation of melanotic pseudotumors and reduced survival of invading *Plasmodium berghei*, due to increased ookinete lysis and melanization compared with mosquitoes that had normal levels of SRPN2 (Michel et al., 2005). Recombinant *A. gambiae* SRPN2 can block proPO activation in *M. sexta* plasma and inhibits *M. sexta* PAP-3, suggesting that it is likely to inhibit an *A. gambiae* proPO activating protease (Michel et al., 2006).

Serpins that inhibit proteases upstream of PAP in the activation cascade have been identified in *M. sexta*. Recombinant serpins-4 and -5 block proPO activation in plasma, but they do not inhibit the purified PAPs (Tong and Kanost, 2005). To identify the proteases inhibited by serpins-4 and -5, we used antibodies to serpin-4 and serpin-5 to purify their protease–serpin complexes formed in plasma. Analysis of these complexes by immunoblot, mass spectrometry, and Edman degradation to identify the inhibited proteases indicated that serpin-4 inhibits HP1, HP6, and HP21, and serpin-5 inhibits HP1 and HP6 (Tong et al., 2005). Inhibition of HP21 by serpin-4 likely regulates the production of active PAP-2 and PAP-3 (Fig. 4.2). HP1 and HP6 are candidates for components of a different branch of the proPO activation cascade. As HP21 was detected in serpin-4 complexes formed after treatment of plasma with *Micrococcus luteus* but not after treatment with *Escherichia coli*, there may more than one branch of the cascade leading to PAP activation, with HP21 not required for activation by *E. coli*. A difference in molecular interactions required for activation by *E. coli* and *M. luteus* was also apparent in observations of differential effects of salt concentration on proPO activation by Gram-positive and Gram-negative bacteria (Tong and Kanost, 2005).

Proteins that directly inhibit PO rather than the activating proteases have been identified in a few insect species. A 4 kDa peptide from hemolymph of *Musca domestica* is an efficient PO inhibitor (Tsukamoto et al., 1992). This peptide contains a sequence motif similar to a family of toxins from snails and spiders (Daquinag et al., 1999). It has the unusual property of containing a tyrosine residue that is hydroxylated to form DOPA within the peptide (Daquinag et al., 1995).

A protein from *A. gambiae* that contains five tandem motifs similar to the house-fly PO inhibitor may also function as a PO inhibitor, as RNAi knockdown resulted in greater melanization of wounds compared with controls (Shi et al., 2006). A *M. sexta* protein containing a homologous domain, produced as a recombinant protein, inhibited PO only at very high concentration (Lu and Jiang, 2007), and thus its role remains in question. A 380 kDa PO inhibitor from larval cuticle (Sugumaran and Nellaiappan, 2000) and a non-protein, low molecular weight PO inhibitor from hemolymph (Lu and Jiang, 2007) of *M. sexta* have been isolated but not yet well characterized with regard to structure or physiological function.

Other proteins in hemolymph may modulate the function of PO. Interaction of PO with dopachrome isomerase (Sugumaran et al., 2000a) or quinone isomerase (Sugumaran et al., 2000b) led to decreased PO activity. A 43 kDa protein from hemolymph of *T. molitor* inhibits melanin synthesis *in vitro*, and its downregulation by RNAi resulted in increased melanization (Zhao et al., 2005). A 160 kDa hemolymph protein from the same species, with some sequence similarity to vitellogenin, stimulated melanin synthesis. The mechanism of action and physiological roles of these protein factors are not yet clear.

4.6 CONCLUSIONS AND FUTURE PROSPECTS

Research in recent years has led to better understanding of some of the proteins involved in insect proPO activation and has verified the hypothesis that protease cascades are central to the process. Nevertheless, several unique aspects of the proPO system make this an extremely challenging pursuit. The problems include the fact that hemocytes that produce proPO are fragile and extremely difficult to study in culture; active PO becomes 'sticky' and difficult to manipulate; the proteases of the cascade are present in plasma at low concentration and once active tend to become rapidly degraded or inhibited by serpins; insect genomes contain large families of clip domain proteases, and their degree of sequence conservation between species is often not sufficient for clearly identifying orthologous protease genes (thus, even though PAPs have been identified in several large species conducive to biochemical analysis of plasma, proteases that cleave proPO in *D. melanogaster* have not at this point been demonstrated); it is likely that more than one pathway, or intersecting pathways exist for proPO activation within a species; the products of PO catalysis are unstable and rapidly undergo additional enzymatic and non-enzymatic reactions, ultimately producing insoluble, cross-linked products (in small amounts *in vivo*), presenting difficult problems in analytical chemistry.

New technology in genomics, proteomics, and RNAi is likely to promote more rapid gains in identifying and investigating the function of proteins involved in the proPO system and insect immunity. We can identify several problems and questions in need of future research. There has not yet been a report of recombinant expression of native, activatable proPO. Availability of such material will provide an essential reagent for investigating proPO activation and PO enzymology, particularly for small insects for which purification of large amounts of proPO from hemolymph is not feasible. There is a great need for three-dimensional structural information for an arthropod proPO, which at this point has not yet been reported. Knowledge of such structures is essential for developing an understanding of the conformational changes that lead to activation of the proPO zymogen. Connections between proPO activation and clotting or wound healing must be investigated to gain a basic understanding of how extensively these processes may interact. PO function *in vivo* occurs at surfaces rather than in solution. Methods should be developed to gain a better understanding of the assembly of molecular complexes of plasma proteins that include microbial polysaccharides, pattern recognition proteins, proteases and protease homolog cofactors, and PO and to study the function of such complexes immobilized at surfaces of microbes or wounds.

There are also important gaps in our understanding of the metabolism of tyrosine derivatives to produce PO substrates in insects. What substrates are actually available for oxidation by PO in a given species and how is that metabolism regulated, particularly during immune challenge? There is also a lack of detailed understanding of the chemistry occurring during melanin formation *in vivo* after the production of quinones by PO. What are the intermediates and products and how are they toxic to pathogens and parasites? Much of the current level of understanding in this area is based on studies of mammalian tyrosinase and melanin, and there may be important variations in the chemistry occurring in insects. The physiological significance of the identified naturally occurring PO inhibitors is unclear and needs much further experimentation. Finally, studies will need to be extended beyond a few model species to gain a full understanding of PO function in insect immunity. There is a particular need for more study of hemimetabolous species, which have received much less recent attention at a molecular level. This group may have significant differences from the few dipteran, lepidopteran, and coleopteran species that have served as models for detailed investigation.

ACKNOWLEDGMENTS

Research in the authors' laboratory has been supported by NIH grants GM41247 and AI057815. We thank Nancy Beckage, Jefferey Fabrick, and Xiao-Qiang Yu for providing photographs used in Fig. 4.1.

REFERENCES

Adachi, K., Hirata, T., Nishioka, T., and Sakaguchi, M. (2003). Hemocyte components in crustaceans convert hemocyanin into a phenoloxidase-like enzyme. *Comp. Biochem. Physiol. B. Biochem. Mol. Biol.* **134**, 135–141.

Altincicek, B., Linder, M., Linder, D., Preissner, K. T., and Vilcinskas, A. (2007). Microbial metallo-proteinases mediate sensing of invading pathogens and activate innate immune responses in the lepidopteran model host *Galleria mellonella*. *Infect Immun.* **75**, 175–183.

Arakane, Y., Muthukrishnan, S., Beeman, R. W., Kanost, M. R., and Kramer, K. J. (2005). Laccase 2 is the phenoloxidase gene required for beetle cuticle tanning. *Proc. Natl. Acad. Sci. USA* **102**, 11337–11342.

Asada, N. (1998). Reversible activation of prophenoloxidase with 2-propanol in *Drosophila melanogaster. J. Exp. Zool.* **282**, 28–31.

Asada, N., and Sezaki, H. (1999). Properties of phenoloxidases generated from prophenoloxidase with 2-propanol and the natural activator in *Drosophila melanogaster. Biochem. Genet.* **37**, 149–158.

Asada, N., Fukumitsu, T., Fujimoto, K., and Masuda, K. (1993). Activation of prophenoloxidase with 2-propanol and other organic compounds in *Drosophila melanogaster. Insect Biochem. Mol. Biol.* **23**, 515–520.

Asano, T., and Ashida, M. (2001). Cuticular pro-phenoloxidase of the silkworm, *Bombyx mori*. Purification and demonstration of its transport from hemolymph. *J. Biol. Chem.* **276**, 11100–11112.

Asgari, S. (2006). Venom proteins from polydnavirus-producing endoparasitoids: Their role in host–parasite interactions. *Arch. Insect Biochem. Physiol.* **61**, 146–156.

Ashida, M., and Brey, P. (1997). Recent advances in research on the insect prophenoloxidase cascade. In *Molecular Mechanisms of Immune Responses in Insects* (P. Brey, and D. Hultmark, Eds.), pp. 135–171. Chapman & Hall, London.

Ashida, M., and Dohke, K. (1980). Activation of pro-phenoloxidase by the activating enzyme of the silkworm, *Bombyx mori. Insect Biochem.* **10**, 37–47.

Ashida, M., Dohke, K., and Ohnishi, E. (1974). Activation of prophenoloxidase III. Release of a peptide from prophenoloxidase by the activating enzyme. *Biochem. Biophys. Res. Commun.* **57**, 1089–1095.

Ashida, M., Ishizaki, Y., and Iwahana, H. (1983). Activation of pro-phenoloxidase by bacterial cell walls or beta-1,3-glucans in plasma of the silkworm, *Bombyx mori. Biochem. Biophys. Res. Commun.* **113**, 562–568.

Aso, Y., Kramer, K. J., Hopkins, T. L., and Lookhart, G. L. (1985). Characterization of hemolymph protyrosinase and a cuticular activator from *Manduca sexta* (L.). *Insect Biochem.* **15**, 9–17.

Aso, Y., Yamashita, T., Meno, K., and Murakami, M. (1994). Inhibition of prophenoloxidase-activating enzyme from *Bombyx mori* by endogenous chymotrypsin inhibitors. *Biochem. Mol. Biol. Int.* **33**, 751–758.

Aspán, A., Huang, T. S., Cerenius, L., and Söderhäll, K. (1995). cDNA cloning of prophenoloxidase from the freshwater crayfish *Pacifastacus leniusculus* and its activation. *Proc. Natl. Acad. Sci. USA* **92**, 939–943.

Baker, J., and Fabrick, J. (2002). Unusual responses of Indianmeal moth larvae (Lepidoptera: Pyralidae) to envenomation and parasitization by a braconid ectoparasitoid. *J. Entomol. Sci.* **37**, 370–374.

Barrett, F. M. (1991). Phenoloxidases and the integument. In *Physiology of the Insect Epidermis* (K. Binnington, and A. Retnakaron, Eds.), pp. 195–212. CSIRO Publications, East Melbourne, Victoria, Australia.

Beckage, N. E. (1998). Modulation of immune responses to parasitoids by polydnaviruses. *Parasitology* **116**(Suppl), S57–S64.

Bidla, G., Dushay, M. S., and Theopold, U. (2007). Crystal cell rupture after injury in *Drosophila* requires the JNK pathway, small GTPases and the TNF homolog Eiger. *J. Cell Sci.* **120**, 1209–1215.

Biedermann, W. (1898). Beiträge zur vergleichenden Pysiologie der Verdauung. I. Die Verdauung der Larve von Tenebrio molitor. *Pflüger's Arch.* **72**, 105–162.

Boigegrain, R. A., Mattras, H., Brehelin, M., Paroutaud, P., and Coletti-Previero, M. A. (1992). Insect immunity: Two proteinase inhibitors from hemolymph of *Locusta migratoria. Biochem. Biophys. Res. Commun.* **189**, 790–793.

Brehélin, M., Drif, L., Baud, L., and Boemare, N. (1989). Insect haemolymph: Cooperation between humoral and cellular factors in *Locusta migratoria. Insect Biochem.* **19**, 301–307.

Brehelin, M., Boigegrain, R. A., Drif, L., and Coletti-Previero, M. A. (1991). Purification of a protease inhibitor which controls prophenoloxidase activation in hemolymph of *Locusta migratoria* (insecta). *Biochem. Biophys. Res. Commun.* **179**, 841–846.

Brivio, M. F., Pagani, M., and Restelli, S. (2002). Immune suppression of *Galleria mellonella* (Insecta, Lepidoptera) humoral defenses induced by *Steinernema feltiae* (Nematoda, Rhabditida): Involvement of the parasite cuticle. *Exp. Parasitol.* **101**, 149–156.

Burmester, T. (2001). Molecular evolution of the arthropod hemocyanin superfamily. *Mol. Biol. Evol.* **18**, 184–195.

Burmester, T. (2002). Origin and evolution of arthropod hemocyanins and related proteins. *J. Comp. Physiol. [B]* **172**, 95–107.

Castillejo-Lopez, C., and Hacker, U. (2005). The serine protease Sp7 is expressed in blood cells and regulates the melanization reaction in *Drosophila. Biochem. Biophys. Res. Commun.* **338**, 1075–1082.

Castillo, J. C., Robertson, A. E., and Strand, M. R. (2006). Characterization of hemocytes from the mosquitoes *Anopheles gambiae* and *Aedes aegypti. Insect Biochem. Mol. Biol.* **36**, 891–903.

Cerenius, L., and Söderhäll, K. (2004). The prophenoloxidase-activating system in invertebrates. *Immunol. Rev.* **198**, 116–126.

Chase, M. R., Raina, K., Bruno, J., and Sugumaran, M. (2000). Purification, characterization and molecular cloning of prophenoloxidases from *Sarcophaga bullata. Insect Biochem. Mol. Biol.* **30**, 953–967.

Chen, C., Durrant, H. J., Newton, R. P., and Ratcliffe, N. A. (1995). A study of novel lectins and their involvement in the activation of the prophenoloxidase system in *Blaberus discoidalis. Biochem. J.* **310**(Pt 1), 23–31.

Chen, C., Rowley, A. F., Newton, R. P., and Ratcliffe, N. A. (1999). Identification, purification and properties of a beta-1,3-glucan-specific lectin from the serum of the cockroach, *Blaberus discoidalis* which is implicated in immune defence reactions. *Comp. Biochem. Physiol. B. Biochem. Mol. Biol.* **122**, 309–319.

Chen, C. C., and Chen, C. S. (1995). *Brugia pahangi*: Effects of melanization on the uptake of nutrients by microfilariae *in vitro. Exp. Parasitol.* **81**, 72–78.

Cherqui, A., Duvic, B., Reibel, C., and Brehelin, M. (1998). Cooperation of dopachrome conversion factor with phenoloxidase in the eumelanin pathway in haemolymph of *Locusta migratoria* (Insecta). *Insect Biochem. Mol. Biol.* **28**, 839–848.

Christensen, B. M., Li, J., Chen, C. C., and Nappi, A. J. (2005). Melanization immune responses in mosquito vectors. *Trends Parasitol.* **21**, 192–199.

Czapla, T. H., Hopkins, T. L., and Kramer, K. J. (1990). Catecholamines and related *o*-diphenols in cockroach hemolymph and cuticle during sclerotization and melanization: Comparative studies on the order Dictyoptera. *J. Comp. Physiol. [B]* **160**, 175–181.

Daquinag, A. C., Nakamura, S., Takao, T., Shimonishi, Y., and Tsukamoto, T. (1995). Primary structure of a potent endogenous dopa-containing inhibitor of phenol oxidase from *Musca domestica. Proc. Natl. Acad. Sci. USA* **92**, 2964–2968.

Daquinag, A. C., Sato, T., Koda, H., Takao, T., Fukuda, M., Shimonishi, Y., and Tsukamoto, T. (1999). A novel endogenous inhibitor of phenoloxidase from *Musca domestica* has a cystine motif commonly found in snail and spider toxins. *Biochemistry* **38**, 2179–2188.

De Gregorio, E., Han, S. J., Lee, W. J., Baek, M. J., Osaki, T., Kawabata, S., Lee, B. L., Iwanaga, S., Lemaitre, B., and Brey, P. T. (2002). An immune-responsive serpin regulates the melanization cascade in *Drosophila. Dev. Cell* **3**, 581–592.

Decker, H., and Jaenicke, E. (2004). Recent findings on phenoloxidase activity and antimicrobial activity of hemocyanins. *Dev. Comp. Immunol.* **28**, 673–687.

Decker, H., and Rimke, T. (1998). Tarantula hemocyanin shows phenoloxidase activity. *J. Biol. Chem.* **273**, 25889–25892.

Decker, H., and Terwilliger, N. (2000). Cops and robbers: Putative evolution of copper oxygen-binding proteins. *J. Exp. Biol.* **203**, 1777–1782.

Decker, H., and Tuczek, F. (2000). Tyrosinase/catecholoxidase activity of hemocyanins: Structural basis and molecular mechanism. *Trends Biochem. Sci.* **25**, 392–397.

Decker, H., Ryan, M., Jaenicke, E., and Terwilliger, N. (2001). SDS-induced phenoloxidase activity of hemocyanins from *Limulus polyphemus, Eurypelma californicum,* and *Cancer magister. J. Biol. Chem.* **276**, 17796–17799.

Dittmer, N. T., Suderman, R. J., Jiang, H., Zhu, Y. C., Gorman, M. J., Kramer, K. J., and Kanost, M. R. (2004). Characterization of cDNAs encoding putative laccase-like multicopper oxidases and developmental expression in the tobacco hornworm, *Manduca sexta,* and the malaria mosquito, *Anopheles gambiae. Insect Biochem. Mol. Biol.* **34**, 29–41.

Dohke, K. (1973). Studies on prephenoloxidase-activating enzyme from cuticle of the silkworm *Bombyx mori.* I. Activation reaction by the enzyme. *Arch. Biochem. Biophys.* **157**, 203–209.

Durrant, H. J., Ratcliffe, N. A., Hipkin, C. R., Aspan, A., and Soderhall, K. (1993). Purification of the pro-phenol oxidase enzyme from haemocytes of the cockroach *Blaberus discoidalis. Biochem. J.* **289**(Pt 1), 87–91.

Evans, J. D., Aronstein, K., Chen, Y. P., Hetru, C., Imler, J. L., Jiang, H., Kanost, M., Thompson, G. J., Zou, Z., and Hultmark, D. (2006). Immune pathways and defence mechanisms in honey bees *Apis mellifera. Insect Mol. Biol.* **15**, 645–656.

Fabrick, J. A., Baker, J. E., and Kanost, M. R. (2003). cDNA cloning, purification, properties, and function of a beta-1,3-glucan recognition protein from a pyralid moth, *Plodia interpunctella. Insect Biochem. Mol. Biol.* **33**, 579–594.

Fujimoto, K., Okino, N., Kawabata, S., Iwanaga, S., and Ohnishi, E. (1995). Nucleotide sequence of the cDNA encoding the proenzyme of phenol oxidase A1 of *Drosophila melanogaster. Proc. Natl. Acad. Sci. USA* **92**, 7769–7773.

Garcia-Borron, J. C., and Solano, F. (2002). Molecular anatomy of tyrosinase and its related proteins: Beyond the histidine-bound metal catalytic center. *Pigment Cell Res.* **15**, 162–173.

Gettins, P. G. (2002). Serpin structure, mechanism, and function. *Chem. Rev.* **102**, 4751–4804.

Gillespie, J. P., Kanost, M. R., and Trenczek, T. (1997). Biological mediators of insect immunity. *Annu. Rev. Entomol.* **42**, 611–643.

Gorman, M. J., An, C., and Kanost, M. R. (2007). Characterization of tyrosine hydroxylase from *Manduca sexta. Insect Biochem. Mol. Biol.* (in press).

Gorman, M. J., Wang, Y., Jiang, H., and Kanost, M. R. (2007). *Manduca sexta* hemolymph proteinase 21 activates prophenoloxidase-activating proteinase 3 in an insect innate immune response proteinase cascade. *J. Biol. Chem.* **282**, 11742–11749.

Götz, P. (1986). Mechanisms of encapsulation in dipteran hosts. *Symp. Zool. Soc. Lond.* **56**, 1–19.

Götz, P., and Vey, A. (1974). Humoral encapsulation in Diptera (Insecta): Defense reactions of *Chironomus* larvae against fungi. *Parasitology* **68**, 193–205.

Gupta, S., Wang, Y., and Jiang, H. B. (2005a). *Manduca sexta* prophenoloxidase (proPO) activation requires proPO-activating proteinase (PAP) and serine proteinase homologs (SPHs) simultaneously. *Insect Biochem. Mol. Biol.* **35**, 241–248.

Gupta, S., Wang, Y., and Jiang, H. B. (2005b). Purification and characterization of *Manduca sexta* prophenoloxidase-activating proteinase-1, an enzyme involved in insect immune responses. *Protein Expr. Purif.* **39**, 261–268.

Hall, M., Scott, T., Sugumaran, M., Soderhall, K., and Law, J. H. (1995). Proenzyme of *Manduca sexta* phenol oxidase: Purification, activation, substrate specificity of the active enzyme, and molecular cloning. *Proc. Natl. Acad. Sci. USA* **92**, 7764–7768.

Hergenhahn, H., Aspan, A., and Söderhäll, K. (1987). Purification and characterization of a high-Mr proteinase inhibitor of pro-phenol oxidase activation from crayfish plasma. *Biochem. J.* **248**, 223–228.

Hernandez, S., Lanz, H., Rodriguez, M. H., Torres, J. A., Martinez-Palomo, A., and Tsutsumi, V. (1999). Morphological and cytochemical characterization of female *Anopheles albimanus* (Diptera: Culicidae) hemocytes. *J. Med. Entomol.* **36**, 426–434.

Hillyer, J. F., and Christensen, B. M. (2002). Characterization of hemocytes from the yellow fever mosquito, *Aedes aegypti. Histochem. Cell Biol.* **117**, 431–440.

Hillyer, J. F., Schmidt, S. L., and Christensen, B. M. (2003). Hemocyte-mediated phagocytosis and melanization in the mosquito *Armigeres subalbatus* following immune challenge by bacteria. *Cell Tissue Res.* **313**, 117–127.

Hoffmann, J. A., Porte, A., and Joly, P. (1970). Physiologie des insectes. Sur la localisation díune activité phénoloxydasique dans les coagulocytes de *Locusta migratoria* L. (Orthoptére). *C.R. Acad. Sci. Paris* **270**(Série D), 629–631.

Hopkins, T. L., Morgan, T. D., and Kramer, K. J. (1984). Catecholamines in haemolymph and cuticle during larval, pupal and adult development of *Manduca sexta* (L.). *Insect Biochem.* **14**, 533–540.

Iwama, R., and Ashida, M. (1986). Biosynthesis of prophenoloxidase in hemocytes of larval hemolymph of the silkworm, *Bombyx mori. Insect Biochem.* **16**, 547–555.

Jaenicke, E., and Decker, H. (2003). Tyrosinases from crustaceans form hexamers. *Biochem. J.* **371**, 515–523.

Ji, C. Y., Wang, Y., Guo, X. P., Hartson, S., and Jiang, H. B. (2004). A pattern recognition serine proteinase triggers the prophenoloxidase activation cascade in the tobacco hornworm, *Manduca sexta. J. Biol. Chem.* **279**, 34101–34106.

Jiang, H., and Kanost, M. R. (2000). The clip-domain family of serine proteinases in arthropods. *Insect Biochem. Mol. Biol.* **30**, 95–105.

Jiang, H., Wang, Y., Ma, C., and Kanost, M. R. (1997). Subunit composition of pro-phenol oxidase from *Manduca sexta*: Molecular cloning of subunit ProPO-P1. *Insect Biochem. Mol. Biol.* **27**, 835–850.

Jiang, H., Wang, Y., and Kanost, M. R. (1998). Pro-phenol oxidase activating proteinase from an insect, *Manduca sexta*: A bacteria-inducible protein similar to *Drosophila* easter. *Proc. Natl. Acad. Sci. USA* **95**, 12220–12225.

Jiang, H., Wang, Y., and Kanost, M. R. (1999). Four serine proteinases expressed in *Manduca sexta* haemocytes. *Insect Mol. Biol.* **8**, 39–53.

Jiang, H., Wang, Y., Yu, X. Q., and Kanost, M. R. (2003a). Prophenoloxidase-activating proteinase-2 from hemolymph of *Manduca sexta*. A bacteria-inducible serine proteinase containing two clip domains. *J. Biol. Chem.* **278**, 3552–3561.

Jiang, H., Wang, Y., Yu, X. Q., Zhu, Y., and Kanost, M. (2003b). Prophenoloxidase-activating proteinase-3 (PAP-3) from *Manduca sexta* hemolymph: A clip-domain serine proteinase regulated by serpin-1J and serine proteinase homologs. *Insect Biochem. Mol. Biol.* **33**, 1049–1060.

Jiang, H., Wang, Y., Gu, Y., Guo, X., Zou, Z., Scholz, F., Trenczek, T. E., and Kanost, M. R. (2005). Molecular identification of a bevy of serine proteinases in *Manduca sexta* hemolymph. *Insect Biochem. Mol. Biol.* **35**, 931–943.

Kanost, M. R. (1999). Serine proteinase inhibitors in arthropod immunity. *Dev. Comp. Immunol.* **23**, 291–301.

Kanost, M. R. (2007). Serpins in a Lepidopteran insect, *Manduca sexta*. In *Molecular and Cellular Aspects of the Serpinopathies and Disorder in Serpin Activity* (G. A. Silverman, and D. A. Lomas, Eds.), pp. 229–242. World Scientific Publishing Co., Hackensack, New Jersey.

Kanost, M. R., and Jiang, H. (1996). Proteinase inhibitors in invertebrate immunity. In *New Directions in Invertebrate Immunology* (K. Söderhäll, S. Iwanaga, and G. Vanta, Eds.), pp. 155–173. SOS Publications, Fair Haven, NJ.

Kanost, M. R., Jiang, H., Wang, Y., Yu, X. Q., Ma, C., and Zhu, Y. (2001). Hemolymph proteinases in immune responses of *Manduca sexta. Adv. Exp. Med. Biol.* **484**, 319–328.

Kanost, M. R., Jiang, H., and Yu, X. Q. (2004). Innate immune responses of a lepidopteran insect, *Manduca sexta. Immunol. Rev.* **198**, 97–105.

Karlsson, C., Korayem, A. M., Scherfer, C., Loseva, O., Dushay, M. S., and Theopold, U. (2004). Proteomic analysis of the *Drosophila* larval hemolymph clot. *J. Biol. Chem.* **279**, 52033–52041.

Kawabata, T., Yasuhara, Y., Ochiai, M., Matsuura, S., and Ashida, M. (1995). Molecular cloning of insect pro-phenol oxidase: A copper-containing protein homologous to arthropod hemocyanin. *Proc. Natl. Acad. Sci. USA* **92**, 7774–7778.

Kellenberger, C., and Roussel, A. (2005). Structure–activity relationship within the serine protease inhibitors of the pacifastin family. *Protein Pept. Lett.* **12**, 409–414.

Kim, M. H., Joo, C. H., Cho, M. Y., Kwon, T. H., Lee, K. M., Natori, S., Lee, T. H., and Lee, B. L. (2000). Bacterial-injection-induced syntheses of N-beta-alanyldopamine and dopa decarboxylase in the hemolymph of coleopteran insect, *Tenebrio molitor* larvae. *Eur. J. Biochem.* **267**, 2599–2608.

Kim, M. S., Baek, M. J., Lee, M. H., Park, J. W., Lee, S. Y., Soderhall, K., and Lee, B. L. (2002). A new easter-type serine protease cleaves a masquerade-like protein during prophenoloxidase activation in *Holotrichia diomphalia* larvae. *J. Biol. Chem.* **277**, 39999–40004.

Koizumi, N., Imamura, M., Kadotani, T., Yaoi, K., Iwahana, H., and Sato, R. (1999). The lipopolysaccharide-binding protein participating in hemocyte nodule formation in the silkworm *Bombyx mori* is a novel member of the C-type lectin superfamily with two different tandem carbohydrate-recognition domains. *FEBS Lett.* **443**, 139–143.

Korner, A., and Pawelek, J. (1982). Mammalian tyrosinase catalyzes three reactions in the biosynthesis of melanin. *Science* **217**, 1163–1165.

Kramer, K. J., and Hopkins, T. L. (1987). Tyrosine metabolism for insect cuticle tanning. *Arch. Insect Biochem. Physiol.* **6**, 279–301.

Kurata, S. (2004). Recognition of infectious non-self and activation of immune responses by peptidoglycan recognition protein (PGRP)-family members in *Drosophila. Dev. Comp. Immunol.* **28**, 89–95.

Kwon, T. H., Kim, M. S., Choi, H. W., Joo, C. H., Cho, M. Y., and Lee, B. L. (2000). A masquerade-like serine proteinase homologue is necessary for phenoloxidase activity in the coleopteran insect, *Holotrichia diomphalia* larvae. *Eur. J. Biochem.* **267**, 6188–6196.

Lai, S. C., Chen, C. C., and Hou, R. F. (2002). Immunolocalization of prophenoloxidase in the process of wound healing in the mosquito *Armigeres subalbatus* (Diptera: Culicidae). *J. Med. Entomol.* **39**, 266–274.

Lavine, M. D., and Strand, M. R. (2002). Insect hemocytes and their role in immunity. *Insect Biochem. Mol. Biol.* **32**, 1295–1309.

Leclerc, V., Pelte, N., El Chamy, L., Martinelli, C., Ligoxygakis, P., Hoffmann, J. A., and Reichhart, J. M. (2006). Prophenoloxidase activation is not required for survival to microbial infections in *Drosophila. EMBO Rep.* **7**, 231–235.

Lee, S. Y., Cho, M. Y., Hyun, J. H., Lee, K. M., Homma, K. I., Natori, S., Kawabata, S. I., Iwanaga, S., and Lee, B. L. (1998a). Molecular cloning of cDNA for pro-phenol-oxidase-activating factor I, a serine protease is induced by lipopolysaccharide or 1,3-beta-glucan in coleopteran insect, *Holotrichia diomphalia* larvae. *Eur. J. Biochem.* **257**, 615–621.

Lee, S. Y., Kwon, T. H., Hyun, J. H., Choi, J. S., Kawabata, S., Iwanaga, S., and Lee, B. L. (1998b). *In vitro* activation of pro-phenol-oxidase by two kinds of pro-phenol-oxidase-activating factors isolated from hemolymph of coleopteran, *Holotrichia diomphalia* larvae. *Eur. J. Biochem.* **254**, 50–57.

Lee, K. Y., Zhang, R., Kim, M. S., Park, J. W., Park, H. Y., Kawabata, S., and Lee, B. L. (2002). A zymogen form of masquerade-like serine proteinase homologue is cleaved during pro-phenoloxidase activation by Ca2+ in coleopteran and *Tenebrio molitor* larvae. *Eur. J. Biochem.* **269**, 4375–4383.

Lee, M. H., Osaki, T., Lee, J. Y., Baek, M. J., Zhang, R., Park, J. W., Kawabata, S., Soderhall, K., and Lee, B. L. (2004). Peptidoglycan recognition proteins involved in 1,3-beta-D-glucan-dependent prophenoloxidase activation system of insect. *J. Biol. Chem.* **279**, 3218–3227.

Lee, S. Y., Lee, B. L., and Söderhäll, K. (2004). Processing of crayfish hemocyanin subunits into phenoloxidase. *Biochem. Biophys. Res. Commun.* **322**, 490–496.

Lemaitre, B., and Hoffmann, J. (2007). The host defense of *Drosophila melanogaster. Annu. Rev. Immunol.* **25**, 697–743.

Leonard, C., Söderhäll, K., and Ratcliffe, N. A. (1985). Studies on prophenoloxidase and protease activity of *Blaberus craniifer* haemocytes. *Insect Biochem.* **15**, 803–810.

Levi, M., de Jonge, E., and van der Poll, T. (2003). Sepsis and disseminated intravascular coagulation. *J. Thromb. Thrombolysis.* **16**, 43–47.

Li, D., Scherfer, C., Korayem, A. M., Zhao, Z., Schmidt, O., and Theopold, U. (2002). Insect hemolymph clotting: Evidence for interaction between the coagulation system and the prophenoloxidase activating cascade. *Insect Biochem. Mol. Biol.* **32**, 919–928.

Liang, Z., Sottrup-Jensen, L., Aspan, A., Hall, M., and Soderhall, K. (1997). Pacifastin, a novel 155-kDa heterodimeric proteinase inhibitor containing a unique transferrin chain. *Proc. Natl. Acad. Sci. USA* **94**, 6682–6687.

Ligoxygakis, P., Pelte, N., Ji, C., Leclerc, V., Duvic, B., Belvin, M., Jiang, H., Hoffmann, J. A., and Reichhart, J. M. (2002). A serpin mutant links Toll activation to melanization in the host defence of *Drosophila. EMBO J.* **21**, 6330–6337.

Liu, C. T., Hou, R. F., Ashida, M., and Chen, C. C. (1997). Effects of inhibitors of serine protease, phenoloxidase and dopa decarboxylase on the melanization of *Dirofilaria immitis* microfilariae with *Armigeres subalbatus* haemolymph *in vitro. Parasitology* **115**(Pt 1), 57–68.

Lu, Z., and Jiang, H. (2007). Regulation of phenoloxidase activity by high- and low-molecular-weight inhibitors from the larval hemolymph of *Manduca sexta. Insect Biochem. Mol. Biol.* **37**, 478–485.

Ma, C., and Kanost, M. R. (2000). A beta1,3-glucan recognition protein from an insect, *Manduca sexta*, agglutinates microorganisms and activates the phenoloxidase cascade. *J. Biol. Chem.* **275**, 7505–7514.

McDermid, H., and Locke, M. (1983). Tyrosine storage vacuoles in insect fat body. *Tissue Cell* **15**, 137–158.

Michel, K., and Kafatos, F. C. (2005). Mosquito immunity against *Plasmodium. Insect Biochem. Mol. Biol.* **35**, 677–689.

Michel, K., Budd, A., Pinto, S., Gibson, T. J., and Kafatos, F. C. (2005). *Anopheles gambiae* SRPN2 facilitates midgut invasion by the malaria parasite *Plasmodium berghei. EMBO Rep.* **6**, 891–897.

Michel, K., Suwanchaichinda, C., Morlais, I., Lambrechts, L., Cohuet, A., Awono-Ambene, P. H., Simard, F., Fontenille, D., Kanost, M. R., and Kafatos, F. C. (2006). Increased melanizing activity in *Anopheles gambiae* does not affect development of *Plasmodium falciparum. Proc. Natl. Acad. Sci. USA* **103**, 16858–16863.

Minakhina, S., and Steward, R. (2006). Melanotic mutants in *Drosophila*: Pathways and phenotypes. *Genetics* **174**, 253–263.

Molina, F. G., Munoz, J. L., Varon, R., Lopez, J. N., Canovas, F. G., and Tudela, J. (2007). An approximate analytical solution to the lag period of monophenolase activity of tyrosinase. *Int. J. Biochem. Cell Biol.* **39**, 238–252.

Munkirs, D. D., Christensen, B. M., and Tracy, J. W. (1990). High pressure liquid chromatographic analysis of hemolymph plasma catecholamines in immune-reactive *Aedes aegypti. J. Invertebr. Pathol.* **56**, 267–279.

Nagai, T., Osaki, T., and Kawabata, S. (2001). Functional conversion of hemocyanin to phenoloxidase by horseshoe crab antimicrobial peptides. *J. Biol. Chem.* **276**, 27166–27170.

Nappi, A. J. (1974). The role of melanization in the immune reaction of larvae of *Drosophila algonquin* against *Pseudeucoila bochei. Parasitology* **66**, 23–32.

Nappi, A. J., and Christensen, B. M. (2005). Melanogenesis and associated cytotoxic reactions: Applications to insect innate immunity. *Insect Biochem. Mol. Biol.* **35**, 443–459.

Nappi, A. J., and Sugumaran, M. (1993). Some biochemical aspects of eumelanin formation in insect immunity. In *Insect Immunity* (J. P. N. Pathak, Ed.), pp. 131–148. Kluwer Academic, Boston.

Nappi, A. J., and Vass, E. (1993). Melanogenesis and the generation of cytotoxic molecules during insect cellular immune reactions. *Pigment Cell Res.* **6**, 117–126.

Nappi, A. J., and Vass, E. (2001). Cytotoxic reactions associated with insect immunity. *Adv. Exp. Med. Biol.* **484**, 329–348.

Nappi, A. J., Carton, Y., and Frey, F. (1991). Parasite-induced enhancement of hemolymph tyrosinase activity in a selected immune reactive strain of *Drosophila melanogaster. Arch. Insect Biochem. Physiol.* **18**, 159–168.

Nappi, A. J., Carton, Y., and Vass, E. (1992). Reduced cellular immune competence of a temperature-sensitive dopa decarboxylase mutant strain of *Drosophila melanogaster* against the parasite *Leptopilina boulardi. Comp. Biochem. Physiol. B* **101**, 453–460.

Nappi, A. J., Frey, F., and Carton, Y. (2005). *Drosophila* serpin 27A is a likely target for immune suppression of the blood cell-mediated melanotic encapsulation response. *J. Insect Physiol.* **51**, 197–205.

Ochiai, M., and Ashida, M. (1988). Purification of a beta-1,3-glucan recognition protein in the prophenoloxidase activating system from hemolymph of the silkworm, *Bombyx mori. J. Biol. Chem.* **263**, 12056–12062.

Ochiai, M., and Ashida, M. (1999). A pattern recognition protein for peptidoglycan – Cloning the cDNA and the gene of the silkworm, *Bombyx mori. J. Biol. Chem.* **274**, 11854–11858.

Ochiai, M., and Ashida, M. (2000). A pattern-recognition protein for beta-1,3-glucan – The binding domain and the cDNA cloning of beta-1,3-glucan recognition protein from the silkworm, *Bombyx mori. J. Biol. Chem.* **275**, 4995–5002.

Ohnishi, E., Dohke, K., and Ashida, M. (1970). Activation of prophenoloxidase II. Activation by α-chymotrypsin. *Arch. Biochem. Biophys.* **139**, 143–148.

Okun, M. R. (1996). The role of peroxidase in mammalian melanogenesis: A review. *Physiol. Chem. Phys. Med. NMR* **28**, 91–100; discussion 100–101.

Ourth, D. D., and Renis, H. E. (1993). Antiviral melanization reaction of *Heliothis virescens* hemolymph against DNA and RNA viruses *in vitro. Comp. Biochem. Physiol. B* **105**, 719–723.

Park, D. S., Shin, S. W., Hong, S. D., and Park, H. Y. (2000). Immunological detection of serpin in the fall webworm, *Hyphantria cunea* and its inhibitory activity on the prophenoloxidase system. *Mol. Cells* **10**, 186–192.

Paskewitz, S., and Riehle, M. A. (1994). Response of Plasmodium refractory and susceptible strains of *Anopheles gambiae* to inoculated Sephadex beads. *Dev. Comp. Immunol.* **18**, 369–375.

Paskewitz, S. M., Brown, M. R., Collins, F. H., and Lea, A. O. (1989). Ultrastructural localization of phenoloxidase in the midgut of refractory *Anopheles gambiae* and association of the enzyme with encapsulated *Plasmodium cynomolgi. J. Parasitol.* **75**, 594–600.

Paskewitz, S. M., Andreev, O., and Shi, L. (2006). Gene silencing of serine proteases affects melanization of Sephadex beads in *Anopheles gambiae. Insect Biochem. Mol. Biol.* **36**, 701–711.

Pau, R. N., and Kelly, C. (1975). The hydroxylation of tyrosine by an enzyme from third-instar larvae of the blowfly *Calliphora erythrocephala. Biochem. J.* **147**, 565–573.

Piao, S., Song, Y. L., Kim, J. H., Park, S. Y., Park, J. W., Lee, B. L., Oh, B. H., and Ha, N. C. (2005). Crystal structure of a clip-domain serine protease and functional roles of the clip domains. *EMBO J.* **24**, 4404–4414.

Pless, D. D., Aguilar, M. B., Falcon, A., Lozano-Alvarez, E., and Heimer de la Cotera, E. P. (2003). Latent phenoloxidase activity and *N*-terminal amino acid sequence of hemocyanin from *Bathynomus giganteus*, a primitive crustacean. *Arch. Biochem. Biophys.* **409**, 402–410.

Polanowski, A., and Wilusz, T. (1996). Serine proteinase inhibitors from insect hemolymph. *Acta. Biochim. Pol.* **43**, 445–453.

Pomerantz, S. H. (1966). The tyrosine hydroxylase activity of mammalian tyrosinase. *J. Biol. Chem.* **241**, 161–168.

Pye, A. E. (1974). Microbial activation of prophenoloxidase from immune insect larvae. *Nature* **251**, 610–613.

Ramesh, N., Sugumaran, M., and Mole, J. E. (1988). Purification and characterization of two trypsin inhibitors from the hemolymph of *Manduca sexta* larvae. *J. Biol. Chem.* **263**, 11523–11527.

Ratcliffe, N. A., and Gagen, S. J. (1977). Studies on the *in vivo* cellular reactions of insects: An ultrastructural analysis of nodule formation in *Galleria mellonella. Tissue Cell* **9**, 73–85.

Ratcliffe, N. A., Brookman, J. L., and Rowley, A. F. (1991). Activation of the prophenoloxidase cascade and initiation of nodule formation in locusts by bacterial lipopolysaccharides. *Dev. Comp. Immunol.* **15**, 33–39.

Rizki, T. M., Rizki, R. M., and Bellotti, R. A. (1985). Genetics of a *Drosophila* phenoloxidase. *Mol. Gen. Genet.* **201**, 7–13.

Rodriguez-Lopez, J. N., Tudela, J., Varon, R., Garcia-Carmona, F., and Garcia-Canovas, F. (1992). Analysis of a kinetic model for melanin biosynthesis pathway. *J. Biol. Chem.* **267**, 3801–3810.

Ross, J., Jiang, H., Kanost, M. R., and Wang, Y. (2003). Serine proteases and their homologs in the *Drosophila melanogaster* genome: An initial analysis of sequence conservation and phylogenetic relationships. *Gene* **304**, 117–131.

Satoh, D., Horii, A., Ochiai, M., and Ashida, M. (1999). Prophenoloxidase-activating enzyme of the silkworm, *Bombyx mori* – Purification, characterization, and cDNA cloning. *J. Biol. Chem.* **274**, 7441–7453.

Saul, S. J., and Sugumaran, M. (1986). Protease inhibitor controls prophenoloxidase activation in *Manduca sexta. FEBS Lett.* **208**, 113–116.

Saul, S. J., Bin, L., and Sugumaran, M. (1987). The majority of prophenoloxidase in the hemolymph of *Manduca sexta* is present in the plasma and not in the hemocytes. *Dev. Comp. Immunol.* **11**, 479–485.

Sezaki, H., Kawamoto, N., and Asada, N. (2001). Effect of ionic concentration on the higher-order structure of prophenol oxidase in *Drosophila melanogaster. Biochem. Genet.* **39**, 83–92.

Shelby, K. S., and Popham, H. J. R. (2006). Plasma phenoloxidase of the larval tobacco budworm, *Heliothis virescens*, is virucidal. *J. Insect Sci.* **6**, article 13.

Shelby, K. S., Adeyeye, O. A., Okot-Kotber, B. M., and Webb, B. A. (2000). Parasitism-linked block of host plasma melanization. *J. Invertebr. Pathol.* **75**, 218–225.

Shi, L., Li, B., and Paskewitz, S. M. (2006). Cloning and characterization of a putative inhibitor of melanization from *Anopheles gambiae. Insect Mol. Biol.* **15**, 313–320.

Shiao, S. H., Higgs, S., Adelman, Z., Christensen, B. M., Liu, S. H., and Chen, C. C. (2001). Effect of prophenoloxidase expression knockout on the melanization of microfilariae in the mosquito *Armigeres subalbatus. Insect Mol. Biol.* **10**, 315–321.

Silverman, G. A., Bird, P. I., Carrell, R. W., Church, F. C., Coughlin, P. B., Gettins, P. G., Irving, J. A., Lomas, D. A., Luke, C. J., Moyer, R. W., Pemberton, P. A., Remold-O'Donnell, E., Salvesen, G. S., Travis, J., and Whisstock, J. C. (2001). The serpins are an expanding superfamily of structurally similar but functionally diverse proteins. Evolution, mechanism of inhibition, novel functions, and a revised nomenclature. *J. Biol. Chem.* **276**, 33293–33296.

Simonet, G., Claeys, I., and Broeck, J. V. (2002a). Structural and functional properties of a novel serine protease inhibiting peptide family in arthropods. *Comp. Biochem. Physiol. B. Biochem. Mol. Biol.* **132**, 247–255.

Simonet, G., Claeys, I., Vanderperren, H., November, T., De Loof, A., and Vanden Broeck, J. (2002b). cDNA cloning of two different serine protease inhibitor precursors in the migratory locust, *Locusta migratoria. Insect Mol. Biol.* **11**, 249–256.

Simonet, G., Claeys, I., Franssens, V., De Loof, A., and Broeck, J. V. (2003). Genomics, evolution and biological functions of the pacifastin peptide family: A conserved serine protease inhibitor family in arthropods. *Peptides* **24**, 1633–1644.

Simonet, G., Breugelmans, B., Proost, P., Claeys, I., Van Damme, J., De Loof, A., and Vanden Broeck, J. (2005). Characterization of two novel pacifastin-like peptide precursor isoforms in the desert locust (*Schistocerca gregaria*): cDNA cloning, functional analysis and real-time RT-PCR gene expression studies. *Biochem. J.* **388**, 281–289.

Söderhäll, K., and Cerenius, L. (1998). Role of the prophenoloxidase-activating system in invertebrate immunity. *Curr. Opin. Immunol.* **10**, 23–28.

Stanley, D. W., Miller, J. S., and Howard, R. W. (1998). The influence of bacterial species and intensity of infections on nodule formation in insects. *J. Insect Physiol.* **44**, 157–164.

Stoltz, D. B., and Cook, D. I. (1983). Inhibition of host phenoloxidase activity by parasitoid hymenoptera. *Experientia* **39**, 1022–1024.

Sugumaran, M. (2002). Comparative biochemistry of eumelanogenesis and the protective roles of phenoloxidase and melanin in insects. *Pigment Cell Res.* **15**, 2–9.

Sugumaran, M., and Nellaiappan, K. (1991). Lysolecithin – A potent activator of prophenoloxidase from hemolymph of the lobster, *Homarus americanas. Biochem. Biophys. Res. Commun.* **176**, 1371–1376.

Sugumaran, M., and Kanost, M. (1993). Regulation of insect hemolymph phenoloxidases. In *Parasites and Pathogens of Insects* (N. Beckage, S. Thompson, and B. Federic, Eds.), pp. 317–342. Academic Press, San Diego, California.

Sugumaran, M., and Nellaiappan, K. (2000). Characterization of a new phenoloxidase inhibitor from the cuticle of *Manduca sexta. Biochem. Biophys. Res. Commun.* **268**, 379–383.

Sugumaran, M., Saul, S. J., and Ramesh, N. (1985). Endogenous protease inhibitors prevent undesired activation of prophenolase in insect hemolymph. *Biochem. Biophys. Res. Commun.* **132,** 1124–1129.

Sugumaran, M., Nellaiappan, K., Amaratunga, C., Cardinale, S., and Scott, T. (2000a). Insect melanogenesis. III. Metabolon formation in the melanogenic pathway-regulation of phenoloxidase activity by endogenous dopachrome isomerase (decarboxylating) from *Manduca sexta. Arch. Biochem. Biophys.* **378,** 393–403.

Sugumaran, M., Nellaiappan, K., and Valivittan, K. (2000b). A new mechanism for the control of phenoloxidase activity: Inhibition and complex formation with quinone isomerase. *Arch. Biochem. Biophys.* **379,** 252–260.

Sussman, A. (1949). The functions of tyrosinase in insects. *Q. Rev. Biol.* **24,** 328–341.

Tamang, D., Tseng, S. M., Huang, C. Y., Tsao, I. Y., Chou, S. Z., Higgs, S., Christensen, B. M., and Chen, C. C. (2004). The use of a double subgenomic Sindbis virus expression system to study mosquito gene function: Effects of antisense nucleotide number and duration of viral infection on gene silencing efficiency. *Insect Mol. Biol.* **13,** 595–602.

Tang, H., Kambris, Z., Lemaitre, B., and Hashimoto, C. (2006). Two proteases defining a melanization cascade in the immune system of *Drosophila. J. Biol. Chem.* **281,** 28097–28104.

Taylor, R. L. (1969). A suggested role for the polyphenol-phenoloxidase system in invertebrate immunity. *J. Invertebr. Pathol.* **14,** 427–428.

Thomas, B. R., Yonekura, M., Morgan, T. D., Czapla, T. H., Hopkins, T. L., and Kramer, K. J. (1989). A trypsin-solubilized laccase from pharate pupal integument of the tobacco hornworm, *Manduca sexta. Insect Biochem.* **19,** 611–622.

Tong, Y., and Kanost, M. R. (2005). *Manduca sexta* serpin-4 and serpin-5 inhibit the prophenol oxidase activation pathway: cDNA cloning, protein expression, and characterization. *J. Biol. Chem.* **280,** 14923–14931.

Tong, Y., Jiang, H., and Kanost, M. R. (2005). Identification of plasma proteases inhibited by *Manduca sexta* serpin-4 and serpin-5 and their association with components of the prophenol oxidase activation pathway. *J. Biol. Chem.* **280,** 14932–14942.

Trudeau, D., Washburn, J. O., and Volkman, L. E. (2001). Central role of hemocytes in *Autographa californica M* nucleopolyhedrovirus pathogenesis in *Heliothis virescens* and *Helicoverpa zea. J. Virol.* **75,** 996–1003.

Tsukamoto, T., Ichimaru, Y., Kanegae, N., Watanabe, K., Yamaura, I., Katsura, Y., and Funatsu, M. (1992). Identification and isolation of endogenous insect phenoloxidase inhibitors. *Biochem. Biophys. Res. Commun.* **184,** 86–92.

van Holde, K. E., Miller, K. I., and Decker, H. (2001). Hemocyanins and invertebrate evolution. *J. Biol. Chem.* **276,** 15563–15566.

Vie, A., Cigna, M., Toci, R., and Birman, S. (1999). Differential regulation of *Drosophila* tyrosine hydroxylase isoforms by dopamine binding and cAMP-dependent phosphorylation. *J. Biol. Chem.* **274,** 16788–16795.

Voit, R., and Feldmaier-Fuchs, G. (1990). Arthropod hemocyanins. Molecular cloning and sequencing of cDNAs encoding the tarantula hemocyanin subunits a and e. *J. Biol. Chem.* **265,** 19447–19452.

Volz, J., Osta, M. A., Kafatos, F. C., and Muller, H. M. (2005). The roles of two clip domain serine proteases in innate immune responses of the malaria vector *Anopheles gambiae. J. Biol. Chem.* **280,** 40161–40168.

Volz, J., Muller, H. M., Zdanowicz, A., Kafatos, F. C., and Osta, M. A. (2006). A genetic module regulates the melanization response of *Anopheles* to *Plasmodium. Cell Microbiol.* **8,** 1392–1405.

Wang, Y., and Jiang, H. (2004a). Purification and characterization of *Manduca sexta* serpin-6: A serine proteinase inhibitor that selectively inhibits prophenoloxidase-activating proteinase-3. *Insect Biochem. Mol. Biol.* **34,** 387–395.

Wang, Y., and Jiang, H. B. (2004b). Prophenoloxidase (proPO) activation in *Manduca sexta*: An analysis of molecular interactions among proPO, proPO-activating proteinase-3, and a cofactor. *Insect Biochem. Mol. Biol.* **34,** 731–742.

Wang, Y., and Jiang, H. (2006). Interaction of beta-1,3-glucan with its recognition protein activates hemolymph proteinase 14, an initiation enzyme of the prophenoloxidase activation system in *Manduca sexta*. *J. Biol. Chem.* **281**, 9271–9278.

Wang, Y., and Jiang, H. (2007). Reconstitution of a branch of the *Manduca sexta* prophenoloxidase activation cascade *in vitro*: Snake-like hemolymph proteinase 21 (HP21) cleaved by HP14 activates prophenoloxidase-activating proteinase-2 precursor. *Insect Biochem. Mol. Biol.* **37**, 1015–1025.

Wang, R., Lee, S. Y., Cerenius, L., and Söderhäll, K. (2001). Properties of the prophenoloxidase activating enzyme of the freshwater crayfish, *Pacifastacus leniusculus*. *Eur. J. Biochem.* **268**, 895–902.

Wang, X., Rocheleau, T. A., Fuchs, J. F., Hillyer, J. F., Chen, C. C., and Christensen, B. M. (2004). A novel lectin with a fibrinogen-like domain and its potential involvement in the innate immune response of *Armigeres subalbatus* against bacteria. *Insect Mol. Biol.* **13**, 273–282.

Wang, X. G., Fuchs, J. F., Infanger, L. C., Rocheleau, T. A., Hillyer, J. F., Chen, C. C., and Christensen, B. M. (2005). Mosquito innate immunity: Involvement of beta 1,3-glucan recognition protein in melanotic encapsulation immune responses in *Armigeres subalbatus*. *Mol. Biochem. Parasitol.* **139**, 65–73.

Waterhouse, R. M., Kriventseva, E. V., Meister, S., Xi, Z., Alvarez, K. S., Bartholomay, L. C., Barillas-Mury, C., Bian, G., Blandin, S., Christensen, B. M., Dong, Y., Jiang, H., Kanost, M. R., Koutsos, A. C., Levashina, E. A., Li, J., Ligoxygakis, P., Maccallum, R. M., Mayhew, G. F., Mendes, A., Michel, K., Osta, M. A., Paskewitz, S., Shin, S. W., Vlachou, D., Wang, L., Wei, W., Zheng, L., Zou, Z., Severson, D. W., Raikhel, A. S., Kafatos, F. C., Dimopoulos, G., Zdobnov, E. M., and Christophides, G. K. (2007). Evolutionary dynamics of immune-related genes and pathways in disease-vector mosquitoes. *Science* **316**, 1738–1743.

Williams, M. J. (2007). *Drosophila* hemopoiesis and cellular immunity. *J. Immunol.* **178**, 4711–4716.

Wyatt, G. (1961). The biochemistry of insect hemolymph. *Annu. Rev. Entomol.* **6**, 75–102.

Xie, J. J., Chen, Q. X., Wang, Q., Song, K. K., and Qiu, L. (2007). Activation kinetics of cetylpyridinium chloride on the prophenol oxidase from pupae of blowfly (*Sarcophaga bullata*). *Pestic. Biochem. Physiol.* **87**, 9–13.

Yasuhara, Y., Koizumi, Y., Katagiri, C., and Ashida, M. (1995). Reexamination of properties of prophenoloxidase isolated from larval hemolymph of the silkworm *Bombyx mori*. *Arch. Biochem. Biophys.* **320**, 14–23.

Yoshida, H., Ochiai, M., and Ashida, M. (1986). Beta-1,3-glucan receptor and peptidoglycan receptor are present as separate entities within insect prophenoloxidase activating system. *Biochem. Biophys. Res. Commun.* **141**, 1177–1184.

Yoshida, H., Kinoshita, K., and Ashida, M. (1996). Purification of a peptidoglycan recognition protein from hemolymph of the silkworm, *Bombyx mori*. *J. Biol. Chem.* **271**, 13854–13860.

Yu, X. Q., and Kanost, M. R. (2004). Immulectin-2, a pattern recognition receptor that stimulates hemocyte encapsulation and melanization in the tobacco hornworm, *Manduca sexta*. *Dev. Comp. Immunol.* **28**, 891–900.

Yu, X. Q., Zhu, Y. F., Ma, C., Fabrick, J. A., and Kanost, M. R. (2002). Pattern recognition proteins in *Manduca sexta* plasma. *Insect Biochem. Mol. Biol.* **32**, 1287–1293.

Yu, X. Q., Jiang, H., Wang, Y., and Kanost, M. R. (2003). Nonproteolytic serine proteinase homologs are involved in prophenoloxidase activation in the tobacco hornworm, *Manduca sexta*. *Insect Biochem. Mol. Biol.* **33**, 197–208.

Zhang, R., Cho, H. Y., Kim, H. S., Ma, Y. G., Osaki, T., Kawabata, S., Soderhall, K., and Lee, B. L. (2003). Characterization and properties of a 1,3-beta-D-glucan pattern recognition protein of *Tenebrio molitor* larvae that is specifically degraded by serine protease during prophenoloxidase activation. *J. Biol. Chem.* **278**, 42072–42079.

Zhang, G., Lu, Z. Q., Jiang, H., and Asgari, S. (2004). Negative regulation of prophenoloxidase (proPO) activation by a clip-domain serine proteinase homolog (SPH) from endoparasitoid venom. *Insect Biochem. Mol. Biol.* **34**, 477–483.

Zhao, X., Ferdig, M. T., Li, J., and Christensen, B. M. (1995). Biochemical pathway of melanotic encapsulation of *Brugia malayi* in the mosquito, *Armigeres subalbatus*. *Dev. Comp. Immunol.* **19**, 205–215.

Zhao, M., Soderhall, I., Park, J. W., Ma, Y. G., Osaki, T., Ha, N. C., Wu, C. F., Soderhall, K., and Lee, B. L. (2005). A novel 43-kDa protein as a negative regulatory component of phenoloxidase-induced melanin synthesis. *J. Biol. Chem.* **280**, 24744–24751.

Zhu, Y., Johnson, T. J., Myers, A. A., and Kanost, M. R. (2003a). Identification by subtractive suppression hybridization of bacteria-induced genes expressed in *Manduca sexta* fat body. *Insect Biochem. Mol. Biol.* **33**, 541–559.

Zhu, Y., Wang, Y., Gorman, M. J., Jiang, H., and Kanost, M. R. (2003b). *Manduca sexta* serpin-3 regulates prophenoloxidase activation in response to infection by inhibiting prophenoloxidase-activating proteinases. *J. Biol. Chem.* **278**, 46556–46564.

Zlateva, T., Di Muro, P., Salvato, B., and Beltramini, M. (1996). The *o*-diphenol oxidase activity of arthropod hemocyanin. *FEBS Lett.* **384**, 251–254.

Zou, Z., and Jiang, H. (2005). Gene structure and expression profile of *Manduca sexta* prophenoloxidase-activating proteinase-3 (PAP-3), an immune protein containing two clip domains. *Insect Mol. Biol.* **14**, 433–442.

Zou, Z., Wang, Y., and Jiang, H. (2005). *Manduca sexta* prophenoloxidase activating proteinase-1 (PAP-1) gene: Organization, expression, and regulation by immune and hormonal signals. *Insect Biochem. Mol. Biol.* **35**, 627–636.

Zou, Z., Lopez, D. L., Kanost, M. R., Evans, J. D., and Jiang, H. (2006). Comparative analysis of serine protease-related genes in the honey bee genome: Possible involvement in embryonic development and innate immunity. *Insect Mol. Biol.* **15**, 603–614.

5

EVIDENCE FOR SPECIFICITY AND MEMORY IN THE INSECT INNATE IMMUNE RESPONSE

LINH N. PHAM AND DAVID S. SCHNEIDER

Department of Microbiology and Immunology, Stanford University,
Stanford, CA 94305-5124, USA

ABSTRACT: Insects, like other invertebrates, rely solely on their innate immune response to fight invading microbes. Typically, innate immunity is characterized as lacking memory and specificity. However, experimental evidence suggests that the insect immune response does not respond identically to repeat challenges and can exhibit a high degree of specificity. After a priming dose of bacteria, some insects exhibit a prolonged activation of the immune response that nonspecifically protects against subsequent challenges. In some cases, microbial challenge of mothers results in the production of more resistant offspring. In addition to these two examples of insect memory lacking specificity, transplantation and bacterial priming experiments have demonstrated both specificity and memory in the insect innate immune response. Here we review the current literature and discuss potential mechanisms underlying specific memory in insect immunity.

Abbreviations:

AMP	=	antimicrobial peptide
Dscam	=	Down syndrome cell adhesion molecule
FREP	=	fibrinogen-related protein
JAK/STAT	=	Janus kinase/signal transducer and activator of transcription
JNK	=	jun N-terminal kinase
LPS	=	lipopolysaccharide
MAMPs	=	microbial-associated molecular patterns
PGRP	=	peptidoglycan recognition protein
PO	=	phenoloxidase
proPO	=	prophenoloxidase
RAG	=	recombination-activating gene
TEP	=	thioester-containing protein
TNF	=	tumor necrosis factor

5.1　INTRODUCTION

Immune responses are historically divided into two categories, adaptive and innate. For the purposes of this review, we will define a classically adaptive immune response as recombination-activating gene (RAG)-dependent immunity that involves somatic recombination in T and B lymphocytes. RAG-dependent immunity is unique to vertebrates and is described as possessing the characteristics of specificity and memory. Upon primary exposure to an immune challenge, an activation period of 2 weeks is required to achieve a maximal immune response (Janeway, 2005). However, upon secondary exposure, the immune response is more rapid (on the order of hours) and stronger. Memory is the ability to raise an enhanced second response, and this ability persists for the life of the organism. Specificity refers to the fact that the enhanced response is only triggered by the same elicitor encountered during the first exposure and is directed only against that elicitor. In contrast, innate immune responses are conserved among all organisms and these responses are activated immediately upon encountering an elicitor, on the order of seconds. Typically, innate immune responses are not characterized as possessing memory – i.e. the innate immune response is thought to respond identically to repeated challenges (Hoffmann, 2003; Janeway, 2005). Moreover, innate immunity is usually assumed to exhibit only broad specificity, such as the ability to distinguish between different classes of microbes (Hoffmann, 2003; Janeway, 2005; Lemaitre and Hoffmann, 2007).

Recent evidence demonstrates that invertebrates are able to modulate their innate immune response after repeated challenges (as reviewed in Kurtz, 2005; Little et al., 2005). We divide these 'adaptive' innate immune responses into two different classes. The first class involves two categories of nonspecific memory: prolonged nonspecific activation and transgenerational memory. The second class includes transplantation and bacterial priming experiments that demonstrate the insect innate immune response can exhibit both memory and greater specificity than would be anticipated (i.e. cross-protection is not observed between different microbial challenges). After discussion of these two classes, we close with speculation concerning potential molecular mechanisms that may contribute to specificity and memory.

To understand these experiments, it is important to be familiar with the basic foundations of the insect immune response; therefore, we begin with an overview of insect immunity. Later in this chapter, we will present evidence concerning how these branches of the immune response can be modulated after immune challenges.

5.2   BACKGROUND ON INSECT IMMUNITY

In this section we present a general characterization of the humoral and cellular immune responses of insects. We then discuss immune responses that are characteristic of insects including melanization, behavioral fever, and social immune responses. Last, we will discuss what is known about conserved signaling pathways that mediate these processes.

5.2.1   HUMORAL IMMUNITY

The best-characterized immune response of insect, humoral immunity involves the secretion of soluble factors into the open circulatory system of the animal. Many of these factors are secreted by the fat body, the insect homolog of the mammalian liver. The majority of literature focuses on the induction of antimicrobial peptides (AMPs), but data has emerged more recently concerning complement-like factors.

AMPs are small peptides with antimicrobial properties. The number and type vary from insect to insect, although some AMPs are conserved among many species (such as cecropin and defensin) (Bulet et al., 1999). For example, in the fruit fly *Drosophila melanogaster*, genome analysis has predicted at least 30 different AMPs (Adams et al., 2000). Experiments with transgenic flies demonstrated that expression of AMPs in an immunocompromised mutant fly is sufficient to protect against some infections (Tzou et al., 2002). Most insect AMPs are thought to function by disrupting bacterial membranes (Lemaitre and Hoffmann, 2007). In addition to antimicrobial activities, there are examples of vertebrate AMPs acting as mitogens, chemoattractants, and microbial-associated molecular patterns (MAMPS) (Oppenheim and Yang, 2005). It is unknown whether insect AMPs possess properties in addition to microbial killing.

Injection of microbes rapidly induces AMP transcription that typically peaks within 6 h and is downregulated by 3 days post-injection (Lemaitre et al., 1997). However, some AMPs like drosomycin have been shown to be elevated transcriptionally for 1 week and can persist in the hemolymph for 1 month (Uttenweiler-Joseph et al., 1997; Uttenweiler-Joseph et al., 1998). Flies injected with persistent pathogenic bacteria like *Salmonella typhimurium* exhibit elevated AMP induction until they die due to infection (Brandt and Schneider, 2007). AMPs are also induced by wounding alone, but the amplitude of the response is lower than what is observed after septic injury (Lemaitre et al., 1997).

The majority of AMPs detectable in the hemolymph after wounding are produced by the fat body, although hemocytes also produce AMPs (Lemaitre and Hoffmann, 2007). It is unclear whether hemocyte-produced AMPs are only antimicrobial, or if they serve a signaling function. During infections in the absence of wounding, the barrier epithelia are also able to produce AMPs. For example, oral challenge with *Erwinia carotovora carotovora* induces AMP production by gut epithelial cells (Basset et al., 2000). It is not clear whether the AMPs are secreted into the gut lumen or the hemocoel. The role of the Toll and Imd pathways in mediating AMP induction has been well characterized and will be discussed in the signaling section below.

Insects possess members of the α_2-macroglobulin/complement family of proteins such as α_2-macroglobulin, Mcr, and thioester-containing proteins (TEPs). α_2-macroglobulin is a broadly conserved complement-like protein (Nonaka and Yoshizaki, 2004). In other organisms, this protein binds to secreted proteases and mediates their uptake and inactivation by host cells, although it has not been functionally studied for its role in the insect immune response. Mcr is another member of this family that is secreted by hemocytes. Studies in *Drosophila* demonstrate that Mcr binds to the surface of the fungus *Candida albicans* and promotes its subsequent phagocytosis (Stroschein-Stevenson et al., 2006). The *Drosophila* genome possesses four functional TEP homologs that are transcriptionally induced in the fat body and hemocytes after bacterial challenge (Lagueux et al., 2000). Using RNAi, Stroschein-Stevenson et al. (2006) demonstrated that TEP II is required for efficient phagocytosis of the Gram-negative bacteria *E. coli* and TEP III in necessary for efficient phagocytosis of the Gram-positive bacteria *S. aureus*. Similar results were reported in the mosquito *Anopheles gambiae* (Levashina et al., 2001). Mosquito TEP I binds to the surface of the parasite *Plasmodium berghei* and is implicated in limiting the number of oocysts that develop in the mosquito midgut. It is evident that different TEPS opsonize the surface of different microbes, mediating phagocyte recognition and killing.

5.2.2 CELLULAR IMMUNE RESPONSE

Hemocytes are cells that circulate in the hemolymph (see chapter 2). In *Drosophila* larvae, hemocytes circulate freely in the hemolymph, but can adhere to the body wall after wounding or cluster at sites of infection. Hemocytes in adult flies appear to be attached and sessile. Although hematopoiesis has been studied in embryos and larvae (reviewed in Crozatier and Meister, 2007), hemocyte proliferation has not been described in adult flies. The Toll (Qiu et al., 1998) and Janus kinase/signal transducer and activator of transcription (JAK/STAT) pathways (Harrison et al., 1995; Luo et al., 1995) have been implicated in hemocyte activation, but how these pathways regulate hemocyte activation remains uncharacterized. Hemocyte-specific antibodies have been generated, but there are no markers that differentially stain activated cells (Goto et al., 2001; Kurucz et al., 2003).

Hemocytes possess a variety of immune functions. The most obvious is phagocytosis that involves the killing of ingested microbes. Injection of polystyrene beads inhibits subsequent phagocytosis, and using bead injection experiments, it is possible to assess the relative contribution of phagocyte-mediated killing to the immune response (Elrod-Erickson et al., 2000). The clearest example is in the context of *S. pneumoniae* infections where bead inhibition renders the fly incapable of killing a normally sublethal dose of 20 bacteria (Pham et al., 2007).

Hemocytes also secrete a variety of opsonins and AMPs. In addition to mediating bacterial killing, it is also possible that secreted molecules may be involved in coordinating the immune response with the fat body. For example, hemocyte-secreted upd3 activates JAK/STAT signaling in the fat body (Agaisse et al., 2003).

Encapsulation is a process whereby hemocytes are activated and form a multilayered capsule around a foreign object, such as a parasitic wasp egg. The activated hemocytes produce melanin and reactive oxygen species (ROS) that kill the parasite within (Nappi et al., 1995). Nodulation is a process where hemocytes aggregate around foci of bacteria.

5.2.3 MELANIZATION

Melanization is an immediate localized blackening reaction caused by phenoloxidase (PO) activity that occurs at wound sites and around foreign objects like parasitic wasp eggs (see chapter 4). Some bacteria such as *Listeria monocytogenes* and *Salmonella typhimurium* elicit a melanotic response at foci of infection (Brandt and Schneider, 2007; Mansfield et al., 2003). It is thought that melanization contributes to the immune response by walling off cuticular breaches and sequestering microbes. ROS are also concentrated at sites of melanin production and mediate direct killing of microbes (Nappi et al., 1995).

Melanization is initiated by proteases that cleave the zymogen prophenoloxidase (proPO) into active PO, which in turn oxidizes tyrosine derivatives into quinones that polymerize to form melanin. In *Drosophila*, crystal cells play a role in activating the melanotic response. Recent work has demonstrated that jun N-terminal kinase (JNK) signaling and the tumor necrosis factor (TNF) homolog eiger are involved in crystal cell activation (Bidla et al., 2007). Intriguingly, Bidla et al. (2007) also demonstrated that in addition to microbial products, endogenous signals from dying hemocytes play a role in activating the melanotic response in fruit flies. Because other insects lack crystal cells, it will be interesting to determine whether JNK and/or eiger signaling play a role in activating melanization across species.

In some insects, melanin and circulating proPO levels are indicators of immune status. For example, the mealworm beetle *Tenebrio molitor* (Barnes and Siva-Jothy, 2000), the noctuid *Spodoptera exempta* (Reeson et al., 1998), and the desert locust *Schistocerca gregaria* (Elliot et al., 2003) exhibit variation in cuticle coloration. In general, darker insects (indicative of more melanin in the cuticle) are more resistant to infections. Additionally, bumble bees (*Bombus terrestris*) and field crickets (*Gryllus campestris*) with higher levels of circulating proPO are more resistant to infections than related cohorts (Brown et al., 2003; Jacot et al., 2005).

5.2.4 BEHAVIORAL FEVER

Many larger insects can regulate their temperature. However, insects such as the desert locust *Schistocerca gregaria*, two species of grasshopper (*Camnula pellucida* and *Melanoplus sanguinipes*), and two species of cricket (*Acheta domesticus* and *Gryllus bimaculatus*) are also able to thermoregulate by basking in the sun. Such regulation occurs on two levels: passive and active. First, these insects prefer locations where their internal temperature can reach 28–33°C regardless of their infection state (Adamo, 1998; Carruthers et al., 1992; Elliot et al., 2002; Inglis et al., 1996; Louis et al., 1986). These elevated temperatures are correlated with limiting fungal (*Metarhizium anisopliae* or *Entomophaga grylli*) proliferation (Carruthers et al., 1992; Elliot et al., 2002). Second, it has also been demonstrated that upon infection, both crickets and locusts actively migrate to warmer locations where their internal temperature can reach a higher mean value between 33°C and 37°C (Adamo, 1998; Inglis et al., 1996; Louis et al., 1986). At these higher temperatures, proliferation of natural pathogens such as *Metarhizium anisopliae*, *Beauveria bassiana*, and *Rickettsiella grylli* is severely limited. The molecular signals underlying this behavior mechanism remain unknown. However, it is clear that the immune response is able to detect an infection and communicate with the nervous system to cause differential thermoregulatory behavior (see chapter 6). Fevers can be generated socially; bee hive temperatures are tightly regulated and usually maintained at approximately 35°C even though the ambient temperature is colder in the winter and hotter in the summer. If the hive becomes infected with a fungal (*Ascosphaera apis*) pathogen, hive temperature increases half a degree (Starks et al., 2000).

5.2.5 SOCIAL IMMUNITY

The last immune response unique to insects involves group facilitation of disease resistance among social insects. A variety of insects exhibit density-dependent prophylaxis whereby individuals from crowded populations are more resistant to infection than individuals maintained at low density. Insects such as the mealworm beetle *Tenebrio molitor* and the noctuid moth *Spodoptera exempta* adjust their immune response according to the increased risk from pathogens when at high population densities (Barnes and Siva-Jothy, 2000; Reeson et al., 1998). Termites also exhibit an interesting array of social immune behaviors (Traniello et al., 2002). When a termite mound becomes infected, there is communication throughout the nest. Termites begin producing antibiotic secretions from their exocrine glands that help limit microbial growth (Rosengaus et al., 2000). Many studies have focused on allogrooming and have shown that the extent of mutual grooming is increased in infected nests (Rosengaus et al., 1998). Pathogen alarm behavior is a particularly striking response; the infection status of a nest can be communicated to the rest of the termite colony (Rosengaus et al., 1999). For example, after detecting the presence of pathogenic spores, termites in direct contact make strong vibrations along the tunnel substrate. These vibrations serve as a signal to unexposed nestmates in the vicinity and cause them to flee the contaminated site. The mechanism behind this process is uncharacterized. It will be interesting to determine whether other insects are also capable of raising similar complex responses.

5.2.6   CONSERVED SIGNALING PATHWAYS

Due to the abundance of genetic tools present in *Drosophila*, the field of insect immunity is rich with information about signaling pathways that are conserved among many organisms. The Toll and Imd pathways were first characterized within the context of AMP induction.

The Toll receptor is conserved among multicellular organisms. In the fly, Toll was first identified for its contribution to dorsoventral patterning of the embryo. The immune functions of Toll signaling were first discovered in the fly (Lemaitre et al., 1996), and since then, much work in vertebrate innate immunity has focused on characterizing this family of pattern recognition receptors (Trinchieri and Sher, 2007). Vertebrates possess multiple Toll-like receptors that are activated by a variety of MAMPs (Trinchieri and Sher, 2007). In the fly, Toll is not activated directly by a MAMP; instead the ligand for Toll is the protein spaetzle. Proteolytic cleavage of spaetzle is required to produce an active ligand (Jang et al., 2006). There are nine Toll homologs in the fly, only one Toll has been shown to have an immune function in the fly (De Gregorio et al., 2001). The Toll pathway is activated by fungal (β1,3-glucan) and bacterial compounds (Lys-peptidoglycan), resulting in signaling through the NF-kappa B homologs Dif and Dorsal (Filipe et al., 2005; Gottar et al., 2006). This signaling pathway has been extensively described using AMPs as a readout. Toll has also been implicated in hemocyte division or activation because mutant flies with constitutively active alleles exhibit elevated hemocyte numbers (Qiu et al., 1998).

The Imd pathway also results in the activation of an NF-kappa B-like transcription factor, Relish. This pathway is activated by the bacterial product DAP-peptidoglycan (Kaneko and Silverman, 2005; Kaneko et al., 2004), and the downstream signaling components resemble the canonical TNF signaling pathway in vertebrates that terminates in NF-kappa B activation. Activation of the Imd pathway results in AMP induction, and its role in hemocyte activation remains uncharacterized. In vertebrates, the

TNF receptor signals through both NF-kappa B and JNK. Imd also signals through the JNK pathway (Delaney et al., 2006; Silverman et al., 2003).

Drosophila encodes a TNF homolog, eiger, which was first characterized as being able to induce apoptosis when expressed ectopically (Moreno et al., 2002). Whereas the downstream signaling components of the Imd pathway consist of NF-kappa B-like TNF signaling, eiger has only been described to signal through the JNK pathway. Eiger mutants have immune phenotypes: mutant flies lacking eiger are sensitive to extracellular infections, suggesting that TNF activation helps fight these diseases (Schneider et al., 2007). However, eiger mutants are protected against two intracellular pathogens, *Mycobacterium marinum* and *Salmonella typhimurium* (Brandt et al., 2004; Schneider et al., 2007). Similar to vertebrates, persistent activation of eiger appears to have pathological consequences in the context of *S. typhimurium* infections (Brandt et al., 2004; Schneider et al., 2007). Potential interactions between eiger and the downstream components of the Imd pathway remain uncharacterized. Eiger has also been shown to be involved in melanization and clot formation (Bidla et al., 2007).

The role of JNK signaling in *Drosophila* immunity is still under consideration. Characterizing immune functions of JNK has been difficult due to developmental defects and lethality in mutants fly lines. Recent work in larvae demonstrated that JNK signaling is required for wound healing (Galko and Krasnow, 2004). Earlier reports have suggested that AP-1 like transcription factors, which are downstream of the JNK signaling pathway, may regulate AMP expression (Meister et al., 1994). However, this hypothesis has yet to be confirmed. Recently, JNK has also been shown to be involved in crystal cell rupture, resulting in the activation of the melanotic response (Bidla et al., 2007).

JAK/STAT signaling has been shown to be involved in communication from the hemocyte to the fat body. There are three potential cytokines that may bind to the receptor domeless: upd, upd2, and upd3. Only upd3 has been demonstrated to have an immune function. There is genetic evidence that hemocytes secrete upd3 which activates JAK/STAT signaling in the fat body, resulting in transcription of turandot A (Agaisse et al., 2003; Dostert et al., 2005). This gene is used as a read-out for JAK/STAT activation, but its function remains unknown. JAK/STAT activation has also been implicated in fighting viral infections because mutants exhibit sensitivity to viral infection (Dostert et al., 2005).

5.3   ADAPTIVE ASPECTS OF INSECT INNATE IMMUNITY

Because insects are invertebrates that lack RAG-dependent immunity, by definition, insects possess only an innate immune response (Hoffmann, 2003). Innate immunity is typically characterized as acting nonspecifically and identically to repeat challenges. However, it is clear that insects can modulate their immune response after multiple immune challenges (summarized in Table 5.1). We will

TABLE 5.1 Summary of Adaptive Components of Innate Immunity in Insects

Insect species	Priming dose	Challenge	Specificity	Memory*	Effect on offspring	Survival	Microbial killing	AMP	Hemocyte	Melanotic response	Reference
Gryllus campestris	LPS	M. anisopliae			.			+		+	Jacot (2005)
Tenebrio molitor	LPS	M. anisopliae				+		+		+	Moret (2003)
Manduca sexta	E. coli	P. luminescens	-			+		+	-		Eleftherianos (2006)
Bombus terrestris	C. bombi				Male offspring					+	Brown (2003)
	LPS			+ tg				-		+	Moret (2001)
	P. fluorescens	P. fluorescens	+	+		+		+			Sadd (2006)
Galleria mellonella	P. aeruginosa	P. aeruginosa	(-)	(-)		+		+		+	Boman (1972)
Drosophila melanogaster	E. cloacae	E. cloacae	(-)			+		+			
	S. pneumoniae	S. pneumoniae	+	+		+	+	-	+		Pham (2007)
	B. bassiana	B. bassiana	+	+		+			+	-	Pham (2007)
Zootermopsis angusticollis	P. aeruginosa	P. aeruginosa		(+)		+					Rosengaus (1999)
	M. anisopliae	M. anisopliae		+		+					Rosengaus (1999)
Schistocerca gregaria	M. anisopliae	M. anisopliae		+ tg	Solitary locusts	+					Elliot (2003)
Ephestia kuehniella	B. thuringiensis toxin	B. thuringiensis toxin		+ tg	Resistant offspring	+				+	Rahman (2004)
Blaberus craniifer	B. orientalis cuticular transplant	B. orientalis cuticular transplant	+	(+)		Rejection			+		Dularay (1987)
Periplaneta Americana	P. americana cuticular transplant	P. americana cuticular transplant	+	(-)		rejection					Hartman (1989)
	P. aeruginosa	P. aeruginosa	(+)	(+)		+					Faulhaber (1992)

The following notations are used throughout: '+' indicates supporting evidence; '-' means the study tested and did not find supporting evidence; (+) or (-) in brackets denotes limited evidence; '+ tg' indicates support for transgenerational memory.

*Memory is defined as an enhanced immune response upon second challenge and specificity is not required.

discuss three ways in which the insect immune response changes. The first two modes are examples of nonspecific memory: prolonged activation of the immune response and transgenerational effects. Last, we will present data that insect immune responses exhibit memory and a greater degree of specificity than expected for the innate immune response, suggesting that there are adaptive aspects of innate immunity.

5.3.1   NONSPECIFIC MEMORY

To discuss memory, it is best to begin with a functional definition that will be valid independent of mechanism. The simplest definition of immune memory is that a host will raise an enhanced immune response following an initial encounter with an elicitor. A straightforward demonstration of memory is to determine whether a priming dose of a microbial agent changes the immune response to a second challenge. Enhancement of the immune response can be assessed mechanistically by measuring activation states of immune effectors, or can be assayed functionally by determining whether there is a protective effect against a second challenge (Fig. 5.1). In this section we discuss responses that are broad and lack specificity.

5.3.1.1   Prolonged Nonspecific Activation: Examination of Immune Effectors

Schmid-Hempel's group examined the effects of the noninvasive gut trypanosome *Crithidia bombi* on the immune response of the bumble bee, *Bombus terrestris* (Brown et al., 2003). *C. bombi* has a high prevalence in bumble bees and is transmitted via ingestion of contaminated bee feces. Although it is normally a commensal, under starvation conditions background mortality of *C. bombi*-infected bumble bees can increase 50%. Brown et al. (2003) found that infected bumble bees have twice as much circulating proPO levels present in the hemolymph compared to uninfected bees. This difference in proPO levels persisted to the last time point that the researchers assayed, 14 days post-infection. Thus, in response to trypanosome infection, the bees have constitutively upregulated their potential to produce a melanotic response. Notably, *C. bombi* is noninvasive and cannot leave the intestinal tract whereas the change in the melanotic response occurs in the hemolymph. Somehow the gut is able to influence the genes that regulate circulating proPO levels in the hemolymph. Identifying the signaling process by which the intestinal tract can communicate with the immune response will be an interesting avenue of study. There are at least two interesting possibilities worthy of further investigation. First, a host signal could be transmitted from the gut to the hemolymph to regulate proPO production. Alternatively, trypanosome infection may induce gut pathology that leads to activation of proPO production. Perhaps trypanosomes act like *Bacillus thuringiensis* toxin and damage the gut, resulting in gut microbes entering the hemolymph (Broderick et al., 2006).

Brinkhof's laboratory found that in the field cricket *Gryllus campestris*, lipopolysaccharide (LPS) stimulation of nymphs resulted in adults with increased

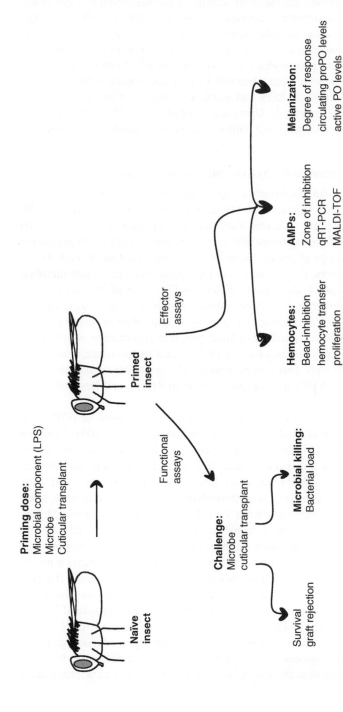

FIGURE 5.1   Methods to test for specificity and memory in the insect immune response.

Assaying for memory in the insect immune response involves testing for enhancement of the immune response after an initial encounter with an elicitor. Effector functions can be compared between naïve and primed insects. Functional assays demonstrate improved survival, faster transplant rejection, or increased microbial killing. Assessing specificity involves performing the same matrix of experiments and comparing the results from a variety of priming doses and a variety of challenges.

Abbreviations: qRT-PCR, quantitative reverse transcription PCR; MALDI-TOF, Matrix-assisted laser desorption/ionization – time of flight mass spectrometry; LPS, lipopolysaccharide; PO, phenoloxidase.

proPO levels and increased antibacterial activity in the hemolymph (Jacot et al., 2005). The increased antibacterial activity is most likely due to AMPs, although as mentioned before, other soluble factors may contribute to bacterial killing. Jacot et al. (2005) found that active PO levels were not upregulated in challenged insects, only the inactive proenzyme levels were increased. As mentioned above, elevated proPO levels were also observed in Crithidia-infected bumble bees. PO activation produces toxic ROS and *quinones* which can cause nonspecific damage to the host (see chapter 4). Upregulation of proPO synthesis increases the sensitivity of the melanotic response without the damage caused by systemic PO activation.

5.3.1.2   Prolonged Nonspecific Activation: Survival Studies

In 1972 Hans Boman published a paper discussing *Enterococcus cloacae* infections in *Drosophila*. Boman et al. (1972) found that priming fruit flies with heat-killed *E. cloacae* protects against a second challenge of *E. cloacae* delivered up to 4 days later. *E. cloacae*-primed flies contained fewer bacteria when injected with a second challenge of either *Pseudomonas aeruginosa* or *E. coli*. Using zone-of-inhibition assays, they demonstrated the persistence of an antimicrobial activity in hemolymph 4 days post-challenge. The zone-of-inhibition assay involves placing hemolymph in defined spots on a poured bacterial agar suspension (Boman et al., 1972). Antibacterial activity is quantified by measuring the area of bacterial clearance around the defined spot. It is important to note that this assay does not distinguish between bacterial killing and bacteriostatic compounds present in the hemolymph. Our current understanding of *Drosophila* immunity supports the idea that AMPs are present in the hemolymph and responsible for enhanced survival.

In response to bacteria such as *E. coli* and *M. luteus*, the peak of AMP transcription is usually around 6 h post-injection (De Gregorio et al., 2001; Lemaitre et al., 1997). Although AMP transcriptional induction is usually used as a proxy for AMP activity, it is important to note that transcriptional changes are not necessarily indicative of AMP levels in the hemolymph. Even though AMP transcription was downregulated by 3 days post-challenge, MALDI-TOF mass spectrometry analysis of hemolymph from *E. coli* and *M. luteus*-challenged flies demonstrated that AMP levels may persist for weeks in flies (Uttenweiler-Joseph et al., 1997). To fully assess permanent changes to AMP activity, it is important to examine both transcriptional changes and activity levels. The mechanism underlying elevation of antimicrobial compounds remains unknown (Fig. 5.2). It is possible that the AMPs are regulated at the level of transcription, translation, or post-translation (secretion or processing) such that after a priming dose of *E. cloacae*, fruit flies persistently produce AMPs. Alternatively, elevated AMP activity could result from a delay in the rate at which AMPs are degraded. Another possibility is that the AMP response is enhanced during a second challenge. It will be interesting to determine how circulating antimicrobial activity can be regulated at different levels.

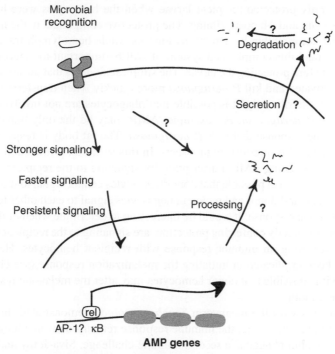

FIGURE 5.2 Potential modes of AMP regulation.
Experimental evidence suggests that the kinetics of AMP induction can be altered after multiple wounding events, and that AMPs can be persistently induced in the context of certain infections. Alterations in upstream signaling events may produce these situations. It is also possible that AMPs are regulated at translational and post-translational levels. Question marks indicate modes of regulation that may exist but have not yet been documented.

Rosengaus et al. (1999) found that priming doses of glutaraldehyde-fixed *P. aeruginosa* protect the dampwood termite *Zootermopsis angusticollis* against a second lethal dose. This protective effect persists up to 1 week and diminishes thereafter. Rosengaus et al. (1999) also demonstrated that a sublethal dose of the fungus *M. anisopliae* protects termites for 3 weeks. Specificity of the response remains to be determined. For example, it is unknown whether a priming dose of *P. aeruginosa* can protect against *M. anisopliae* and vice versa.

Formalin-killed *P. aeruginosa* or LPS derived from *P. aeruginosa* also protects the wax moth *Galleria mellonella* from a second challenge of *P. aeruginosa* (Deverno et al., 1983). This protective effect persisted between 3 and 18 h post-immunization. Deverno et al. (1983) performed hemolymph transfers to determine the nature of the protective effect. Whole hemolymph protected recipient larvae when collected from primed donor larvae between 3 and 40 h post-immunization. The protective factor could be the original elicitor, or an effector like AMPs. In either case, this activity persists for 40 h. In contrast, transfer of hemocytes only protected recipient larvae when the hemocytes were harvested between 30 min and 4 h post-priming. The protective component in the hemocyte fraction did not maintain its activity as long as whole hemolymph fractions. To fully understand this complicated system, it will be important to determine how the hemocytes are protecting the larvae. The simplest model is that the donor hemocytes are activated and kill *P. aeruginosa* more quickly when transferred into the recipient fly. Alternatively, it is possible that phagocytes are not involved in protection from *P. aeruginosa*. For example, AMPs may be the only branch of the insect immune response that kills *P. aeruginosa*. The fat body is responsible for the majority of AMP production in insects. In this scenario, primed donor hemocytes may be inducing AMP transcription by signaling to the recipient fat body. Alternatively, it is also possible that donor hemocytes cannot signal to the fat body in the recipient, and donor and recipient hemocytes signal to each other to coordinate the immune response. It would be interesting to determine whether the donor hemocytes are directly conferring protection, are signaling to the recipient fat body, or are coordinating an immune response with recipient hemocytes. Hemocytes have also been implicated in initiating the melanization response (see chapters 2 and 4), so it is possible that donor hemocytes may alter the melanotic response of the recipient moth.

Experiments with the mealworm *Tenebrio molitor* demonstrated that after a primary immune challenge, the immune response remains activated long enough to protect the animal against a second immune challenge. Siva-Jothy and his colleagues call this elevated activation state 'responsive mode prophylaxis'. Moret et al. (2003) simulated a primary immune insult by introducing the bacterial membrane component LPS with a pin prick and looked for protection against a subsequent infection with the broad fungal entomopathogen *Metarhizium anisopliae*. Immune-stimulated mealworms were protected against *M. anisopliae* infections for up to 1 week. The difference in survival was likely due to increased levels of antimicrobial activity present in the hemolymph of primed insects. The researchers compared hemolymph from LPS-stimulated and mock-treated larvae for its ability to inhibit bacterial growth in a zone-of-inhibition plating assay. The antimicrobial activity is presumed to be due to AMPs secreted in the hemolymph, although it is possible that other secreted compounds contribute to bacterial killing and/or growth inhibition. It is important to note that in the fruit fly, injection of purified LPS is not sufficient to induce an immune response (Kaneko et al., 2004). Rather, another bacterial membrane component, peptidoglycan, is the primary elicitor that activates immune signaling pathways. Kaneko et al. (2004) demonstrated most LPS fractions were contaminated with sufficient levels of peptidoglycan to activate the immune response. It is possible that in other insects, the relevant immune elicitor is also peptidoglycan contaminants rather than LPS.

A similar response has also been shown in *Manduca sexta* caterpillars. Eleftherianos et al. (2006) demonstrated that priming with *E. coli* protects the caterpillars against a second challenge of *Photorhabdus luminescens*. In these experiments,

E. coli is probably acting as a general immune elicitor like peptidoglycan. Similar to the work with the mealworm, hemolymph from *E. coli*-challenged *M. sexta* caterpillars inhibited *P. luminescens* growth in a zone-of-inhibition assay. This increased antibacterial activity present in the hemolymph of immune-stimulated caterpillars could be responsible for increased resistance. This study showed that *E. coli* transcriptionally induces microbial recognition genes (hemolin, immunolectin 2, and peptidoglycan recognition proteins (PGRPs)) as well as antimicrobial effector genes (attacin, cecropin, lebocin, lysozyme, and moricin). PO activity and phagocytosis were not assayed.

Thus far, prolonged immune activation has been tested up to 3 weeks, which is a significant fraction of many insects' lifespans; however, it will also be interesting to determine whether these changes persist for the entire life of the animal. For example, does the immune response remain elevated in adult mealworm beetles and *M. sexta* moths? It would be interesting to know whether prolonged immune activation can persist through metamorphosis. Can immune challenges of larvae cause alterations in the development of the adult insect innate immune response?

For all of these infection examples it will be important to ascertain the relative contribution of hemocytes, AMPs, and melanization. Combining the results from priming experiments that analyze alterations in effectors versus priming experiments that examine survival as an output, the results implicate increased levels of circulating proPO, AMPs, and hemocyte activity. However, increased levels of immune effectors may not always correlate with efficacy of the immune response. It will be important to connect the functional output of survival with immune activation phenotypes.

The mechanisms whereby the innate immune response remains activated in these insects merit further study. Is it at the level of increased production of proPO and AMPs via transcription or translation? Or are these effectors posttranslationally regulated at the level of degradation and turnover? It is conceivable that in addition to elevating proPO levels, the proteases that convert proPO into active PO are elevated in response to a secondary immune challenge (Fig. 5.3). For both of these scenarios, it will be required to determine how the signaling circuitry is altered to mediate these changes and to determine how long these changes persist.

Notably, although researchers have observed increased levels of AMPs and proPO levels, prolonged activation of hemocytes has yet to be described (Fig. 5.4). Can immune stimulation alter phagocyte number or increase the phagocytic rate? Whether hemocytes can be activated to kill microbes more quickly remains to be elucidated. Faster killing rates could be due to intracellular killing within the phagosome, increased production of ROS, or coordination of the AMP and melanotic response. It is possible that prolonged activation of hemocytes has been selected against because it causes extensive nonspecific damage. Alternatively, it is possible that prolonged hemocyte activation has yet to be assayed carefully.

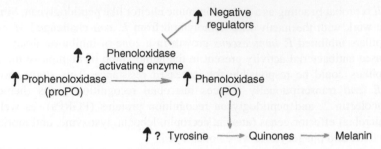

FIGURE 5.3   Melanization cascade.
Melanization involves the zymogen prophenoloxidase (proPO) being converted into phenoloxidase (PO) by PO activating enzymes. Increased proPO levels and increased PO activity have been reported to be differentially activated. Black arrows indicate components that may be at higher or more active levels after immune stimulation to generate a stronger melanotic response. Question marks indicate levels of regulation that could contribute to the response but have not yet been described.

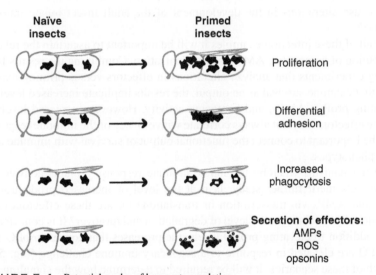

FIGURE 5.4   Potential modes of hemocyte regulation.
The specific primed response in Drosophila melanogaster is dependent on phagocytosis, but hemocyte involvement in other insect immune responses has yet to be addressed. The diagrams depict potential mechanisms that may be differentially activated in the hemocytes of primed insects. These different modes of hemocyte activation have yet to be described in insects, particularly in the context of conferring specificity or memory.
Abbreviations: AMPs, antimicrobial peptides; ROS, reactive oxygen species.

Prolonged immune activation may be detrimental to the host either at the level of resource allocation or nonspecific damage to the host. One example of resource allocation effects has been demonstrated in field cricket *Gryllus campestris*. Brinkhofin's group found that total protein levels were decreased in

LPS-challenged crickets (Jacot et al., 2005). Protein levels were used as a general indicator of the physiological metabolic state of an organism. Schneider's research laboratory also demonstrated that *Mycobacterium marinum*-infected *Drosophila* exhibit a wasting-like disease whereby flies progressively lose metabolic stores in the form of fat and glycogen (Dionne et al., 2006). There are also numerous examples that prolonged immune activation is detrimental to the host. Schmid-Hempel and colleagues found that starved bees exhibit increased mortality after LPS or bead challenge (Moret and Schmid-Hempel, 2000). Siva-Jothy's group performed transplantation experiments in the beetle *Tenebrio molitor* and found that the melanotic response damaged Malpighian tubules near the transplantation site (Sadd and Siva-Jothy, 2006). Mutant fruit flies with constitutively active alleles of Toll or JAK (hopscotch) develop melanotic tumors and have a decreased lifespan (Harrison et al., 1995; Luo et al., 1995; Qiu et al., 1998). Brandt et al. (2004) also implicated eiger, the *Drosophila* TNF homolog, in damaging the host during a *Salmonella typhimurium* infection.

Downregulation of the immune response is the simplest way of limiting damage, but provides no selective advantage if the host is challenged a second time. There are two mechanistic scenarios that could result in enhanced killing – prolonged activation of the immune response and increased immune activation after a second challenge. It is possible that insects have changed the fundamental circuitry of signaling so that in response to a second challenge, the immune response is stronger and/or faster. Between immune challenges, the immune response may return to its basal state, but upon a second challenge, the insect immune response could alter the rate or the amplitude of the response. One example involves the increased proPO levels that have been reported in many insects. Elevated levels of inactive zymogen result in higher levels of active PO being produced more rapidly after an immune challenge. This type of immune enhancement has also been observed when comparing AMP transcription in flies that have been wounded once versus flies that have been wounded twice (Pham et al., 2007). Pham et al. (2007) found that the kinetics of AMP induction is more rapid in doubly wounded flies; the peak of the AMP transcriptional response shifts from 6 to 2 h in flies that have been wounded twice. The amplitude of AMP transcriptional induction was also greater in flies that have been wounded twice. Experiments addressing the physiological relevance of this response have yet to be performed. The effects of multiple wounding events on the cellular response have not been characterized.

The route of infection may also play a role in regulating the nature of a prolonged immune response. Lemaitre's laboratory has been characterizing *Drosophila* oral infections using two bacteria, *Pseudomonas entomophila* and *Erwinia carotovora carotovora*. Replacement of larval food with a *P. entomophila* pellet mixed with banana kills fly larvae (Liehl et al., 2006). In contrast, a similar feeding regime with *E. carotovora carotovora* induces localized AMP production by the larval gut (Basset et al., 2000). It is important to note that the *E. carotovora carotovora* infections are never fatal, even in immunocompromised larvae (Basset et al., 2000).

To distinguish between the contribution of locally and systemically produced AMPs to gut infections, the authors compared the protective effect of a priming dose delivered orally and systemically. Liehl et al. (2006) found that a priming dose of *E. carotovora carotovora* could only protect against oral infection of *P. entomophila* if the priming dose was delivered orally. Introduction of *E. carotovora carotovora* into the hemocoel by a pinprick does not protect against a subsequent oral dose of *P. entomophila*. These experiments demonstrate that the local gut immune response, but not the systemic immune response, is involved in fighting *P. entomophila* infections in the gut. It also suggests that the immune response differs depending on the route of infection. Limiting the site of prolonged immune activation may be another way to minimize the extent of nonspecific damage to the host.

5.3.1.3 Transgenerational Memory

Immune memory is typically discussed as it occurs within a given individual: priming a naïve individual results in a stronger immune response in that individual. Transmission of IgA in mammalian milk might be considered to be a type of transgenerational immune memory, and there are more concrete examples of this type of protective effect in other organisms. For example, maternally induced defense has been demonstrated in plants (*Raphanus raphanistrum*) and the water-flea *Daphnia cucullata* (Agrawal et al., 1999). In insects it is also evident that challenging adults can result in offspring that possess an elevated immune response.

The desert locust *Schistocerca gregaria* exhibits density-dependent polyphenism, meaning that group-living gregaria exhibit different morphologies, coloration, and behavior from low density solitaria (Elliot et al., 2003). A locust's phenotype is determined by the parental state and crowding with respect to other locusts prior to oviposition. Gregarious locusts have been shown to be more resistant to the fungal entomopathogen *M. anisopliae* and possess more AMP activity in the hemolymph. This finding is consistent with the density-dependent prophylaxis hypothesis that organisms living at higher densities are at greater risk of infection (Elliot et al., 2002; Reeson et al., 1998). Due to this intersection of transgenerational effects and infection, Elliot et al. (2003) tested whether immune challenge could affect the locust phase state. They found that *M. anisopliae* infection results in increased production of solitary locusts, even if the mothers had visual and olfactory cues from other gregaria. Interestingly, raising the temperature to mimic a behavioral fever (and therefore mimicking infection without an immune challenge) was sufficient to skew the offspring toward the solitaria phenotype. Although this was not the expected result, the authors reason that solitary locusts disaggregate and avoid other individuals, resulting in an overall decreased risk of infection. In this fashion, locusts are able to modulate their immune responses based on the likelihood of encountering a pathogen. Gregarious locusts have a more elevated immune response due to the increased probability of encountering a pathogen. However, after experiencing an infection, mothers generate solitary locusts that will leave the group, and therefore the infection site.

Schmid-Hempel and collegues also found a maternal effect in the bumble bee *Bombus terrestris* (Moret and Schmid-Hempel, 2001). Social insects are interesting organisms to test for transgenerational effects because some offspring remain in the nest while others leave, and the two populations are therefore subject to different selective pressures. In the case of bumble bees, worker bees will remain with the nest whereas sexuals (daughter queens and males) must leave the nest. Unlike daughter queens, male bees do not hibernate, and their reproductive success depends on their survival once they leave the parental colony. Male bees are more likely to encounter pathogens once they leave the colony because bees maintain a sterile nest. Due to the difference in probability of encountering a pathogen, Moret et al. (2001) wanted to determine whether immune challenge of the parental colony could modulate the immune response of the male bees. Using a split colony design, 70–80% of the workers were injected with LPS or a control solution. Moret et al. (2001) found that the male offspring from LPS-challenged colonies exhibited increased PO activity in their hemolymph, although the level of AMP production remained unaltered.

An intriguing example of maternal transmission of resistance was reported by Rahman et al. (2004) using the flour moth *Ephestia kuehniella*. Sublethal doses of *Bacillus thuringiensis* toxin delivered to a larva will protect that larva against a second lethal dose of toxin. This protective effect is at least partially specific because LPS and zymosan (dead yeast) will not protect larvae. This protective effect can be maternally transferred from a primed adult moth to the offspring. By crossing naïve susceptible moths with toxin-primed resistant moths, Rahman et al. (2004) found that toxin-primed mothers will always produce resistant offspring. Because the degree of toxin resistance varied across a range of values, the authors suggest that the protective effect is mediated by more than one gene. The most parsimonious explanation for this maternal effect is that a factor is incorporated into the egg that results in toxin resistance. However, until the molecular mechanism of *B. thuringiensis* toxin resistance is identified, it will be difficult to ascertain how the toxin-primed mother is able to pass resistance to her offspring. Lipophorin, a hemolymph protein involved in lipid transport between tissues in insects, is an attractive candidate. Schmidt's research laboratory found that lipophorin is also present in the gut lumen of larvae and forms a complex with *B. thuringiensis* toxin (Mahbubur Rahman et al., 2007). Resistant moths also exhibit increased PO activity, but melanization is not responsible for toxin resistance (Mahbubur Rahman et al., 2007). Recent work in the gypsy moth *Lymantria dispar Linnaeus* demonstrated that the fatal septicemia resulting from *B. thuringiensis* infection is caused by the native gut microbiota entering the hemolymph after gut damage (Broderick et al., 2006). Perhaps in response to a sublethal dose of toxin, the native flora of the moths changes. This opens the possibility that the microbes transferred from mother to egg may be altered by *B. thuringiensis* toxin treatment, resulting in resistance.

It is evident that immune challenge not only alters an individual's immune response, but can also influence development of the offspring's immune response. Do other invertebrates exhibit this inherited response? Of primary importance is determining the mechanisms underlying maternal transmission of the protective effects. In the case of the desert locust *Schistocerca gregaria*, transgenerational memory resulted in a permanent phenotype change. What are the molecular alterations that mediate this permanent change? The darker cuticles of solitary locusts are due to increased melanin deposition. How does this increased melanin production relate to increased circulating proPO levels?

With regards to the bumble bee *Bombus terrestris* and the flour moth *Ephestia kuehniella*, how long does this transgenerational protective effect persist? At least for flour moths, the protect effect is not passed to the F2 generation. Because the probability of pathogen encounter varies greatly from generation to generation, and the fact that raising an immune response is costly (as discussed above), it seems reasonable that maternal protection would only persist for one generation. How is this process limited to only one generation? The simplest explanation is that the protective factor is not present, but it is also possible that there is a difference in the signaling pathways that mediate this phenomenon.

The experiments with transgenerational memory demonstrate that the immune response of insects is altered based on the probability of encountering a pathogen. Social immunity experiments have shown that insects can communicate infection states to the rest of the colony (Traniello et al., 2002). Taken together, these data raise the possibility that adaptive social immunity may exist. Rosengaus and colleagues demonstrated that in response to an infection in the colony, termites exhibit more allogrooming and increase secretion of antibacterial compounds from exocrine glands (Rosengaus et al., 1998, 2000). Can other immune changes be triggered in social insects? For example, uninfected individual termites in an infected colony may exhibit increased circulating AMP or proPO levels.

5.3.2 SPECIFIC MEMORY

Up to this point, we discussed two kinds of immune memory; prolonged immune activation and transgenerational effects. In both of these cases, primary exposure to an elicitor enhances the immune responses and confers protection against a second immune challenge. For the majority of these experiments, specificity was not tested and the results suggest that protection was due to general immune activation that would be effective against many kinds of microbes. Demonstration of specificity requires an additional test: one must compare the effects of various priming agents to different kinds of immune challenges. Specificity is inherently an operational definition because it cannot be addressed by looking at AMP levels or the degree of the melanization response. Specificity must be assayed by measuring an enhanced immune response to a variety of microbes, either in terms of survival or increased microbial killing. We present experimental evidence demonstrating that the insect immune response exhibits more specificity than was anticipated based on existing generalizations about the innate response.

5.3.2.1 Transplantation Studies

Cockroach tissue transplantation studies conducted in the late 1980s provide an interesting example of short-term specific memory. Cuticular transplantation experiments were conducted by Lackie (1983). In these experiments, the cuticle, underlying epidermis, and basement membrane were transplanted between last-instar nymphs of donor and recipient cockroaches. After molting, the graft site of the recipient is examined for donor-like cuticle with respect to pattern and coloration. If donor epidermis is not rejected, then it will join seamlessly with the recipient cuticle and the graft site will have cuticle that resembles both the donor and the recipient.

Cuticular transplants between two closely related cockroach species, *Blaberus craniifer* and *Blatta orientalis*, produced interesting results (Dularay and Lackie, 1987). If a recipient cockroach receives one cuticular transplantation, the graft is accepted. However, if the recipient receives a second cuticular transplant between 1 and 4 days after the primary transplant, the second graft is rejected. This data implies that the immune response is activated for 4 days after the primary graft, and this immune activation results in rejection of the second transplant. To test this hypothesis, Dularay and Lackie (1987) injected sepharose beads to activate the cellular immune response. Sepharose beads are encapsulated by hemocytes after injection. Sepharose bead-injected cockroaches will reject a cuticular transplant as well, although to a lesser extent than if they had received two transplants. Together with the fact that hemocytes aggregate at the transplant site, these data suggest that activated hemocytes play a role in rejecting cuticular transplants between closely related species.

Karp's research group performed allografts and autografts in the American cockroach, *Periplaneta americana* (Hartman and Karp, 1989). Allografts involve transplants of tissue between individuals of the same species whereas autografts are tissue transplants from one part of the body to the other in the same individual. Hartman et al. (1989) found that in the cockroach, primary allografts are rejected within 7 days post-transplantation whereas autografts are accepted for the life of the animal. Notably, second set allografts from the same donor are rejected more rapidly whereas allografts from a third party donor are rejected at the same rate as a primary allograft. However, this more rapid rejection of a second set allograft is a short-term effect. Second set allograft rejection can be detected as early as 1–2 days after a primary allograft, although maximal rejection rates are observed between 3 and 7 days post-transplantation. After 7 days, there is no difference in the rejection rate between second set allografts and third party allografts. This result suggests that there is specificity in the cockroach immune response that mediates rejection even though the transplants are within the same species. Interestingly, primary exposure to filter paper containing donor hemocytes was sufficient to mediate second set allograft rejection, implying that hemocytes are being recognized by the recipient immune response. The molecular basis for recognition and rejection between individuals of the same species remains uncharacterized in insects.

Copepods are another type of invertebrate that are able to distinguish between parasites of related sibships (Kurtz and Franz, 2003). In the copepod model, it is unclear whether rejection is mediated by hemocytes as well. Identifying this type of response in genetically tractable invertebrates will be critical to elucidate the mechanisms underlying this degree of specificity and may explain why the specific effect is short term.

5.3.2.2   Bacterial Challenges

Karp's laboratory also tested whether the cockroach immune response exhibits specific memory to challenges with pathogenic bacteria. Faulhaber and Karp (1992) primed cockroaches with glutaraldehyde-killed *P. aeruginosa* and compared survivorship with saline-injected cockroaches after a second lethal challenge of *P. aeruginosa* (Faulhaber and Karp, 1992). They found that a priming dose of *P. aeruginosa* protected against a second lethal challenge of *P. aeruginosa* as early as 1 day post-priming, and the protective effect persisted for up to 14 days post-priming. The researchers carried out the experiment for 49 days and found that after 14 days, *P. aeruginosa*-primed cockroaches died at the same rate as saline controls. Faulhaber and Karp (1992) then tested whether the cockroach immune response exhibits specificity comparing priming with four other bacteria. They used two Gram-positive bacteria, *Streptococcus lactis* and *Micrococcus lysodeikticus*, and two Gram-negative bacteria, *Serratia marcescens* and *Enterobacter cloacae*. Within the first 3 days post-priming, all four bacteria provide cross-protection against a second lethal challenge of *P. aeruginosa*. After 3 days, priming with either *S. lactis* or *M. lysodeikticus* no longer protects against *P. aeruginosa*. Notably, there is a partial specificity because priming doses of the Gram-negative bacteria *S. marcescens* or *E. cloacae* protect against *P. aeruginosa* for up to 6 days post-priming. However, by 7 days post-priming, only a priming dose of *P. aeruginosa* protects against itself. This result is particularly interesting because it demonstrates that the cockroach immune response is mobilized in three phases. Between 1 and 3 days post-priming, the immune response can be nonspecifically activated by either Gram-positive or Gram-negative bacteria. However, by 3–6 days post-priming the immune response is modulated to be slightly more specific in that only Gram-negative bacteria will elicit a protective response against challenge with the Gram-negative bacteria *P. aeruginosa*. By 1 week post-priming, only cockroaches primed with *P. aeruginosa* are protected, showing that the cockroach immune response has been altered to be highly specific. The nature of this specificity or how the immune response changes has yet to be determined.

A similar response has also been described in the bumble bee *Bombus terrestris*. Sadd and Schmid-Hempel (2006) examined the protective effects of priming doses of one Gram-negative bacteria, *Pseudomonas fluorescens*, and two closely related Gram-positive bacteria, *Paenibacillus alvei* and *Paenibacillus larvae*. These bacteria were chosen because sublethal and lethal doses were identified in bumble bees. The authors compared the protective effect of these bacteria at 8 and 22 days post-priming. Priming doses of *P. fluorescens* and *P. larvae* protected bees against a lethal dose of *P. fluorescens* delivered 8 days later. However, greater specificity was observed at the 22 day time point because a priming dose of *P. fluorescens* was the only priming dose that protected against a second lethal challenge of *P. fluorescens*. Thus, the bee innate immune response is similar to that of cockroaches in that there is an early nonspecific phase, and a late specific primed response. The authors also quantified bacterial numbers in hemolymph and found that enhanced survival correlated with the proportion of individuals that had cleared bacteria from their hemolymph. Enhanced survival is therefore correlated with enhanced bacterial killing by the immune response and not increased stress resistance. Notably, using the zone-of-inhibition assay, they were unable to observe an antibacterial activity persisting in the hemolymph for 22 days. The specific protective effect they observe is therefore not due to the persistence of AMPs in the hemolymph. However, it remains possible that AMPs are differentially upregulated after a second challenge, but it is unclear how the high degree of specificity could be observed with AMPs. It seems more likely that the hemocytes are specifically activated, or that differential activation of melanization occurs.

Experiments must be conducted that address which branch of the immune response is critical for fighting these microbes. The cellular response, AMPs, or melanization may be involved in all, or only a subset of the infection phases. By identifying which immune effector mechanism is active at each stage of infection, it will be possible to target candidate pathways. The early nonspecific phase, later specific stage, and the fact that the protective effect diminishes in all of the insects tested imply that differential regulation of the immune response may be involved. It is possible that during the infection a more specific set of AMPs are induced, or that hemocytes alter recognition mechanisms. What are the signaling changes that result in a more specific immune response being produced over the course of infection? It will be also interesting to examine wax moths and termites at later time points to see if they also exhibit greater specificity at some other time interval.

Fruit flies possess a specific primed response that is dependent on phagocytes. Schneider's laboratory found that heat-killed *S. pneumoniae* protects against a second lethal challenge of *S. pneumoniae* (Pham et al., 2007). Enhanced survival is correlated with reduced bacterial load, demonstrating that the immune response is activated to kill bacteria more quickly. This protective effect can be elicited as early as 1 day and persists 14 days post-priming, which is as long as this experiment could be performed because by this time point, the flies are actually 4 weeks old and will die nonspecifically due to the stress of wounding alone. Four other microbes were tested for their ability to elicit a primed response including the Gram-positive bacteria *Listeria monocytogenes*, the Gram-negative bacteria *Salmonella typhimurium*, the fungal entomopathogen *Beauveria bassiana*, and *Mycobacterium marinum*. Only *S. pneumoniae* and *B. bassiana* were able to elicit a protective effect, but it was highly specific. Only a priming dose of *S. pneumoniae* is able to protect against a lethal dose of *S. pneumoniae*, and only a priming dose of *B. bassiana* confers protection against *B. bassiana* infections. No cross-protection was observed between these two microbes. Pham et al. (2007) then examined the relative contributions of the branches of the immune response. Melanization was discounted because no differences in melanization were detected between naïve and primed flies. Using quantitative reverse transcription PCR (qRT-PCR), the authors found that AMP transcription did not remain elevated throughout the course of the experiment, and the kinetics of AMP induction after a second challenge was identical between primed and naïve flies. Notably, AMP induction using a mixture of strong elicitors was not sufficient to protect the fly against a lethal challenge of *S. pneumoniae*, demonstrating that AMPs are not involved in the specific primed response of *Drosophila*. To address the contribution of phagocytes, the authors used a bead inhibition assay whereby injection of polystyrene beads into the hemocoel inhibits hemocyte phagocytosis (Elrod-Erickson et al., 2000). These polystyrene beads are not encapsulated by hemocytes, and the hemocyte numbers are unaltered. Bead-injected flies were very sensitive to *S. pneumoniae* and were unable to clear a dose of *S. pneumoniae* that should be sublethal. This result demonstrates that the effector of the primed response is the hemocytes because phagocytosis is required to kill *S. pneumoniae*. Accordingly, bead-inhibited flies were not protected by a priming dose of *S. pneumoniae*. The critical involvement of phagocytes may explain why the authors did not observe a primed response for the other bacteria tested. *M. marinum*, *L. monocytogenes*, and *S. typhimurium* can live within phagocytes and possess mechanisms to evade intracellular killing. In terms of molecular mechanism, the authors tested the relative contribution of the Toll and Imd pathways. While the Imd pathway was dispensable for the primed response, the Toll pathway was necessary, but not sufficient to elicit a primed response. Toll mutants cannot be primed, but Toll activation using a mixture of strong elicitors was not sufficient to protect the fly. Pham et al. (2007) propose a model whereby the Toll pathway may be required for detection of the microbe, and Toll activation turns on another pathway critical for the specific primed response. Using the molecular and genetic tools available in the fruit fly, it may be possible to identify the relevant signaling pathways involved in this response.

In light of these data, it will be interesting to revisit the bumble bee and cockroach experiments. Are hemocytes involved in specific priming in both of these species? If so, they may be useful models because both animals have significantly more hemocytes than fruit flies and it may be possible to perform a biochemical analysis on the hemolymph and hemocytes. It will also be interesting to determine which bacterial components are sufficient to elicit a primed response.

5.4 POTENTIAL MECHANISMS UNDERLYING SPECIFICITY

Evidence regarding a specific primed response that is dependent on phagocytes raises at least two questions. First, specific recognition of a variety of elicitors requires a diversity of recognition mechanisms. What is the molecular basis for this diversity? Second, diverse nonoverlapping effectors must be produced if there is limited cross-protection in this system. How are immune effectors able to specifically recognize and target the microbe that was encountered in the priming dose?

A few features can be anticipated for the mechanism that underlies the specific primed response. First, molecules involved in recognition and conferring protection should be induced by immune challenge. Induction may occur at the transcriptional, translational, or post-translational level. Second, to account for the specificity observed in this process, we expect some of the molecules to exhibit diversity. A one-to-one level of diversity between elicitor and molecule is not necessary; it is possible that the molecules may work combinatorially to mediate specificity. Third, some sort of selection process is required to account for specificity in the immune response. There must be some mechanism for initiating nonoverlapping effectors that mediate enhanced microbe killing. Last, at least for the specific primed response in flies that is dependent on phagocytes, the mechanism must mediate specific recognition and faster killing by hemocytes.

It is difficult to imagine a scenario whereby AMPs or melanization could be the effector of the specific primed response observed in flies and bees because both mechanisms act against a broad range of microbes. However, it is possible that sequestration or localized activation may contribute a partial degree of specificity. Hemocytes are a more attractive candidate effector, and at least in flies, have been shown to be essential for the specific primed response. Some type of opsonin or receptor must be involved to mediate faster recognition by phagocytes. Either opsonin, the receptor, or both molecules may possess the degree of diversity required to account for the specificity observed in this response. As discussed above, this opsonin or receptor should be induced by immune challenges, and must undergo some type of selection that could be transcriptional or post-translational. For example, an elicitor may induce certain types of molecules, or the molecule that binds to the elicitor could be selected after it is expressed by the insect. In this section, we will discuss two candidate molecules that may contribute diversity to immune recognition mechanisms.

5.4.1   DOWN SYNDROME CELL ADHESION MOLECULE

The Down syndrome cell adhesion molecule (Dscam) is a member of the immunoglobulin superfamily that was originally characterized for its role in neuronal wiring specificity in *Drosophila*. Dscam is composed of 10 immunoglobulin-like domains, 6 fibronectin type II domains, and a transmembrane domain. Constant exons flank variable exons with alternative splice sites. Alternative splicing of the Ig2, Ig3, and Ig7 domains results in more than 38,000 potential isoforms being expressed. In the brain, Dscam is involved in a high degree of recognition diversity and is involved in isoform-specific homophilic binding. Chess's group examined expression of Dscam splice variants in different *Drosophila* tissues throughout development (Neves et al., 2004). Notably, different repertoires of Dscam splice variants were expressed during different developmental stages. Neves et al. (2004) also found that different tissues such as hemocytes and neurons express different pools of splice variants. Together these data suggest there is regulation of alternative splicing of Dscam. Additionally, using single cell PCR, Neves et al. (2004) found that individual hemocytes express up to 14 different Dscam isoforms. Schmucker's laboratory demonstrated that in addition to the brain and hemocytes, Dscam is also expressed in the fat body (Watson et al., 2005). Using custom microarrays that contained specific oligo probes for all alternative spliced exons, Watson et al. (2005) found that more than 18,000 different isoforms are expressed in hemocyte populations. Due to its expression in immunocompetent cells and its involvement in isoform-specific homophilic binding, Dscam is an attractive molecule for generating receptor diversity in the insect immune response. Intriguingly, Watson et al. (2005) found that secreted isoforms were also detected in the hemolymph of flies. Inhibition of the Dscam receptor with a neutralizing antibody resulted in decreased phagocytosis of bacteria, implying that Dscam may be involved in specific binding and recognition of microbes.

Indeed, Dimopoulos' research laboratory found that in the mosquito *Anopheles gambiae*, different pools of splice variants are induced in response to different immune challenges (Dong et al., 2006). Dong et al. (2006) found that pathogen-specific splice repertoires are induced by *S. aureus* versus *E. coli*. By using RNAi to knock down specific splice variants, the authors were able to induce specific sensitivities in mosquitoes. For example, when expression of *E. coli*-specific splice variants is eliminated, mosquitoes are sensitized to *E. coli* infection, but not *S. aureus* infection. These data suggest that Dscam could contribute the necessary diversity required to generate specific immune responses. Neves et al. (2004) found that individual unchallenged fly hemocytes can express multiple Dscam splice variants. It will be interesting to determine whether immune challenge results in a change in the number of receptors expressed by an individual hemocyte.

Using the functional assays developed in flies, bees, and cockroaches, it will be interesting to determine whether Dscam is involved in specific priming. An attractive hypothesis is that microbial challenge induces alternative splicing of Dscam, thereby generating specific receptors that can recognize the particular microbe. These receptors can also be cleaved off the surface to act as an opsonin that persists in a hemolymph. Upon second challenge, the opsonin would bind to the surface of the microbe, mediating more rapid recognition and phagocytosis by hemocytes. Importantly, induction of microbe-specific splice repertoires is not sufficient to prove that Dscam is involved in specific priming. It will be necessary to determine whether inhibition of a specific splice variant repertoire eliminates the protective effect.

5.4.2 FIBRINOGEN-RELATED PROTEINS

The Loker laboratory characterized the fibrinogen-related proteins (FREPs) in the freshwater snail *Biomphalaria glabrata* (Zhang et al., 2004). FREPs are another member of the immunoglobulin superfamily and are very diverse in snails. This protein contains either one or two immunoglobulin domains and a fibrinogen domain. FREPs are induced by infection in snails and act as an opsonin to coat the surface of an immune elicitor. Zhang et al. (2004) sequenced FREPs from a single snail and observed a greater sequence diversity than predicted from germline sequence. The authors found evidence supporting the hypothesis that germline sequences are diversified in somatic cells at a low frequency through mutational and recombinatorial processes, although they stress that the mechanism underlying immunoglobulin diversification is probably not the same as somatic recombination of vertebrate immunoglobulin domains. FREP homologs have been identified in insects as well. Christenson's group found 53 homologs in the mosquito *Anopheles gambiae* and 20 copies in the genome of *Drosophila melanogaster* (Wang et al., 2005). FREPs are another class of molecules with a high degree of diversity that may account for the specificity present in the insect immune response.

5.5 CONCLUDING REMARKS

It has been generally assumed that due to the short lifespan of insects, the ability to mount a specific immune response would not confer a selective advantage. Experimental immunology now argues otherwise. In the post-genomic era, it will be critical to define the molecular mechanisms underlying these adaptive responses. Research to date has provided much insight to the molecular mechanisms involved in microbial recognition and signaling. In light of the functional experiments demonstrating adaptive aspects of innate immunity, we can anticipate further studies regarding the effectors of the immune response. Other than transcriptional induction of AMPs, many questions remain to be answered in this area. How are the cellular, melanotic, and AMP responses coordinated in response to a pathogenic infection? Elucidating the mechanisms that contribute to specificity and memory will be critical. It may be a far stretch to assume that these responses are conserved in vertebrates, but until the mechanism is known, there is also no reason to suggest that it is not conserved. Identifying the molecules involved in the process is the only way to determine which hypothesis is correct.

On the one hand, adaptive immunity is usually thought to occur within a single individual. However, a survey of insect immunity across species demonstrates the existence of transgenerational, behavioral, and social aspects of the immune response. Considering these responses in an adaptive context could reveal a novel area of immunity. Historically, performing experiments in insects has allowed researchers to clearly test the interaction between basic processes. For example, interactions between immunity and basic biological processes like metabolism, fecundity, and circadian rhythm have all been reported (Brandt and Schneider, 2007; Dionne et al., 2006; Shirasu-Hiza et al., 2007). Exploring whether there are adaptive aspects to social and behavioral immunity remains to be determined.

ACKNOWLEDGMENTS

That authors wish to thank Joe Dan Dunn, Junaid Ziauddin, and members of the Schneider laboratory for comments on the manuscript. This work was supported by grant R01AI053080 from the National Institutes of Health (DSS) and the National Science Foundation (LNP).

REFERENCES

Adamo, S. A. (1998). The specificity of behavioral fever in the cricket *Acheta domesticus*. *J. Parasitol.* **84**, 529–533.

Adams, M. D., Celniker, S. E., Holt, R. A., Evans, C. A., Gocayne, J. D. et al. (2000). The genome sequence of *Drosophila melanogaster*. *Science* **287**, 2185–2195.

Agaisse, H., Petersen, U. M., Boutros, M., Mathey-Prevot, B., and Perrimon, N. (2003). Signaling role of hemocytes in *Drosophila* JAK/STAT-dependent response to septic injury. *Dev. Cell* **5**, 441–450.

Agrawal, A. A., Laforsch, C., and Tollrian, R. (1999). Transgenerational induction of defences in animals and plants. *Nature* **401**, 60–63.

Barnes, A. I., and Siva-Jothy, M. T. (2000). Density-dependent prophylaxis in the mealworm beetle *Tenebrio molitor* L. (Coleoptera: Tenebrionidae): Cuticular melanization is an indicator of investment in immunity. *Proc. Biol. Sci.* **267**, 177–182.

Basset, A., Khush, R. S., Braun, A., Gardan, L., Boccard, F., Hoffmann, J. A., and Lemaitre, B. (2000). The phytopathogenic bacteria *Erwinia carotovora* infects *Drosophila* and activates an immune response. *Proc. Natl. Acad. Sci. USA* **97**, 3376–3381.

Bidla, G., Dushay, M. S., and Theopold, U. (2007). Crystal cell rupture after injury in *Drosophila* requires the JNK pathway, small GTPases and the TNF homolog eiger. *J. Cell. Sci.* **120**, 1209–1215.

Boman, H. G., Nilsson, I., and Rasmuson, B. (1972). Inducible antibacterial defence system in *Drosophila*. *Nature* **237**, 232–235.

Brandt, S. M., Dionne, M. S., Khush, R. S., Pham, L. N., Vigdal, T. J., and Schneider, D. S. (2004). Secreted bacterial effectors and host-produced eiger/TNF drive death in a *Salmonella*-infected fruit fly. *PLoS Biol.* **2**, e418.

Brandt, S. M., and Schneider, D. S. (2007). Bacterial infection of fly ovaries reduces egg production and induces local hemocyte activation. *Dev. Comp. Immunol.* 2007 Mar 15 [Epub ahead of print].

Broderick, N. A., Raffa, K. F., and Handelsman, J. (2006). Midgut bacteria required for *Bacillus thuringiensis* insecticidal activity. *Proc. Natl. Acad. Sci. USA* **103**, 15196–15199.

Brown, M. J., Moret, Y., and Schmid-Hempel, P. (2003). Activation of host constitutive immune defence by an intestinal trypanosome parasite of bumble bees. *Parasitology* **126**, 253–260.

Bulet, P., Hetru, C., Dimarcq, J. L., and Hoffmann, D. (1999). Antimicrobial peptides in insects: Structure and function. *Dev. Comp. Immunol.* **23**, 329–344.

Carruthers, R. I., Larkin, T. S., Firstencel, H., and Feng, Z. D. (1992). Influence of thermal ecology on the mycosis of a rangeland grasshopper. *Ecology* **73**, 190–204.

Crozatier, M., and Meister, M. (2007). *Drosophila* haematopoiesis. *Cell Microbiol.* **9**, 1117–1126.

De Gregorio, E., Spellman, P. T., Rubin, G. M., and Lemaitre, B. (2001). Genome-wide analysis of the *Drosophila* immune response by using oligonucleotide microarrays. *Proc. Natl. Acad. Sci. USA* **98**, 12590–12595.

Delaney, J. R., Stoven, S., Uvell, H., Anderson, K. V., Engstrom, Y., and Mlodzik, M. (2006). Cooperative control of *Drosophila* immune responses by the JNK and NF-kappaB signaling pathways. *Embo. J.* **25**, 3068–3077.

Deverno, P. J., Aston, W. P., and Chadwick, J. S. (1983). Transfer of immunity against *Pseudomonas aeruginosa* P11-1 in *Galleria mellonella* larvae. *Dev. Comp. Immunol.* **7**, 423–434.

Dionne, M. S., Pham, L. N., Shirasu-Hiza, M., and Schneider, D. S. (2006). Akt and FOXO dysregulation contribute to infection-induced wasting in *Drosophila*. *Curr. Biol.* **16**, 1977–1985.

Dong, Y., Taylor, H. E., and Dimopoulos, G. (2006). AgDscam, a hypervariable immunoglobulin domain-containing receptor of the *Anopheles gambiae* innate immune system. *PLoS Biol.* **4**, e229.

Dostert, C., Jouanguy, E., Irving, P., Troxler, L., Galiana-Arnoux, D., Hetru, C., Hoffmann, J. A., and Imler, J. L. (2005). The Jak-STAT signaling pathway is required but not sufficient for the antiviral response of *Drosophila*. *Nat. Immunol.* **6**, 946–953.

Dularay, B., and Lackie, A. M. (1987). The effect of biotic and abiotic implants on the recognition of *Blatta orientalis* cuticular transplants by the cockroach *Periplaneta americana*. *Dev. Comp. Immunol.* **11**, 69–77.

Eleftherianos, I., Marokhazi, J., Millichap, P. J., Hodgkinson, A. J., Sriboonlert, A., ffrench-Constant, R. H., and Reynolds, S. E. (2006). Prior infection of *Manduca sexta* with non-pathogenic *Escherichia coli* elicits immunity to pathogenic *Photorhabdus luminescens*: Roles of immune-related proteins shown by RNA interference. *Insect Biochem. Mol. Biol.* **36**, 517–525.

Elliot, S. L., Blanford, S., and Thomas, M. B. (2002). Host-pathogen interactions in a varying environment: temperature, behavioural fever and fitness. *Proc. of the Royal Society of London Series B-Biological Sciences* **269**, 1599–1607.

Elliot, S. L., Blanford, S., Horton, C. M., and Thomas, M. B. (2003). Fever and phenotype: Transgenerational effect of disease on desert locust phase state. *Ecol. Lett.* **6**, 830–836.

Elrod-Erickson, M., Mishra, S., and Schneider, D. (2000). Interactions between the cellular and humoral immune responses in *Drosophila*. *Curr. Biol.* **10**, 781–784.

Faulhaber, L. M., and Karp, R. D. (1992). A diphasic immune response against bacteria in the American cockroach. *Immunology* **75**, 378–381.

Filipe, S. R., Tomasz, A., and Ligoxygakis, P. (2005). Requirements of peptidoglycan structure that allow detection by the *Drosophila* Toll pathway. *EMBO Rep.* **6**, 327–333.

Galko, M. J., and Krasnow, M. A. (2004). Cellular and genetic analysis of wound healing in *Drosophila* larvae. *PLoS Biol* **2**, e239.

Goto, A., Kumagai, T., Kumagai, C., Hirose, J., Narita, H., Mori, H., Kadowaki, T., Beck, K., and Kitagawa, Y. (2001). A *Drosophila* haemocyte-specific protein, hemolectin, similar to human von Willebrand factor. *Biochem. J.* **359**, 99–108.

Gottar, M., Gobert, V., Matskevich, A. A., Reichhart, J. M., Wang, C., Butt, T. M., Belvin, M., Hoffmann, J. A., and Ferrandon, D. (2006). Dual detection of fungal infections in *Drosophila* via recognition of glucans and sensing of virulence factors. *Cell* **127**, 1425–1437.

Harrison, D. A., Binari, R., Nahreini, T. S., Gilman, M., and Perrimon, N. (1995). Activation of a *Drosophila* Janus kinase (JAK) causes hematopoietic neoplasia and developmental defects. *Embo. J.* **14**, 2857–2865.

Hartman, R. S., and Karp, R. D. (1989). Short-term immunologic memory in the allograft response of the American cockroach, *Periplaneta americana*. *Transplantation* **47**, 920–922.

Hoffmann, J. A. (2003). The immune response of *Drosophila*. *Nature* **426**, 33–38.

Inglis, G. D., Johnson, D. L. and Goettel, M. S. (1996). Effects of temperature and thermoregulation on mycosis by *Beauveria bassiana* in grasshoppers. *Biol. Cont.* **7**, 131–139.

Jacot, A., Scheuber, H., Kurtz, J., and Brinkhof, M. W. (2005). Juvenile immune system activation induces a costly upregulation of adult immunity in field crickets *Gryllus campestris*. *Proc. Biol. Sci.* **272**, 63–69.

Janeway, C. (2005). *Immunobiology: The Immune System in Health and Disease*. Garland Science, New York.

Jang, I. H., Chosa, N., Kim, S. H., Nam, H. J., Lemaitre, B., Ochiai, M., Kambris, Z., Brun, S., Hashimoto, C., Ashida, M., Brey, P. T., and Lee, W. J. (2006). A Spatzle-processing enzyme required for toll signaling activation in *Drosophila* innate immunity. *Dev. Cell* **10**, 45–55.

Kaneko, T., Goldman, W. E., Mellroth, P., Steiner, H., Fukase, K., Kusumoto, S., Harley, W., Fox, A., Golenbock, D., and Silverman, N. (2004). Monomeric and polymeric Gram-negative peptidoglycan but not purified LPS stimulate the *Drosophila* IMD pathway. *Immunity* **20**, 637–649.

Kaneko, T., and Silverman, N. (2005). Bacterial recognition and signalling by the *Drosophila* IMD pathway. *Cell Microbiol.* **7**, 461–469.

Kurtz, J. (2005). Specific memory within innate immune systems. *Trends Immunol.* **26**, 186–192.

Kurtz, J., and Franz, K. (2003). Innate defence: Evidence for memory in invertebrate immunity. *Nature* **425**, 37–38.

Kurucz, E., Zettervall, C. J., Sinka, R., Vilmos, P., Pivarcsi, A., Ekengren, S., Hegedus, Z., Ando, I., and Hultmark, D. (2003). Hemese, a hemocyte-specific transmembrane protein, affects the cellular immune response in *Drosophila*. *Proc. Natl. Acad. Sci. USA* **100**, 2622–2627.

Lackie, A. M. (1983). Immunological recognition of cuticular transplants in insects. *Dev. Comp. Immunol.* **7**, 41–50.

Lagueux, M., Perrodou, E., Levashina, E. A., Capovilla, M., and Hoffmann, J. A. (2000). Constitutive expression of a complement-like protein in toll and JAK gain-of-function mutants of *Drosophila*. *Proc. Natl. Acad. Sci. USA* **97**, 11427–11432.

Lemaitre, B., and Hoffmann, J. (2007). The host defense of *Drosophila melanogaster*. *Annu. Rev. Immunol.* **25**, 697–743.

Lemaitre, B., Nicolas, E., Michaut, L., Reichhart, J., and Hoffmann, J. (1996). The dorsoventral regulatory gene cassette spaetzle/Toll/cactus controls the potent antifungal response in *Drosophila* adults. *Cell* **86**, 973–983.

Lemaitre, B., Reichhart, J., and Hoffmann, J. (1997). *Drosophila* host defense: Differential induction of antimicrobial peptide genes after infection by various classes of microorganisms. *Proc. Natl. Acad. Sci. USA* **94**, 14614–14619.

Levashina, E. A., Moita, L. F., Blandin, S., Vriend, G., Lagueux, M., and Kafatos, F. C. (2001). Conserved role of a complement-like protein in phagocytosis revealed by dsRNA knockout in cultured cells of the mosquito, *Anopheles gambiae*. *Cell* **104**, 709–718.

Liehl, P., Blight, M., Vodovar, N., Boccard, F., and Lemaitre, B. (2006). Prevalence of local immune response against oral infection in a *Drosophila/Pseudomonas* infection model. *PLoS Pathog.* **2**, e56.

Little, T. J., Hultmark, D., and Read, A. F. (2005). Invertebrate immunity and the limits of mechanistic immunology. *Nat. Immunol.* **6**, 651–654.

Louis, C., Jourdan, M., and Cabanac, M. (1986). Behavioral fever and therapy in a rickettsia-infected orthoptera. *Amer. J. Physiol.* **250**, R991–R995.

Luo, H., Hanratty, W. P., and Dearolf, C. R. (1995). An amino acid substitution in the *Drosophila* hopTum-l Jak kinase causes leukemia-like hematopoietic defects. *Embo. J.* **14**, 1412–1420.

Mahbubur Rahman, M., Roberts, H. L., and Schmidt, O. (2007). Tolerance to *Bacillus thuringiensis* endotoxin in immune-suppressed larvae of the flour moth *Ephestia kuehniella*. *J. Invertebr. Pathol.* **96**, 125–132.

Mansfield, B. E., Dionne, M. S., Schneider, D. S., and Freitag, N. E. (2003). Exploration of host–pathogen interactions using Listeria monocytogenes and *Drosophila melanogaster*. *Cell Microbiol.* **5**, 901–911.

Meister, M., Braun, A., Kappler, C., Reichhart, J., and Hoffmann, J. (1994). Insect immunity. A transgenic analysis in *Drosophila* defines several functional domains in the diptericin promoter. *Embo. J.* **13**, 5958–5966.

Moreno, E., Yan, M., and Basler, K. (2002). Evolution of TNF signaling mechanisms: JNK-dependent apoptosis triggered by eiger, the *Drosophila* homolog of the TNF superfamily. *Curr. Biol.* **12**, 1263–2368.

Moret, Y., and Schmid-Hempel, P. (2000). Survival for immunity: The price of immune system activation for bumblebee workers. *Science* **290**, 1166–1168.

Moret, Y., and Schmid-Hempel, P. (2001). Immune defence in bumble-bee offspring. *Nature* **414**, 506.

Moret, Y., and Siva-Jothy, M. T. (2003). Adaptive innate immunity? Responsive-mode prophylaxis in the mealworm beetle, Tenebrio molitor. *Proc. Biol. Sci.* **270**, 2475–2480.

Nappi, A. J., Vass, E., Frey, F., and Carton, Y. (1995). Superoxide anion generation in *Drosophila* during melanotic encapsulation of parasites. *Eur. J. Cell. Biol.* **68**, 450–456.

Neves, G., Zucker, J., Daly, M., and Chess, A. (2004). Stochastic yet biased expression of multiple Dscam splice variants by individual cells. *Nat. Genet.* **36**, 240–246.

Nonaka, M., and Yoshizaki, F. (2004). Primitive complement system of invertebrates. *Immunol. Rev.* **198**, 203–215.

Oppenheim, J. J., and Yang, D. (2005). Alarmins: Chemotactic activators of immune responses. *Curr. Opin. Immunol.* **17**, 359–365.

Pham, L. N., Dionne, M. S., Shirasu-Hiza, M., and Schneider, D. S. (2007). A specific primed immune response in *Drosophila* is dependent on phagocytes. *PLoS Pathog.* **3**, e26.

Qiu, P., Pan, P., and Govind, S. (1998). A role for the *Drosophila* Toll/Cactus pathway in larval hematopoiesis. *Development* **125**, 1909–1920.

Rahman, M. M., Roberts, H. L., Sarjan, M., Asgari, S., and Schmidt, O. (2004). Induction and transmission of *Bacillus thuringiensis* tolerance in the flour moth Ephestia kuehniella. *Proc. Natl. Acad. Sci. USA* **101**, 2696–2699.

Reeson, A. F., Wilson, K., Gunn, A., Hails, R. S., and Goulson, D. (1998). Baculovirus resistance in the noctuid *Spodoptera exempta* is phenotypically plastic and responds to population density. *Proc. Royal Society of London Series B-Biological Sciences* **265**, 1787–1791.

Rosengaus, R. B., Maxmen, A. B., Coates, L. E., and Traniello, J. F. A. (1998). Disease resistance: A benefit of sociality in the dampwood termite *Zootermopsis angusticollis* (Isoptera: Termopsidae). *Behav. Ecol. Sociobiol.* **44**, 125–134.

Rosengaus, R. B., Jordan, C., Lefebvre, M. L., and Traniello, J. F. (1999). Pathogen alarm behavior in a termite: A new form of communication in social insects. *Naturwissenschaften* **86**, 544–548.

Rosengaus, R. B., Lefebvre, M. L., and Traniello, J. F. A. (2000). Inhibition of fungal spore germination by Nasutitermes: Evidence for a possible antiseptic role of soldier defensive secretions. *J. Chem. Ecol.* **26**, 21–39.

Sadd, B. M., and Schmid-Hempel, P. (2006). Insect immunity shows specificity in protection upon secondary pathogen exposure. *Curr. Biol.* **16**, 1206–1210.

Sadd, B. M., and Siva-Jothy, M. T. (2006). Self-harm caused by an insect's innate immunity. *Proc. Biol. Sci.* **273**, 2571–2574.

Schneider, D. S., Ayres, J. S., Brandt, S. M., Costa, A., Dionne, M. S., Gordon, M. D., Mabery, E. M., Moule, M. G., Pham, L. N., and Shirasu-Hiza, M. M. (2007). *Drosophila* eiger mutants are sensitive to extracellular pathogens. *PLoS Pathog.* **3**, e41.

Shirasu-Hiza, M., Dionne, M. S., Pham, L. N., Ayres, J. S., and Schneider, D. S. (2007). Interactions between circadian rhythm and immunity in *Drosophila melanogaster*. *Curr. Biol.* **17**, R1–R2.

Silverman, N., Zhou, R., Erlich, R. L., Hunter, M., Bernstein, E., Schneider, D., and Maniatis, T. (2003). Immune activation of NF-kappaB and JNK requires *Drosophila* TAK1. *J. Biol. Chem.* **278**, 48928–48934.

Starks, P. T., Blackie, C. A., and Seeley, T. D. (2000). Fever in honeybee colonies. *Naturwissenschaften* **87**, 229–231.

Stroschein-Stevenson, S. L., Foley, E., O'Farrell P, H., and Johnson, A. D. (2006). Identification of *Drosophila* gene products required for phagocytosis of *Candida albicans*. *PLoS Biol.* **4**, e4.

Traniello, J. F., Rosengaus, R. B., and Savoie, K. (2002). The development of immunity in a social insect: Evidence for the group facilitation of disease resistance. *Proc. Natl. Acad. Sci. USA* **99**, 6838–6842.

Trinchieri, G., and Sher, A. (2007). Cooperation of Toll-like receptor signals in innate immune defence. *Nat. Rev. Immunol.* **7**, 179–190.

Tzou, P., Reichhart, J. M., and Lemaitre, B. (2002). Constitutive expression of a single antimicrobial peptide can restore wild-type resistance to infection in immunodeficient *Drosophila* mutants. *Proc. Natl. Acad. Sci. USA* **99**, 2152–2157.

Uttenweiler-Joseph, S., Moniatte, M., Lambert, J., Van Dorsselaer, A., and Bulet, P. (1997). A matrix-assisted laser desorption ionization time-of-flight mass spectrometry approach to identify the origin of the glycan heterogeneity of diptericin, an *O*-glycosylated antibacterial peptide from insects. *Anal. Biochem.* **247**, 366–375.

Uttenweiler-Joseph, S., Moniatte, M., Lagueux, M., Van, D. A., Hoffmann, J., and Bulet, P. (1998). Differential display of peptides induced during the immune response of *Drosophila*: a matrix-assisted laser desorption ionization time-of-flight mass spectrometry study. *Proc. Natl. Acad. Sci. USA* **95**, 11342–11347.

Wang, X., Zhao, Q., and Christensen, B. M. (2005). Identification and characterization of the fibrinogen-like domain of fibrinogen-related proteins in the mosquito, *Anopheles gambiae*, and the fruitfly, *Drosophila melanogaster*, genomes. *BMC Genomics* **6**, 114.

Watson, F. L., Puttmann-Holgado, R., Thomas, F., Lamar, D. L., Hughes, M., Kondo, M., Rebel, V. I., and Schmucker, D. (2005). Extensive diversity of Ig-superfamily proteins in the immune system of insects. *Science* **309**, 1874–1878.

Zhang, S. M., Adema, C. M., Kepler, T. B., and Loker, E. S. (2004). Diversification of Ig superfamily genes in an invertebrate. *Science* **305**, 251–254.

6

BIDIRECTIONAL CONNECTIONS BETWEEN THE IMMUNE SYSTEM AND THE NERVOUS SYSTEM IN INSECTS

SHELLEY ANNE ADAMO

*Department of Psychology and Neuroscience, Dalhousie University,
Halifax, NS, Canada*

ABSTRACT: Bidirectional connections between the immune system and the nervous system are well established in mammals, and there is solid evidence for their existence in mollusks. However, the issue has been relatively neglected in insects. There is indirect evidence suggesting that immune systems and nervous systems interact in insects. For example, injection of bacterial products such as lipopolysaccharides results in reproducible changes in behavior similar to the changes observed in vertebrates in response to bacterial exposure. Insects exposed to bacterial products may exhibit behavioral fever, illness-induced anorexia, decreased learning ability and increased egg laying, suggesting that immune activity can alter neural function. As in vertebrates, the expression of certain behaviors, such as flight-or-fight, results in transient immunosuppression, suggesting that neural activity influences immune function. Indirect evidence hints at the identity of compounds mediating immune–neural connections. For example, neurohormonally released octopamine appears to alter immune function. Growth-blocking peptide produced by immune cells increases dopamine production within the central nervous system. In mammals, neural–immune bidirectional connections are critical for the proper functioning of the immune system. Our poor understanding of neural–immune interactions in insects may limit our ability to comprehend insect immune systems.

6.1   INTRODUCTION

Insects rank as one of the most successful animal groups in terms of total biomass and geographic distribution (Romoser and Stoffolano, 1998). The bidirectional connections that exist between the central nervous system (CNS) and a variety of other physiological systems (Nation, 2002) contribute to their success. These interconnections allow the nervous system to integrate information from both internal and external sources and produce the most adaptive behavior and physiological set points. However, we still do not understand how two of the most complex and important physiological systems, the nervous system and the immune system, interact. In both vertebrates (Steinman, 2004; Sternberg, 2006) and mollusks (Ottaviani et al., 1997; Stefano et al., 2002), these two vital physiological systems are interconnected. Interestingly, some of the molecules linking the two systems are the same in both phyla, suggesting the possibility that connections between the nervous system and the immune system are ancient (Ottaviani and Franceschi, 1996). If so, then these connections should exist in insects as well.

There are good adaptive reasons why we should expect bidirectional connections between the immune system and the nervous system in any animal (e.g. see Sternberg, 2006). For example, in vertebrates, the activated immune system induces sickness behavior, a coordinated series of behavioral changes (Dantzer, 2004; Hart, 1988). These changes in behavior increase the chance that the animal will survive its infection (Dantzer, 2004).

This chapter reviews evidence in insects for: (1) modulation of the nervous system by the immune system, (2) modulation of the immune system by the nervous system and (3) the identity of the molecules that may link them.

6.2   EVIDENCE THAT THE IMMUNE SYSTEM CAN INFLUENCE NERVOUS SYSTEM FUNCTION

To demonstrate an unequivocal connection between the immune system and the nervous system, factors released by cells of the immune system (e.g. cytokines) should directly or indirectly alter neuronal function. For example, in mammals (see Dantzer, 2004 for review), bacterial lipopolysaccharide (LPS) binds to a Toll-like receptor found on macrophages. This receptor binding results in the release of cytokines (e.g. interleukin-1, IL-1). Cytokines released by macrophages play an important role in regulating the host's immune response (Roitt et al., 1996). Neurons in some brain areas also have receptors for cytokines (Dantzer, 2004). However, both mammals and insects (Nation, 2002) have a blood–brain barrier. Cytokines, such as IL-1, are relatively large proteins and cannot cross the blood–brain barrier (Dantzer, 2004). Nevertheless, blood-borne cytokines can still influence the mammalian CNS by three methods: (1) using a putative transporter molecule that would carry cytokines into the brain, (2) crossing into the CNS through areas of the brain devoid of the blood–brain barrier and (3) influencing the activity of peripheral sensory neurons (e.g. the vagus nerve) that signal to the CNS (Dantzer, 2004; Kis et al., 2006). After entering the brain, cytokines may either bind with neuronal receptors or by binding to receptors on the brain's own macrophages, the microglia, induce them to secrete compounds such as IL-1 locally into the brain (Dantzer, 2004). Dantzer (2004) suggests that it is this local release of immune-derived factors that has the most significant influence on neuronal activity. How cytokines such as IL-1 influence neuronal firing is still under investigation. In some neurons IL-1 appears to decrease glutaminergic synaptic transmission (Luk et al., 1999). Therefore, there is solid evidence that immune-derived factors such as cytokines influence behavior by altering neuronal function in specific brain areas.

In mollusks, the evidence is less detailed, but still compelling. A particularly elegant example comes from a series of studies examining how a parasite depresses egg laying in its snail host *Lymnaea stagnalis* (see De Jong-Brink, 1995; De Jong-Brink et al., 2001). Cells of the snail's immune system release schistosomin, a cytokine-like compound. Schistosomin is released into the hemolymph in response to a factor produced by the parasite. Most non-cephalopod mollusks are thought to have little, if any, blood–brain barrier (Bundgaard and Abbott, 1992). Therefore, circulating cytokines have direct access to the CNS. Once in the CNS, schistosomin alters the electrophysiological activity of the central neuroendocrine cells (e.g. the caudodorsal cells) that regulate egg laying (Hordijk et al., 1992). Schistosomin reduces the depolarizing after potential usually observed in these cells, decreasing their excitability (Hordijk et al., 1992). This decrease, along with other changes, results in a reduction in egg-laying behavior (De Jong-Brink, 1995). Therefore, in both mollusks and vertebrates, factors released by the immune system alter neuronal activity.

There are few studies on immune–neural interactions in insects. As with mollusks, some of the evidence comes from studies examining how parasites and pathogens influence host behavior (Adamo, 2002, 2005). Changes in host behavior after infection imply a change in neural function, and therefore could indicate an underlying immune–neural connection. The difficulty with this line of evidence is that the change in behavior may be due to some mechanism correlated with infection as opposed to being caused by an immune-derived factor. For example, the host's behavior may be altered by some secretion of the pathogen (e.g. parasitic manipulation) or by pathological changes caused by the pathogen (e.g. damage to an insect host's sensory system) (Adamo, 2002; Moore, 2002). Below I discuss studies that report a behavioral change using a non-pathological immune challenge (e.g. dead bacteria or bacterial components). These studies provide good evidence that immune-derived factors influence neural function. Interestingly, the changes in behavior due to a non-pathogenic immune challenge are similar to the changes in behavior that occur in vertebrates after an immune challenge (Table 6.1).

6.2.1 BEHAVIORAL FEVER

One of the earliest indications that the immune system and nervous system may interact in insects was the observation that an injection of LPS (a component of Gram negative bacterial cell wall) is able to induce behavioral fever in cockroaches (Bronstein and Conner, 1984). In behavioral fever, insects change their temperature preference and migrate to warmer areas. In a variety of insects, migrating to a warmer temperature increases resistance to different types of pathogens, meaning that this putative immune–neural connection could provide the animal with an adaptive advantage (Moore, 2002). How behavioral fever actually decreases mortality is not entirely clear (Thomas and Blanford, 2003). It can increase some immunological functions and can also directly lower pathogen vitality (e.g. Ouedraogo et al., 2003).

Not all immune reactions appear to elicit behavioral fever. In crickets (Adamo, 1998), and possibly locusts (Springate and Thomas, 2005; but see Bundey et al., 2003), behavioral fever is not a general response to any infection, but is pathogen

TABLE 6.1   Sickness Behaviors in Insects and Mammals

Insects[a]		Mammals[b]	
Behavior	Possible mediating factors	Behavior	Major mediating factors
Behavioral fever	Prostaglandins	Fever	Cytokines, prostaglandins
Anorexia	Octopamine	Anorexia	Cytokines
Decreased reproductive behavior	?	Decreased reproductive behavior	Cytokines
Increased reproductive behavior (egg laying)	?		
Decreased learning	?	Decreased learning	Cytokines
		Increased pain sensitivity	Cytokines
		Decreased locomotion	Cytokines
		Increased sleep	Cytokines

[a]See text for references.

[b]Taken from reviews by Hart (1988), Maier (2003) and Dantzer (2004). In mammals, effects of cytokines can be complex, altering some forms of learning but not others, and reducing foraging behavior more than food consumption (see Dantzer, 2004). Only main behavioral effects are listed.

specific. For example, crickets do not exhibit behavioral fever when encapsulating sephadex beads, when infested with the tachinid parasitoid *Ormia ochracea*, when infected with the bacterium *Serratia marcescens* or when infested with gregarines (gut protozoans) (Adamo, 1998). However, crickets do move to warmer temperatures when infected with an intracellular parasite, *Rickettsiella grylli* (Adamo, 1998; Louis et al., 1986). Different pathogens elicit different types of immune responses (Brennan and Anderson, 2004; Gillespie et al., 1997). If we knew the identity of the compounds released by the immune system during different types of immune responses, we might be able to determine which ones are likely to be mediating this putative immune–neural link.

Eicosanoids may be involved in the induction of behavioral fever. Eicosanoids are involved in mediating cellular immune reactions (Stanley and Miller, 2006; Stanley and Miller, chapter 3). The source of immunologically active eicosanoids is not known, but it could be the fat body and/or hemocytes. Both are immune tissues and sites of eicosanoid biosythnesis and there is evidence that hemocytes secrete eicosanoids during cellular immune reactions (Stanley and Miller, 2006). Bundey et al. (2003) blocked the expression of behavioral fever in locusts by inhibiting eicosanoid production using dexamethasone. Dexamethasone inhibits the production of arachidonic acid, a precursor molecule for eicosanoids such as prostaglandins (Stanley, 2000). Injecting animals with arachidonic acid as well as dexamethasone 'rescued' these animals, i.e. animals injected with both dexamethasone and arachidonic acid showed behavioral fever when challenged (Bundey et al., 2003). Other studies suggest that the eicosanoid responsible for inducing behavioral fever may be a prostaglandin. In another arthropod, the scorpions *Buthus occitanus* and *Androctonus australis*, injections of PGE1 resulted in behavioral fever (Cabanac and Le Guelte, 1980). Interestingly, prostaglandins play a role in producing fever in mammals (Blatteis, 2006).

Injection of bacterial cell wall components or other immune activators do not produce an immediate behavioral fever response. Bundey et al. (2003) noticed no change in temperature preference 2.5–5 h post injection, but did note a significant change at the next time point, 24 h later. Similarly, Bronstein and Conner (1984) found that the cockroach (*Gromphadorhina portentosa*) took more than 10 h after an LPS injection to develop a preference for warmer temperatures. Beetles (*Onymacris plana*) given an LPS injection exhibited an elevated temperature preference within 1 h after the injection with the peak effect 5 h later (McClain et al., 1988). However, injections of PGE1 induced behavioral fever in 12 min in scorpions (Cabanac and Le Guelte, 1980). Although the results cited above are from different species, the short delay between the injection of PGE1 and the induction of behavioral fever is consistent with the hypothesis that prostaglandin exerts its effect on the nervous system after being released by an immune response. McClain et al. (1988; Table 6.1) review data of the timing of behavioral fever from a wider range of arthropods. They (McClain et al., 1988) also find that immune activators require a longer time to elicit behavioral fever than do injections of prostaglandins.

Unfortunately, we do not know the neural basis of behavioral fever. In fact, our understanding of temperature control in insects remains sparse. Insects have peripheral thermal receptors, but how this information is integrated within the CNS is unknown (Chapman, 1998). Therefore, although there is strong indirect evidence that behavioral fever is produced by an immune–neural connection, unraveling the mechanistic details will be difficult.

6.2.2 ILLNESS-INDUCED ANOREXIA

Infection is known to reduce feeding in insects (Lacey and Brooks, 1997). Activating the immune system (e.g. with an injection of LPS) is sufficient to elicit illness-induced anorexia in insects (e.g. *Manduca sexta*; Adamo, 2005; Adamo et al., 2007; Bedoyan et al., 1992; Dunn et al., 1994). There is evidence that an immune–neural connection is responsible, at least in part, for the reduction in feeding (Fig. 6.1) in immune-challenged *M. sexta* (Adamo, 2005, 2006; Adamo et al., 2007).

The biogenic amine octopamine may play a role in this connection. Octopamine is a neuroactive compound; neurons have receptors for octopamine (Roeder, 2005)

FIGURE 6.1   *M. sexta* feeding on a tobacco plant.

and the effect of octopamine on neuronal activity has been well documented (Orchard et al., 1993). Octopamine is also an immune modulator (see Adamo, 2006; Roeder, 2005). Hemocytes have receptors for octopamine (e.g. Orr et al., 1985), and octopamine can, for example, enhance phagocytosis (Baines and Downer, 1994; Baines et al., 1992). Octopamine concentrations increase in the hemolymph during an immune challenge (Adamo, 2005; Dunphy and Downer, 1994). The source of this octopamine is unknown, but it could come from hemocytes (Adamo, 2005). High circulating octopamine concentrations disrupt the neural circuit for swallowing located in the frontal ganglion (Fig. 6.2), a part of the insect CNS (Miles and Booker, 2000). Therefore, if octopamine is released by the immune system, it could alter neuronal firing in the frontal ganglion. How octopamine passes through the insect blood–brain barrier remains to be determined. However, the blood–brain barrier has receptors for octopamine (Schofield and Treherne, 1986), and insects do have octopamine transporter molecules, although the ones found to date are typically associated with reuptake into neurons (Donly and Caveney, 2005).

Serotonin, another biogenic amine, may also play a role in producing illness-induced anorexia. Locusts show illness-induced anorexia when injected with the immune-activating agent laminarin (Goldsworthy, personal communication). This behavior can be reversed by mianserin (Goldsworthy, personal communication), an inhibitor of both serotonin and octopamine receptors in locusts (Hiripi et al., 1994; Molaei and Lange, 2003).

6.2.3   CHANGES IN REPRODUCTION

Immune activity and reproduction are often negatively correlated in insects (Lawniczak et al., 2007; Siva-Jothy et al., 2005). The suppressive effect of immune activation on reproduction is usually thought to be due to competition for resources between these two energetically expensive endeavors. There are only a few examples

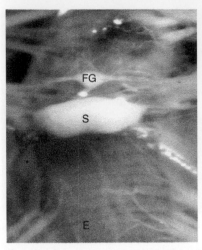

FIGURE 6.2   The frontal ganglion (FG) and supraesophageal ganglion (S) of a 5th-instar larval *M. sexta*. The ribbed structure (E) is the esophagus. The rostral end is toward the top of the photograph, while the caudal end is toward the bottom. *Source*: Photography courtesy of C. Miles.

in which immune system activation leads to a relatively rapid change in reproductive behavior, suggesting that the change may be caused by a direct immune–neural connection.

For example, Jacot et al. (2004) found that LPS injections in male crickets led to a decline in their calling behavior within a day. A decline in reproduction is commonly observed in immune-challenged vertebrates (Table 6.1).

Adamo (1999) found that injections of LPS led to an increase in egg laying in the cricket *Acheta domesticus*. In a related species of cricket, *Gryllus texensis*, bacterial infection also leads to an increase in egg laying (Shoemaker et al., 2006a). Egg laying was not increased by the encapsulation response (induced by the injection of sephadex beads), suggesting that it is produced only by some immune responses (Adamo, 1999). Increased oviposition after infection is thought to be adaptive because it allows female crickets to lay stored eggs that might otherwise perish with the sick female (Adamo, 1999; Shoemaker et al., 2006a). Because crickets have no parental care, the death of the female soon after laying her eggs would not affect the survival of her offspring.

How the immune system might induce increased egg laying is a tempting target for further study. The neural basis of egg-laying behavior is reasonably well understood in some orthopterans (e.g. locusts, Thompson, 1986), providing researchers with potential neural targets with which to test the effects of immune-derived factors.

6.2.4   CHANGES IN LEARNING BEHAVIOR

Immune activation results in a decline in learning ability in honeybees (Mallon et al., 2003). Injections of LPS led to a decline in the ability of honeybees to perform a classical conditioning task, i.e. associating a novel odor with a food reward (Mallon et al., 2003). The decline in learning was measurable 12 min after the injection. However, using a similar classical conditioning paradigm, LPS injections had no effect on learning in bumblebees, unless the bees were deprived of protein (Riddell and Mallon, 2006). Both Mallon et al. (2003) and Riddell and Mallon (2006) discuss the possibility that an immune–neural connection is responsible for the effect. However, why an immune–neural connection would exist in bumblebees only under conditions of protein deprivation is unclear. Possibly the decline in learning in hymenoptera during an immune reaction is not due to an immune–neural connection. In both vertebrates and mollusks, immune–neural connections do not require food deprivation for their activation. Both Mallon et al. (2003) and Riddell and Mallon (2006) suggest alternative explanations for the decline in learning, such as a competition for resources between the mechanisms underlying learning and those underlying an immune response.

6.3   EVIDENCE FOR NEURAL INFLUENCES ON IMMUNE FUNCTION

In vertebrates, the nervous system regulates immune function (Sternberg, 2006). For example, organs important for immune function, such as lymph nodes, are innervated by the sympathetic nervous system (Steinman, 2004). Neurons of the sympathetic nervous system release norepinephrine that then binds to receptors on immune cells leading to a change in their function (Webster et al., 2002). More recently, it has been shown that some vertebrate neuropeptides have antimicrobial properties (Brogden et al., 2005). Neurons may function as part of the immune system in some locations (e.g. the oral cavity in mammals) releasing antimicrobial compounds directly into infected areas (Brogden et al., 2005). Insect neurons have some of the same or related neuropeptides that have antimicrobial properties in vertebrates (e.g. Settembrini et al., 2003; Winther et al., 2006), suggesting that insect neurons could also play a direct role in immune function.

Below are examples suggesting that the nervous system influences the immune system in insects.

6.3.1   NEUROENDOCRINE INFLUENCES ON IMMUNITY

Beckage (chapter 10) and Adamo (2006) review evidence showing that hormones, many of which are directly or indirectly regulated by the CNS (Nijhout, 1994), influence immune function. For example, hormones important for the regulation of development also influence the expression of immune-related genes (Roxström-Lindquist et al., 2005). The ability of hormones to regulate both development and immune function could help the animal sculpt its immune system into the most adaptive configuration for each life stage.

6.3.2 NEURAL INNERVATION OF IMMUNE ORGANS

The fat body, a major organ of the immune system (Gillespie et al., 1997), is innervated in insects (Hazarika and Gupta, 1987). Fat body innervation supports the possibility of direct neural influence on immune function, although such a connection remains to be demonstrated.

6.3.3 CHANGES IN BEHAVIOR THAT RESULT IN CHANGES IN IMMUNE FUNCTION

In mammals, some behaviors correlate with changes in immune function (Glaser and Kiecolt-Glaser, 2005). Research has shown that some of these correlations exist because the neural activity mediating a behavior also activates a neural–immune connection. For example, activating flight-or-fight behavior in mammals results not just in fleeing or fighting, but also in the activation of the mammalian stress response (Sapolsky, 1992). This neural/neuroendocrine circuit can influence immune function using a variety of pathways (Glaser and Kiecolt-Glaser, 2005; Sternberg, 2006; Webster et al., 2002).

6.3.3.1 Reproductive Behavior

In insects, the production of certain behaviors also correlates with changes in immune function. For example, reproductive behavior can correlate with changes in both cellular and humoral immune function in insects (see Adamo, 2006; Lawniczak et al., 2007; Siva-Jothy et al., 2005). However, some of these correlations may occur for reasons other than a neural–immune link. For example, males transfer compounds to females during mating. Some of these compounds have direct effects on the expression of immune-related genes in the female (e.g. McGraw et al., 2004; Peng et al., 2005). Therefore, in females, changes in immune function after mating may be caused by substances and/or pathogens introduced by the male and not by a neural–immune connection. Alternatively, immune function may decline after reproductive behavior begins because of a depletion of resources needed for an immune response once reproduction is underway (Lawniczak et al., 2007). This depletion may not necessarily be energetic, but could be caused by competition for compounds needed by both functions (Siva-Jothy et al., 1998).

6.3.3.2 Stress-Induced Immunosuppression

The observation that stressful stimuli can influence immune function in mammals was one of the first documented examples of neural–immune communication. A wide range of stressors can alter immune function because of the potent effects of the neurally produced stress response on cells of the immune system (Glaser and Kiecolt-Glaser, 2005). Mollusks also show a coordinated stress response resulting in changes in immune function (e.g. Stefano et al., 2002).

Various stressors can influence immune function in insects (Brey, 1994). In crickets, flight-or-fight behavior (Fig. 6.3) leads to a transient decrease in resistance to

FIGURE 6.3   Tethered flight of the cricket *Gryllus texensis.*

bacteria (Adamo and Parsons, 2006). In orthoptera, flight-or-fight also results in an increase in the biogenic amine octopamine in the hemolymph (e.g. crickets, Adamo et al., 1995), presumably from a neural source (Orchard et al., 1993). Neurons containing octopamine send processes out to the periphery and are thought to release octopamine as a neurohormone (Orchard et al., 1993). The release of octopamine during flight-or-fight leads to increased energy supply, enhanced muscle performance and increased sensory perception, similar to the effects of adrenergic activation during mammalian stress (Roeder, 2005). Therefore, the release of octopamine during flight-or-fight has been considered analogous to the activation of the adrenergic system during similar behaviors in mammals (Orchard et al., 1993, Roeder, 2005).

As discussed in Section 6.2.2, octopamine can alter hemocyte function in cockroaches (Baines et al., 1992) and lepidopterans (*Galleria mellonella*, Diehl-Jones et al., 1996). Therefore, octopamine appears to mediate a neural–immune connection in insects, in much the same way that stress hormones play a similar role in mammals (e.g. see Webster et al., 2002). In insects, the 'stress' hormone octopamine appears to enhance rather than depress immune function (see Section 6.2.2; Adamo, 2006). Nevertheless, if crickets are given injections of octopamine, they become immunosuppressed, just as they do after flight (Adamo and Parsons, 2006).

The transient decline in resistance to bacteria after flight appears to be caused, in part, by a conflict between the demands of lipid metabolism and immune surveillance (Adamo and Parsons, 2006, Adamo et al., 2007; Adamo et al., unpublished data). By increasing hemocyte function (e.g. phagocytosis), octopamine may mitigate the negative effects that changes in lipid metabolism have on immunity. This phenomenon may be similar to the reconfiguration of immune resources in response to flight-or-fight in vertebrates (Dhabhar, 2002). However, as Roeder (2005) cautions, we do not know the biological significance of octopaminergic actions on immune function. Nevertheless, the ability of neurohormonal octopamine to alter immune function supplies good evidence for the existence of a neural–immune connection in insects.

6.4   WHAT ARE THE POSSIBLE CHEMICAL SIGNALS BETWEEN THE IMMUNE AND NERVOUS SYSTEMS?

Salzet et al. (2000) point out that molecules used by both the immune and nervous systems could carry messages between the two. In this section I use this criterion, as well as other evidence, to identify compounds that are likely to mediate communication between the immune and nervous systems in insects. However, molecules traveling between the immune system and CNS need to cross the blood–brain barrier. The major insect immune organs, the fat body and hemocytes, exist outside the blood–brain barrier. Mammals, which also have a blood–brain barrier, have specialized immune cells called microglia that reside inside the brain. Microglia are thought to be responsible for many mammalian immune–neural interactions (Dantzer, 2004). Microglia in mollusks (Peruzzi et al., 2004) and annelids (leeches, Vergote et al., 2006) also appear to be involved in immune–neural bidirectional communication. Insect nervous systems contain microglia too (Sonetti et al., 1994), but very little is known about them. Further study of the function of microglia in insects is likely to reveal important information about immune–neural interactions in insects.

6.4.1   BIOGENIC AMINES

The possible role of octopamine has been discussed in Section 6.2.2. Another biogenic amine, serotonin, can also influence hemocyte function (Baines and Downer, 1992; Baines et al., 1992). Like octopamine, serotonin is used within the CNS for interneuronal communication, and neurons have receptors for serotonin (Chapman, 1998). Serotonin, too, is used as a neurohormone in some insects (Nijhout, 1994) and, therefore, could also mediate a neural–immune connection.

Dopamine, another biogenic amine, may also play a role in immune–neural communication, as discussed in the next section.

6.4.2   CYTOKINES

In vertebrates, cytokines are common mediators of immune–neural connections (Maier, 2003). Insects also have cytokines (Agaisse et al., 2003), and there is good evidence that they may play a similar role. For example, two peptides of the ENF family, growth-blocking peptide (GBP) and plasmatocyte spreading peptide (PSP1) are cytokines found in Lepidoptera (Strand et al., 2000). They may also be mediators of immune–neural communication in insects.

GBP is found in both immune cells (e.g. fat body, Hayakawa et al., 1998) and glial cells of the CNS (Hayakawa et al., 2000). In the periphery, GBP is immunomodulatory (Strand et al., 2000). GBP appears to bind with a GBP receptor on plasmatocytes (Watanabe et al., 2006), leading to a change in hemocyte behavior (Strand et al., 2000). GBP also induces the epidermis to increase synthesis of the biogenic amine dopamine, leading to an increase in the concentration of dopamine in the hemolymph (Noguchi et al., 2003). GBP also appears to induce increased dopamine production within the CNS (Noguchi et al., 2003) and to regulate dopamine release from neurons (Hayakawa et al., 2000), suggesting that neurons have receptors for GBP. Because dopamine is a neuromodulator (Chapman, 1998), increases in dopamine content within the CNS are likely to result in changes in neural activity. Moreover, injections of GBP into larval lepidopterans can cause transient paralysis and other behavioral effects, suggesting that it does alter neural function (Hayakawa, 1995, 2006). Therefore, GBP appears to be a mediator of an immune–neural connection either directly or indirectly by its effect on dopamine concentrations within the CNS.

Studies on GBP were not initiated with the aim of uncovering an immune–neural connection. GBP was discovered while exploring how a parasitic wasp, *Cotesia kariyai*, is able to manipulate the physiology of its caterpillar host (*Pseudaletia separata*) (Hayakawa, 2006). GBP increases in the hemolymph of parasitized *P. separata* and plays a role in arresting host development (Hayakawa, 1995). Arresting host development is an important goal for the parasitic wasp because it allows the wasp larvae to emerge from the host during the host's larval stage, instead of being trapped within the sclerotized pupal cuticle (Hayakawa, 1995). Noguchi et al. (2003) argue that the increase in dopamine in the CNS leads to increased juvenile hormone release from the corpora allata, and that this is one of the mechanisms by which the wasp interferes with host development. Unfortunately, we do not fully understand the role GBP plays under normal conditions. Possibly GBP allows the host to slow growth and development after injury or infection, allowing it time to repair itself prior to molting (Lavine and Strand, 2002; Strand et al., 2000). This raises the intriguing possibility that the wasp has evolved the ability to exploit an immune–neural connection in the host for its own ends.

PSP1, another insect cytokine, is expressed in both neural and immune tissue (Lavine and Strand, 2002). PSP1 is thought to bind to a putative 190 kDa receptor, although the molecular identification of the receptor is still ongoing (Clark et al., 2004). After having bound to its receptor, PSP1 activates hemocytes (Strand et al., 2000). Its function within the CNS remains unknown.

Molecules similar to vertebrate cytokines such as IL-1 and tumor necrosis factor (TNF) have been found in insects (Beschin et al., 2001). Vertebrate-like cytokines may act as cytokines in insects as well as they localize to immune cells (e.g. hemocytes, Franchini et al., 1996; Wittwer et al., 1999). Moreover, recently a homolog of a vertebrate cytokine, TNF-alpha was found to be upregulated during infection in *Drosophila melanogaster* (Irving et al., 2005), suggesting that it may play some role in the immune response. Vertebrate-like cytokines have also been found in a range of other animals, such as mollusks (Ottaviani and Francheschi, 1996; Stephano et al., 2002). Similarities in the identity and function of cytokines between mollusks and mammals suggest that compounds such as IL-1 may play an ancient function in connecting the immune and nervous systems (Maier, 2003, however see Beschin et al., 2001). For example, hemocytes of the mollusk *Aplysia* release compounds, such as IL-1-beta, that increase sensory neuron excitability (Clatworthy, 1996; Clatworthy and Grose, 1999). Vertebrate-like cytokines may play a similar role in insects.

6.4.3   OPIOID PEPTIDES AND OPIATE ALKALOIDS

In mollusks and mammals, opioids and opiates are involved in both the immune and nervous systems, and these compounds appear to modulate immune responses during stressful stimuli (Kavaliers, 1991; Lacoste et al., 2001a; Ottaviani and Franceschi, 1996; Ottaviani et al., 1997; Salzet et al., 2000; Stefano et al., 2002). For example, in the oyster *Crassostrea gigas*, pro-opiomelanocortin-derived peptides, such as ACTH (adrenocorticotropic hormone), are able to induce neurosecretory cells located in the heart to release norepinephrine, which in turn can influence hemocyte function (Lacoste et al., 2001b). Unfortunately, there is much less information available about the role these compounds play in insects. Opioids and opiates can alter hemocyte function in insects (Scharrer et al., 1996). There is also evidence suggesting that there are opioid (Stefano and Scharrer, 1981) and opiate (Stefano and Scharrer, 1996) receptors within the CNS of insects. Furthermore, there is indirect behavioral evidence that opioids and/or opiates are involved in CNS function because they appear to play a role in regulating analgesia in cockroaches (Brown et al., 1994; Gritsai et al., 2004).

6.4.4   NITRIC OXIDE

Nitric oxide (NO) is involved in both neuronal signaling (Bicker, 2001) and cellular and humoral immune function (Nappi et al., 2000) in insects. In *Drosophila*, NO plays a role in the production of cytotoxic molecules during encapsulation, and can induce increased antimicrobial peptide production (Nappi et al., 2000). NO is involved in both neuronal signaling and immune function in other invertebrates as well (Ottaviani et al., 1997). In mollusks, NO may carry bidirectional signals between neurons and microglia (Peruzzi et al., 2004).

6.4.5   EICOSANOIDS

The possible role of eicosanoids, especially prostaglandins, in neural–immune interactions has been discussed in Section 6.2.1. A difficulty for prostaglandins as potential mediators of immune–neural interactions is that most of them are unable to cross a blood–brain barrier (Kis et al., 2006). Nevertheless, this barrier does not stop prostaglandins from entering the CNS in mammals (e.g. Eguchi et al., 1988). Once in the mammalian brain, prostaglandins can alter neural activity because neurons have receptors for them (e.g. see Piomelli, 1994). Unfortunately, little is known about prostaglandin receptors in insects (Stanley, 2000; Stanley and Miller, 2006), although receptors for prostaglandins have been found in other arthropods (e.g. Qian et al., 1997). There is indirect evidence that prostaglandins can cross the insect blood–brain barrier and bind with neural receptors. During mating in the cricket *Tellogryllus commodus*, PGE2 moves from the female reproductive tract into the hemolymph (see Stanley, 2000). This prostaglandin activates egg laying in females and it is thought to do so by interacting with neurons in the terminal ganglion (Stanley, 2000).

Interestingly, mating in female crickets results in increased resistance to disease (Shoemaker et al., 2006b) as well as in increased oviposition (Stanley, 2000). Could PGE2 play a role in producing both phenomena? Prostaglandins are released during an immune response (Stanley and Miller, 2006). PGE2 may enhance nodule formation in *M. sexta* (Phelps et al., 2003). Perhaps PGE2 acts as a signal from the immune system to the nervous system resulting in the adaptive increase in egg laying. If this is true, then males may have evolved the ability to manipulate female oviposition by exploiting such an immune–neural connection. Or, viewed from a different perspective, males may give females enzymes for, and/or precursors of, compounds such as prostaglandins as a nuptial gift that results in an increase in both female immune function and oviposition.

6.5   CONCLUSIONS

Our understanding of how immune and nervous systems interact in insects (Fig. 6.4) has lagged far behind our understanding of such interactions in mollusks and mammals. Given the wealth of information on insect nervous systems and immune systems, the fusion of these two fields is likely to prove as successful and important for insect biologists as their combination has proven to be for those studying vertebrates (see Dantzer, 2004; Sternberg, 2006). This fusion will also be of practical importance for topics such as biological control and integrated pest management. For example, behavioral fever in insect pests can limit the effectiveness of some fungi as biological control agents (Blanford and Thomas, 2001).

The enormous number of species within the insects is both a liability and an opportunity when studying immune–neural connections. Usually information about possible immune–neural interactions in insects is known from a few species at most. Although immune–neural links may have an ancient evolutionary basis (Maier, 2003; Ottaviani and Franceschi, 1996; Stephano et al., 2002), they will also be sculpted by the animal's present ecological niche. Therefore, these connections are likely to vary across species. This variability may limit the generalizability of results, but it will also provide insight as to how these connections respond to different evolutionary pressures.

In mammals, the input of the CNS is considered integral to the response of the innate immune system (Sternberg, 2006). Insect immune systems are thought to resemble vertebrate innate immunity in a number of ways (Brennan and Anderson, 2004). Therefore, it may be impossible to understand insect immune systems in depth without a better appreciation of insect immune–neural bidirectional connections.

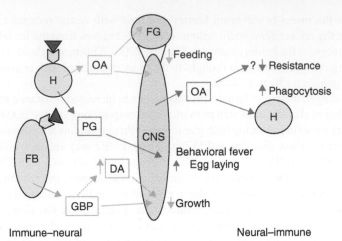

FIGURE 6.4   Possible immune–neural bidirectional connections in insects. The left half of the figure represents possible immune–neural connections. The triangles represent immune activators. Activated hemocytes (H) release octopamine (OA). The OA interacts with neurons in the frontal ganglion (FG) and elsewhere in the central nervous system (CNS) to depress feeding. Activated hemocytes also release eicosanoids such as prostaglandins (PG). PG may enter the CNS resulting in behavioral fever and an increase in egg laying. Immune-activated fat body (FB) releases growth-blocking peptide (GBP). GBP leads to an increase in the dopamine (DA) concentration of the hemolymph and within the CNS. This DA increase could influence CNS functioning. GBP may also enter the CNS and alter neuronal activity resulting in reduced growth. The right side of the figure represents possible neural–immune connections. During some behaviors, the CNS releases OA into the hemolymph. OA can bind with receptors on H leading to an increase in phagocytosis. Nevertheless, increasing OA concentration in the hemolymph leads to a decrease in resistance to bacteria. The mechanism causing the decline is unknown.

REFERENCES

Adamo, S. A. (1998). The specificity of behavioral fever in the cricket *Acheta domesticus*. *J. Parasitol.* **84**, 529–533.

Adamo, S. A. (1999). Evidence for adaptive changes in egg-laying in crickets exposed to bacteria and parasites. *Anim. Behav.* **57**, 117–124.

Adamo, S. A. (2002). Modulating the modulators: Parasites, neuromodulators and host behavioral change. *Brain Behav. Evolut.* **60**, 370–377.

Adamo, S. A. (2005). Parasitic suppression of feeding in the tobacco hornworm, *Manduca sexta*: Parallels with feeding depression after an immune challenge. *Arch. Insect Biochem. Physiol.* **60**, 185–197.

Adamo, S. A. (2006). Comparative psychoneuroimmunology: Evidence from the insects. *Behav. Cognit. Neurosci. Rev.* **5**, 128–140.

Adamo, S. A., and Parsons, N. M. (2006). The emergency life-history stage and immunity in the cricket, *Gryllus texensis*. *Anim. Behav.* **72**, 235–244.

Adamo, S. A., Linn, C. E., and Hoy, R. R. (1995). The role of neurohormonal octopamine during 'fight or flight' behaviour in the field cricket *Gryllus bimaculatus*. *J. Exp. Biol.* **198**, 1691–1700.

Adamo, S. A., Fidler, T. L., and Forestell, C. A. (2007). Illness-induced anorexia and its possible function in the caterpillar, *Manduca sexta*. *Brain Behav. Immun.* **21**, 293–300.

Agaisse, H., Petersen, U. M., Boutros, M., Mathey-Prevot, B., and Perrimon, N. (2003). Signaling role of hemocytes in *Drosophila* Jak/Stat-dependent response to septic injury. *Dev. Cell* **5**, 441–450.

Baines, D., and Downer, R. G. H. (1992). 5-Hydroxytryptamine-sensitive adenylate cyclase affects phagocytosis in cockroach hemocytes. *Arch. Insect Biochem. Physiol.* **21**, 303–316.

Baines, D., and Downer, R. G. H. (1994). Octopamine enhances phagocytosis in cockroach hemocytes: Involvement of inositol trisphosphate. *Arch. Insect Biochem. Physiol.* **26**, 249–261.

Baines, D., DeSantis, T., and Downer, R. (1992). Octopamine and 5-hydroxytryptamine enhance the phagocytic and nodule formation activities of cockroach (*Periplaneta americana*) haemocytes. *J. Insect Physiol.* **38**, 905–914.

Bedoyan, J. K., Patil, C. S., Kyriakides, T. R., and Spence, K. D. (1992). Effect of excess dietary glucose on growth and immune response on *Manduca sexta*. *J. Insect Physiol.* **38**, 525–532.

Beschin, A., Bilej, M., Torreele, E., and de Baetselier, P. (2001). On the existence of cytokines in invertebrates. *Cell. Mol. Life Sci.* **58**, 801–814.

Bicker, G. (2001). Nitric oxide: An unconventional messenger in the nervous system of an orthopteroid insect. *Arch. Insect Biochem. Physiol.* **48**, 100–110.

Blanford, S., and Thomas, M. B. (2001). Adult survival, maturation, and reproduction of the desert locust *Schistocerca gregaria* infected with the fungus *Metarhizium anisopliae* var acridum. *J. Invertebr. Pathol.* **78**, 1–8.

Blatteis, C. M. (2006). Endotoxic fever: New concepts of its regulation suggest new approaches to its management. *Pharmacol. Therapeut.* **111**, 194–223.

Brennan, C. A., and Anderson, K. V. (2004). *Drosophila*: The genetics of innate immune recognition and response. *Annu. Rev. Immunol.* **22**, 457–483.

Brey, P. T. (1994). The impact of stress on insect immunity. *Bull. Inst. Pasteur* **92**, 101–118.

Brogden, K. A., Guthmiller, J. M., Salzet, M., and Zasloff, M. (2005). The nervous system and innate immunity: The neuropeptide connection. *Nat. Immunol.* **6**, 558–564.

Bronstein, S. M., and Conner, W. E. (1984). Endotoxin-induced behavioral fever in the Madagascar cockroach, *Gromphadorhina portentosa*. *J. Insect Physiol.* **30**, 327–330.

Brown, G. E., Anderson, C. L., and Scruggs, J. L. (1994). Shock-induced analgesia in the cockroach (*Periplaneta americana*). *Psychol. Rep.* **74**, 1051–1057.

Bundey, S., Raymond, S., Dean, P., Roberts, S. K., Dillon, R. J., and Charnley, A. K. (2003). Eicosanoid involvement in the regulation of behavioral fever in the desert locust, *Schistocerca gregoria*. *Arch. Insect Biochem. Physiol.* **52**, 183–192.

Bundgaard, M., and Abbott, N. J. (1992). Fine structure of the blood–brain interface in the cuttlefish *Sepia officinalis* (Mollusca, Cephalopoda). *J. Neurocytol.* **21**, 260–275.

Cabanac, M., and Le Guelte, L. (1980). Temperature regulation and prostaglandin E1 fever in scorpions. *J. Physiol.* **303**, 365–370.

Chapman, R. F. (1998). *The Insects: Structure and Function*. Cambridge University Press, Cambridge.

Clark, K. D., Garczynski, S. F., Arora, A., Crim, J. W., and Strand, M. R. (2004). Specific residues in plasmatocyte-spreading peptide are required for receptor binding and functional antagonism of insect immune cells. *J. Biol. Chem.* **279**, 33246–33252.

Clatworthy, A. L. (1996). A simple systems approach to neural–immune communication. *Comparat. Biochem. Physiol. A* **115**, 1–10.

Clatworthy, A. L., and Grose, E. (1999). Immune-mediated alterations in nociceptive sensory function in *Aplysia californica*. *J. Exp. Biol.* **202**, 623–630.

Dantzer, R. (2004). Cytokine-induced sickness behaviour: A neuroimmune response to activation of the innate immunity. *Euro. J. Pharmacol.* **500**, 399–411.

De Jong-Brink, M. (1995). How schistosomes profit from the stress responses they elicit in their hosts. *Adv. Parasitol.* **35**, 177–256.

De Jong-Brink, M., Bergamin-Sassen, M., and Solis Soto, M. (2001). Multiple strategies of schistosomes to meet their requirements in the intermediate snail host. *Parasitology* **123**, S129–S141.

Dhabhar, F. (2002). Stress-induced augmentation of immune function – The role of stress hormones, leuckocyte trafficking, and cytokines. *Brain Behav. Immun.* **16**, 785–798.

Diehl-Jones, W. L., Mandato, C. A., Whent, G., and Downer, R. G. H. (1996). Monoaminergic regulation of hemocyte activity. *J. Insect Physiol.* **42**, 13–19.

Donly, B. C., and Caveney, S. (2005). A transporter for phenolamine uptake in the arthropod CNS. *Arch. Insect Biochem. Physiol.* **59**, 172–183.

Dunn, P. E., Bohnert, T. J., and Russell, V. (1994). Regulation of antibacterial protein synthesis following infection and during metamorphosis of *Manduca sexta. Ann. NY Acad. Sci.* **712**, 117–130.

Dunphy, G. B., and Downer, R. G. H. (1994). Octopamine, a modulator of the haemocytic nodulation response of non-immune *Galleria mellonella. J. Insect Physiol.* **40**, 267–272.

Eguchi, N., Hayashi, H., Urade, Y., Ito, S., and Hayaishi, O. (1988). Central action of prostaglandin E2 and its methyl ester in the induction of hyperthermia after their systemic administration in urethane-anesthetized rats. *J. Pharmacol. Exp. Therapeut.* **247**, 671–679.

Franchini, A., Miyan, J. A., and Ottaviani, E. (1996). Induction of ACTH- and TNF-alpha-like molecules in the hemocytes of *Calliphora vomitoria (Insecta, Diptera). Tissue Cell* **28**, 587–592.

Gillespie, J. P., Kanost, M. R., and Trenczek, T. (1997). Biological mediators of insect immunity. *Annu. Rev. Entomol.* **42**, 611–643.

Glaser, R., and Kiecolt-Glaser, J. K. (2005). Stress-induced immune dysfunction: Implications for health. *Nat. Rev. Immunol.* **5**, 243–251.

Gritsai, O. B., Dubynin, V. A., Pilpenko, V. E., and Petrov, O. P. (2004). Effects of peptide and non-peptide opioids on protective reaction of the cockroach *Periplaneta americana* in the 'hot camera'. *J. Evolut. Biochem. Physiol.* **40**, 153–160.

Hart, B. L. (1988). Biological basis of the behavior of sick animals. *Neurosci. Biobehav. Rev.* **12**, 123–137.

Hayakawa, Y. (1995). Growth-blocking peptide: An insect biogenic peptide that prevents the onset of metamorphosis. *J. Insect Physiol.* **41**, 1–6.

Hayakawa, Y. (2006). Insect cytokine growth-blocking peptide (GBP) regulates insect development. *Appl. Entomol. Zool.* **41**, 545–554.

Hayakawa, Y., Ohnishi, A., and Endo, Y. (1998). Mechanism of parasitism-induced elevation of haemolymph growth-blocking peptide levels in host insect larvae (*Pseudaletia separata*). *J. Insect Physiol.* **44**, 859–866.

Hayakawa, Y., Ohnishi, A., Mizoguchi, A., and Yamashika, C. (2000). Distribution of growth-blocking peptide in the insect central nervous tissue. *Cell Tissue Res.* **300**, 459–464.

Hazarika, H. L., and Gupta, A. P. (1987). Anatomy of the retrocerebral complex of *Blattella germanica* L. (Dictyoptera Blattellidae). *Zoologischer-Anzeiger* **219**, 257–264.

Hiripi, L., Szilveszter, J., and Downer, R. G. H. (1994). Characterization of tyramine and octopamine receptors in the insect (*Locusta migratoria migratorioides*) brain. *Brain Res.* **633**, 119–126.

Hordijk, P. L., De Jong-Brink, M., Ter Maat, A., Pieneman, A. W., Lodder, J. C., and Kits, K. S. (1992). The neuropeptide schistosomin and haemolymph from parasitized snails induce similar changes in excitability in neuroendocrine cells controlling reproduction and growth in a freshwater snail. *Neurosci. Lett.* **136**, 193–197.

Irving, P., Ubeda, J. M., Doucet, D., Troxler, L., Lagueux, M., Zachary, D., Hoffmann, J. A., Hetru, C., and Meister, M. (2005). New insights into *Drosophila* larval hemocyte functions through genome-wide analysis. *Cell. Microbiol.* **7**, 335–350.

Jacot, A., Scheuber, H., and Brinkhof, M. W. G. (2004). Costs of an induced immune response on sexual display and longevity in field crickets. *Evolution* **58**, 2280–2286.

Kavaliers, M. (1991). Opioid peptides, nociception and analgesia in molluscs. In *Comparative Aspects of Neuropeptide Function* (E. Florey, and G. B. Stefano, Eds.), pp. 87–96. Manchester University Press, Manchester.

Kis, B., Isse, T., Snipes, J. A., Chen, L., Yamashita, H., Ueta, Y., and Busija, D. W. (2006). Effects of LPS stimulation on the expression of prostaglandin carriers in the cells of the blood–brain and blood–cerebrospinal fluid barriers. *J. Appl. Physiol.* **100**, 1392–1399.

Lacey, L. A., and Brooks, W. M. (1997). Initial handling and diagnosis of diseased insects. In *Manual of Techniques in Insect Pathology* (L. A. Lacey, Ed.), pp. 1–15. Academic Press, San Diego.

Lacoste, A., Malham, S. K., Cueff, A., Jalabert, F., Gélébart, F., and Poulet, S. A. (2001a). Evidence for a form of adrenergic response to stress in the mollusc *Crassostrea gigas*. *J. Exp. Biol.* **204**, 1247–1255.

Lacoste, A., Malham, S. K., Cueff, A., and Poulet, S. A. (2001b). Noradrenaline modulates oyster hemocyte phagocytosis via a β -adrenergic receptor-cAMP signaling pathway. *Gen. Comparat. Endocrinol.* **122**, 252–259.

Lavine, M. D., and Strand, M. R. (2002). Insect hemocytes and their role in immunity. *Insect Biochem. Mol. Biol.* **32**, 1295–1309.

Lawniczak, M. K. N., Barnes, A. I., Linklater, J. R., Boone, J. M., Wigby, S., and Chapman, T. (2007). Mating and immunity in invertebrates. *Trends Ecol. Evolut* **22**, 48–55.

Louis, C., Jourdan, M., and Cabanac, M. (1986). Behavioral fever and therapy in a rickettsia-infected Orthoptera. *Am. J. Physiol.* **250**, R991–R995.

Luk, W. P., Zhang, Y., White, T. D., Lue, F. A., Wu, C. P., Jiang, C. G., Zhang, L., and Moldofsky, H. (1999). Adenosine: A mediator of interleukin-1 beta induced hippocampal synaptic inhibition. *J. Neurosci.* **19**, 4238–4244.

Maier, S. F. (2003). Bi-directional immune–brain communication: Implications for understanding stress, pain and cognition. *Brain Behav. Immun.* **17**, 69–85.

Mallon, E. B., Brockmann, A., and Schmid-Hempel, P. (2003). Immune response inhibits associative learning in insects. *Proc. Roy. Soc. Lond. B* **270**, 2471–2473.

McClain, E., Magnuson, P., and Warner, S. J. (1988). Behavioural fever in a Namib desert tenebrionid beetle, *Onymacris plana*. *J. Insect Physiol.* **34**, 279–284.

McGraw, L. A., Gibson, G., Clark, A. G., and Wolfner, M. F. (2004). Genes regulated by mating, sperm or seminal proteins in mated female *Drosophila melanogaster*. *Curr. Biol.* **14**, 1509–1514.

Miles, C. I., and Booker, R. (2000). Octopamine mimics the effects of parasitism on the foregut of the tobacco hornworm *Manduca sexta*. *J. Exp. Biol.* **203**, 1689–1700.

Molaei, G., and Lange, A. B. (2003). The association of serotonin with the alimentary canal of the African migratory locust, *Locusta migratoria*: Distribution, physiology and pharmacological profile. *J. Insect Physiol.* **49**, 1073–1082.

Moore, J. (2002). *Parasites and the Behavior of Animals*. Oxford University Press, New York.

Nappi, A. J., Vass, E., Frey, F., and Carton, Y. (2000). Nitric oxide involvement in *Drosophila* immunity. *Nitric Oxide* **4**, 423–430.

Nation, J. L. (2002). *Insect Physiology and Biochemistry*. CRC Press, Boca Raton.

Nijhout, H. F. (1994). *Insect Hormones*. Princeton University Press, Princeton, NJ.

Noguchi, H., Tsuzuki, S., Tanaka, K., Matsumoto, H., Hiruma, K., and Hayakawa, Y. (2003). Isolation and characterization of a dopa decarboxylase cDNA and the induction of its expression by an insect cytokine, growth blocking peptide in *Pseudaletia separata*. *Insect Biochem. Mol. Biol.* **33**, 209–217.

Orchard, I., Ramirez, J. M., and Lange, A. B. (1993). A multifunctional role for octopamine in locust flight. *Annu. Rev. Entomol.* **38**, 227–249.

Orr, G. L., Gole, J. W. D., and Downer, R. G. H. (1985). Characterization of an octopamine-sensitive adenylate cyclase in hemocyte membrane fragments of the American cockroach *Periplaneta americana* L. *Insect Biochem.* **15**, 695–701.

Ottaviani, E., and Franceschi, C. (1996). The neuroimmunology of stress from invertebrates to man. *Prog. Neurobiol.* **48**, 421–440.

Ottaviani, E., Franchini, A., and Franceschi, C. (1997). Pro-opiomelanocortin-derived peptides, cytokines, and nitric oxide in immune responses and stress: An evolutionary approach. *Int. Rev. Cytol.* **170**, 79–141.

Ouedraogo, R. M., Cusson, M., Goettel, M. S., and Brodeur, J. (2003). Inhibition of fungal growth in thermoregulating locusts, *Locusta migratoria*, infected by the fungus *Metarhizium anisopliae* var *acridum*. *J. Invertebr. Pathol.* **82**, 103–109.

Peng, J., Zipperlen, P., and Kubli, E. (2005). *Drosophila* sex-peptide stimulates female innate immune system after mating via the Toll and Imd pathways. *Curr. Biol.* **15**, 1690–1694.

Peruzzi, E., Fontana, G., and Sonetti, D. (2004). Presence and role of nitric oxide in the central nervous system of the freshwater snail *Planorbarius corneus*: Possible implication in neuron–microglia communication. *Brain Res.* **1005**, 9–29.

Phelps, P. K., Miller, J. S., and Stanley, D. W. (2003). Prostaglandins, not lipoxygenase products, mediate insect microaggregation reactions to bacterial challenge in isolated hemocyte preparations. *Comparat. Biochem. Physiol. A* **136**, 409–416.

Piomelli, D. (1994). Eicosanoids in synaptic transmission. *Crit. Rev. Neurobiol.* **8**, 65–83.

Qian, Y., Essenberg, R. C., Dillwith, J. W., Bowman, A. S., and Sauer, J. R. (1997). A specific prostaglandin E2 receptor and its role in modulating salivary secretion in the female tick, *Amblyomma americanum* (L.). *Insect Biochem. Mol. Biol.* **27**, 387–395.

Riddell, C. E., and Mallon, E. B. (2006). Insect psychoneuroimmunology: Immune response reduces learning in protein starved bumblebees (*Bombus terrestris*). *Brain Behav. Immun.* **20**, 135–138.

Roeder, T. (2005). Tyramine and octopamine: Ruling behavior and metabolism. *Annu. Rev. Entomol.* **50**, 447–477.

Roitt, I., Brostoff, J., and Male, D. (1996). *Immunology*. Mosby, London.

Romoser, W. S., and Stoffolano, J. G. (1998). *The Science of Entomology*. McGraw-Hill, Boston.

Roxström-Lindquist, K., Assefaw-Redda, Y., Rosinska, K., and Faye, I. (2005). 20-Hydroxyecdysone indirectly regulates *Hemolin* gene expression in *Hyalophora cecropia*. *Insect Mol. Biol.* **14**, 645–652.

Salzet, M., Vieau, D., and Day, R. (2000). Crosstalk between nervous and immune systems through the animal kingdom: Focus on opioids. *Trends Neurosci.* **23**, 550–555.

Sapolsky, R. (1992). Neuroendocrinology of the stress response. In *Behavioral Endocrinology* (J. Becker, S. Breedlove, and D. Crews, Eds.), pp. 287–324. MIT Press, Cambridge, MA.

Scharrer, B., Paemen, L., Smith, E. M., Hughes, T. K., Liu, Y., Pope, M., and Stefano, G. B. (1996). The presence and effects of mammalian signal molecules in immunocytes of the insect *Leucophaea maderae*. *Cell Tissue Res.* **283**, 93–97.

Schofield, P. K., and Treherne, J. E. (1986). Octopamine sensitivity of the blood–brain-barrier of an insect. *J. Exp. Biol.* **123**, 423–439.

Settembrini, B. P., Nowicki, S., Hokfelt, T., and Villar, M. J. (2003). Distribution of NPY and NPY-Y1 receptor-like immunoreactivities in the central nervous system of Triatoma infestans (Insecta: Heteroptera). *J. Comparat. Neurol.* **460**, 141–154.

Shoemaker, K. L., Parsons, N. M., and Adamo, S. A. (2006a). Egg-laying behaviour following infection in the cricket *Gryllus texensis*. *Can. J. Zool.* **84**, 412–418.

Shoemaker, K. L., Parsons, N. M., and Adamo, S. A. (2006b). Mating enhances parasite resistance in the cricket *Gryllus texensis*. *Anim. Behav.* **71**, 371–380.

Siva-Jothy, M. T., Tsubaki, Y., and Hooper, R. E. (1998). Decreased immune response as a proximate cost of copulation and oviposition in a damselfly. *Physiol. Entomol.* **23**, 274–277.

Siva-Jothy, M. T., Moret, Y., and Rolff, J. (2005). Insect immunity: An evolutionary ecology perspective. *Adv. Insect Physiol.* **32**, 1–48.

Sonetti, D., Ottaviani, E., Bianchi, F., Rodriguez, M., Stefano, M. L., Sharrer, B., and Stefano, G. B. (1994). Microglia in invertebrate ganglia. *Proc. Natl. Acad. Sci. USA* **91**, 9180–9184.

Springate, S., and Thomas, M. B. (2005). Thermal biology of the meadow grasshopper, *Chorthippus parallelus*, and the implications for resistance to disease. *Ecol. Entomol.* **30**, 724–732.

Stanley, D. W. (2000). *Eicosanoids in Invertebrate Signal Transduction Systems*. Princeton University Press, Princeton, NJ.

Stanley, D. W., and Miller, J. S. (2006). Eicosanoid actions in insect cellular immune functions. *Entomologia Experimentalis et Applicata* **119**, 1–13.

Stefano, G. B., and Scharrer, B. (1981). High affinity binding of an enkephalin analog in the cerebral ganglion of the insect *Leucophaea maderae* (Blattaria). *Brain Res.* **225**, 107–114.

Stefano, G. B., and Scharrer, B. (1996). The presence of the mu(3) opiate receptor in invertebrate neural tissues. *Comparat. Biochem. Physiol. C* **113**, 369–373.

Stefano, G. B., Cadet, P., Zhu, W., Rialas, C. M., Mantione, K., Benz, D., Fuentes, R., Casares, F., Fricchione, G. L., Fulop, Z., and Slingsby, B. (2002). The blueprint for stress can be found in invertebrates. *Neuroendocrinol. Lett.* **23**, 85–93.

Steinman, L. (2004). Elaborate interactions between the immune and nervous systems. *Nat. Immunol.* **5**, 575–581.

Sternberg, E. M. (2006). Neural regulation of innate immunity: A coordinated nonspecific host response to pathogens. *Nat. Rev. Immunol.* **6**, 318–328.

Strand, M. R., Hayakawa, Y., and Clark, K. D. (2000). Plasmatocyte spreading peptide (PSP1) and growth blocking peptide (GBP) are multifunctional homologs. *J. Insect Physiol.* **46**, 817–824.

Thomas, M. B., and Blanford, S. (2003). Thermal biology in insect–parasite interactions. *Trends Ecol. Evolut.* **18**, 344–350.

Thompson, K. J. (1986). Oviposition digging in the grasshopper. I. Functional anatomy and the motor programme. *J. Exp. Biol.* **122**, 387–411.

Vergote, D., Macagno, E. R., Salzet, M., and Sautiere, P. E. (2006). Proteome modifications of the medicinal leech nervous system under bacterial challenge. *Proteomics* **6**, 4817–4825.

Watanabe, S., Tada, M., Aizawa, T., Yoshida, M., Sugaya, T., Taguchi, M., Kouno, T., Nakamura, T., Mizuguchi, M., Demura, M., Hayakawa, Y., and Kawano, K. (2006). *N*-terminal mutational analysis of the interaction between growth-blocking peptide (GBP) and receptor of insect immune cells. *Protein Peptide Lett.* **13**, 815–822.

Webster, J. I., Tonelli, L., and Sternberg, E. M. (2002). Neuroendocrine regulation of immunity. *Annu. Rev. Immunol.* **20**, 125–163.

Winther, Å. M. E., Acebes, A., and Ferrus, A. (2006). Tachykinin-related peptides modulate odor perception and locomotor activity in *Drosophila*. *Mol. Cell. Neurosci.* **31**, 399–406.

Wittwer, D., Franchini, A., Ottaviani, E., and Wiesner, A. (1999). Presence of IL-1- and TNF-like molecules in *Galleria mellonella* (Lepidoptera) hemocytes in an insect cell line from *Estigmene acraea* (Lepidoptera). *Cytokine* **11**, 637–642.

7

BLOODFEEDING AS AN INTERFACE OF MAMMALIAN AND ARTHROPOD IMMUNITY

LEYLA AKMAN-ANDERSON*, YORAM VODOVOTZ**,
RUBEN ZAMORA** AND SHIRLEY LUCKHART*

*Department of Medical Microbiology and Immunology, School of Medicine,
University of California, Davis, CA 95616, USA
**Department of Surgery, School of Medicine, University of Pittsburgh,
Pittsburgh, PA 15213, USA

ABSTRACT: Bloodfeeding behavior of disease-transmitting arthropods juxtaposes vertebrate and invertebrate immune systems since blood contains not only nutrients, but also an array of immune cells and molecules as well as potential pathogens and pathogen-associated molecules. Numerous blood-derived factors that are transferred into feeding arthropods remain immunologically active. These factors can activate conserved arthropod signal transduction pathways and interact with arthropod cells to regulate vector physiology and the response to ingested pathogens. Based on recent discoveries, it is likely that additional blood factors that are functional in both the mammal and in the arthropod remain to be identified. To better understand these complex interactions, we suggest that a systems biology approach may be necessary to study the impact of the interacting components of vertebrate and invertebrate immunity on arthropod-borne pathogen transmission.

Abbreviations:

aPKC	=	atypical protein kinase C
BMP	=	bone morphogenetic protein
CPB	=	carboxypeptidase B
ERK	=	extracellular signal-regulated kinase
GALE	=	galectin
GNBP	=	Gram-negative bacteria binding protein
GPI	=	glycosylphosphatidylinositol
HlMIF	=	*Haemaphysalis longicornis* MIF
Hz	=	hemozoin
IFN	=	interferon
IκB	=	inhibitors of κB
IL	=	interleukin
Imd	=	immunodeficiency
iNOS	=	inducible nitric oxide synthase
InR	=	insulin receptor
ISC	=	insulin signaling cascade
JNK	=	c-Jun *N*-terminal kinase
LPS	=	lipopolysaccharide
MAC	=	membrane attack complex
MAPK	=	mitogen-activated proteins kinase
MEK	=	mitogen-activated or extracellular signal-regulated protein kinase kinase
MIF	=	migration inhibitory factor
MIP	=	macrophage inflammatory protein
MyD88	=	myeloid differentiation factor 88
NF-κB	=	nuclear factor kappa B
NO	=	nitric oxide
NOS	=	nitric oxide synthase
PAMP	=	pathogen-associated molecular pattern
PDGF	=	platelet-derived growth factor
PDK1	=	3-phosphoinositide-dependent protein kinase
*Pf*GPI	=	*Plasmodium falciparum* GPI
PGRP	=	peptidoglycan recognition proteins
PI-3K	=	phosphatidylinositol 3-kinase

PKB   = Akt/protein kinase B
PKC   = protein kinase C
PM   = peritrophic matrix
PRR   = pattern recognition receptor
qRT-PCR = quantitative reverse transcription PCR
RBC   = red blood cell
ROS   = reactive oxygen species
STAT   = signal transducers and activators of transcription
TAK1   = TGF-β-activated kinase 1
TEP   = thioester-containing protein
TGF-β1   = transforming growth factor beta 1
TLR   = Toll-like receptor
TNF   = tumor necrosis factor.

7.1   INTRODUCTION

In this chapter, we review older findings and highlight new findings regarding host blood-derived factors that remain immunologically active after ingestion by ticks and anopheline mosquitoes as models of vector arthropods. We do not attempt to review the physiology of blood digestion, as it has been studied extensively in anophelines by Billingsley and colleagues (Billingsley, 1990; Billingsley and Hecker, 1991; Jahan et al., 1999), by Crisanti and colleagues (Muller et al., 1993a, b, 1995) and others (Devenport et al., 2004; Prevot et al., 2003; Vizioli et al., 2001) and reviewed for hard ticks by Coons and colleagues (1986). Rather, our intent here is to illustrate that a variety of host blood-derived factors persist through blood digestion to affect the mosquito response to malaria parasite (*Plasmodium* spp.) development and to affect tick transmission of bacterial agents. We will also not review extensive studies of the immunological interfaces between mosquitoes or ticks and their hosts that are dictated by the actions of myriad bioactive salivary gland proteins. This area was reviewed recently by Billingsley et al. (2006) for anophelines and by Schoeler and Wikel (2001) for hard ticks, and serves as a template for efforts to understand the immunological interface in the arthropod midgut. Salivary gland proteins act on highly conserved mammalian effector systems that regulate clotting, blood flow and anti-pathogen immunity. In an analogous way, we aim to show that host blood-derived factors ingested at bloodfeeding can act on highly conserved arthropod signaling pathways to regulate arthropod physiology and immunity.

7.2   CYTOKINES AND GROWTH FACTORS

Beier et al. (1994) were the first to demonstrate that ingestion of human insulin, a hormone and growth factor with pleiotropic physiological functions, could significantly increase oocyst densities of *Plasmodium falciparum*, the most important human malaria parasite, in the Asian malaria vector *Anopheles stephensi* and in the African malaria vector *Anopheles gambiae* relative to controls. More recently, studies with *A. stephensi* revealed that human insulin can induce nitric oxide (NO) production in cultured cells and in the mosquito midgut epithelium following ingestion of the blood meal (Lim et al., 2005). In *A. stephensi* cells *in vitro*, human insulin induced expression of the enzyme responsible for NO synthesis, *A. stephensi* NO synthase (*NOS*), 2-fold relative to controls at 48 h after treatment. Additionally, provision of human insulin by artificial blood meal of heat-inactivated human serum and washed human red blood cells (RBCs) induced *NOS* expression in the midgut approximately 2-fold at 6 h and 4-fold at 36 h post-feeding relative to controls (Lim et al., 2005). Inducible NO production in the *A. stephensi* midgut limits malaria parasite development (Luckhart et al., 1998) through the formation of inflammatory levels of toxic reactive nitrogen oxides (Luckhart et al., 2003; Peterson et al., 2007) that likely induce parasite apoptosis in the midgut lumen (reviewed in Hurd et al., 2006). These observations indicated that insulin control of NO synthesis and immunity, which has also been observed in mammals (reviewed in Heemskerk et al., 1999), is highly conserved in *Anopheles* mosquitoes.

The findings of Beier et al. (1994) and Lim et al. (2005) are not surprising in light of the fact that the pathway that transduces signaling by insulin, the insulin signaling cascade or ISC, is very highly conserved in aedine and anopheline mosquitoes (reviewed in Luckhart and Riehle, 2007). Two major signaling branches transduce the effects of insulin downstream of the insulin receptor (InR; Fig. 7.1): one signaling branch is characterized by sequential activation of phosphatidylinositol 3-kinase (PI-3K) and Akt/protein kinase B (PKB) and the other by activation of the kinases Ras, Raf, MEK (a mitogen-activated or extracellular signal-regulated protein kinase kinase), and ERK (an extracellular signal-regulated protein kinase; Luckhart and Riehle, 2007). Studies by Riehle and Brown (1999, 2002, 2003) of *Aedes aegypti*, the yellow fever mosquito, revealed that a dose range of $1.7–85\,\mu M$ bovine insulin stimulated ecdysteroid production by ovaries isolated from female mosquitoes. This effect was transduced through the ISC, as determined with the use of inhibitors or activators of the InR, PI-3K and Akt/PKB (Riehle and Brown, 1999; Riehle et al., 2002). Interestingly, the MEK inhibitor PD98059 had no effect on steroidogenesis, suggesting that the PI-3K/Akt branch of the ISC is activated for this physiological effect. Recent studies in *A. stephensi* have demonstrated that human insulin at a concentration as low as $1.7 \times 10^{-5}\,\mu M$, which is identical to fasting levels of insulin in human blood (Darby et al., 2001), can activate ERK and Akt/PKB in the *A. stephensi* midgut epithelium (Kang et al., unpublished). These observations confirm that blood levels of human insulin can have significant physiological effects on feeding mosquitoes.

In mammals, blood insulin levels are up- or down-regulated by malaria parasite infection. Thus, feeding mosquitoes are likely to be subjected to a range of insulin concentrations under natural conditions. In mice (Elased et al., 2004) and in

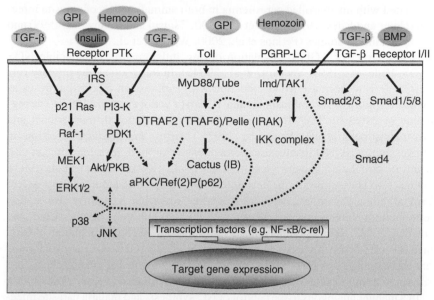

FIGURE 7.1 A model of signal transduction pathways involved in response to blood-derived factors. Sequential activation of conserved proteins involved in insulin, TGF-β and Toll/Imd signal transduction pathways and complex interactions between and among these pathways regulate the expression of target genes such as *NOS*. Blood-derived factors that can activate these pathways include insulin, TGF-β, BMP, GPI and hemozoin. Dotted lines illustrate some examples of cross-talk among signaling proteins of the insulin, Toll/Imd, TGF-β and MAPK-dependent pathways. Mammalian orthologs of proteins involved in Toll pathway are indicated in parentheses. For simplicity, invertebrate orthologs of insulin signaling proteins and Smad proteins are not specified.

humans (White et al., 1983, 1987), parasite infection induces hypoglycemia, which is predictive of severe pathology and fatal outcome. In humans, malaria infection and quinine therapy of infection can also lead to hyperinsulinemia (White et al., 1983; reviewed in Planche et al., 2005). Average insulin levels in hyperinsulinemic malaria patients were $1.6 \times 10^{-4}\,\mu M$, with the highest concentration at $4.7 \times 10^{-4}\,\mu M$ (White et al., 1983), indicating that blood levels of insulin can vary as much as 10- to 25-fold depending on nutrition and disease status. Thus, in the context of the mosquito immune response to malaria parasite infection, ingested insulin likely influences parasite transmission.

Like insulin, the growth factor and cytokine transforming growth factor β1 (TGF-β1) is present in circulating human blood, is pleiotropic in its effects on cell growth, anti-pathogen immunity and inflammation, and plays a critical role in regulating 'immunological balance' during malaria parasite infection (Omer et al., 2000; reviewed in Vodovotz et al., 2004b). Specifically, low levels of TGF-β1 are pro-inflammatory early in infection to promote malaria parasite clearance and high levels are anti-inflammatory later in infection to minimize host immunopathology (Vodovotz et al., 2004b). Alterations of dose and/or timing of TGF-β1 appearance are associated with uncontrolled parasitemia in both animal models of parasite infection (Omer and Riley, 1998; Omer et al., 2000; Tsutsui and Kamiyama, 1999) and in human clinical studies (Perkins et al., 2000; Wenisch et al., 1995).

A key target of TGF-β1 regulation during inflammation is inducible NOS (iNOS; Vodovotz, 1997; Vodovotz et al., 2004b). In this context, the primary role of TGF-β1 is to decrease iNOS induction with suppression reported at doses as low as 100 pg/ml of recombinant human TGF-β1 (Vodovotz et al., 1993). The regulation of iNOS by TGF-β1 can occur at the transcriptional, translational and post-translational levels (Vodovotz, 1997). During *Plasmodium* infection in humans and mice, NO inactivates sporozoites during liver invasion and can also inactivate circulating gametocytes (Mellouk et al., 1994; Naotunne et al., 1993). NO has also been linked to the development of cerebral malaria, the most severe form of human infection, with most studies reporting a positive association between high NO levels and increased pathology in late stage disease (Clark and Cowden, 2003). Not surprisingly, however, studies of NO levels in human malaria have revealed both protective and pathological effects. One likely explanation is that early synthesis is involved in parasite clearance, which if uncontrolled, contributes to immunopathology in late stage disease (reviewed in Riley et al., 2006).

In a manner analogous to the control of NO synthesis and malaria parasite infection in mammals, mammalian TGF-β1 ingested with the blood meal regulates NO synthesis and malaria parasite development in the midgut of *A. stephensi* (Luckhart et al., 2003). Specifically, low levels of human TGF-β1 ($\leq$200 pg/ml) provided in blood inhibited parasite growth in *A. stephensi*, while a higher dose (2000 pg/ml) did not affect parasite growth, suggesting a pro-inflammatory effect at low doses and an anti-inflammatory effect at the highest dose (Luckhart et al., 2003). Like insulin, ingested mammalian TGF-β1 appears to induce *NOS* expression and regulate malaria parasite development through activation of conserved signaling pathways in the cells of the *A. stephensi* midgut epithelium. To understand this regulation, we must first turn to what is known to occur in mammals and the fruit fly, *Drosophila melanogaster*.

In mammals and insects, the TGF-β superfamily is comprised of the bone morphogenetic proteins (BMPs) and the prototypical TGF-βs, which include TGF-β1. In mammalian cells, the Smad proteins comprise the canonical TGF-β signaling pathways, with specific subgroups of Smads that transduce cellular signals from the BMPs and the TGF-βs (Fig. 7.1). However, mammalian BMPs and TGF-βs also signal through a variety of alternative pathways, including those involving Ras, ERK, p38 mitogen-activated proteins kinase (MAPK) and c-Jun *N*-terminal kinases (JNKs) (reviewed by Moustakas and Heldin, 2005; Yue and Mulder, 2000; Fig. 7.1). Given the connections of Ras and ERK with the ISC, these observations would suggest that crosstalk occurs between the insulin and TGF-β signaling pathways in mammals. Indeed, this crosstalk is well known and has significant implications for mammalian physiology (reviewed by Danielpour and Song, 2006). Together with previous studies confirming the involvement of JNK in mosquito immunity (Mizutani et al., 2003), unpublished data from our laboratory also suggest that insulin-TGF-β signaling crosstalk occurs in anopheline mosquito cells and is likely to have significant effects on mosquito physiology.

Returning to the Smad signaling pathway, many studies have demonstrated that homologous Smad proteins in *Drosophila melanogaster* function to transduce signals from endogenous BMPs and the TGF-β activin (reviewed by Raftery and Sutherland, 1999). Specifically, fly BMPs signal through activation of a ligand-binding heteromeric receptor, which activates the cytoplasmic Smad known as Mad. Activated Mad binds to the co-Smad Medea and this complex is translocated to the nucleus to promote gene expression (Raftery and Sutherland, 1999). Similarly, activin binds the corresponding heteromeric receptor, which activates the cytoplasmic Smad known as dSmad2, which binds to Medea and translocates to the nucleus (Raftery and Sutherland, 1999). Like *D. melanogaster*, both *A. gambiae* and *A. stephensi* express a variety of endogenous BMP ligands and the TGF-β ligand activin (Lieber and Luckhart, 2004). In addition, orthologs of the signaling proteins Mad, dSmad2 and Medea are expressed in the *A. stephensi* midgut epithelium, indicating that the signaling architecture is in place to respond to mammalian TGF-β1 ingested at bloodfeeding (Lieber and Luckhart, 2004).

Intriguing recent studies with the soft tick *Ornithodoros moubata* suggest that these arthropods may also be able to respond to ingested mammalian growth factors. Using antibodies to mammalian platelet-derived growth factor (PDGF), to TGF-β1 and to their receptors, Matsuo et al. (2007) identified potential binding sites for the mammalian proteins on tick hemocytes and cross-reacting endogenous proteins that could represent tick orthologs of PDGF and TGF-β1. These observations suggest that mammalian PDGF and TGF-β1, which could cross the midgut epithelium into the hemolymph during digestion, could serve as signals to circulating tick hemocytes to facilitate repair of infection-induced damage of the midgut epithelium (Matsuo et al., 2007).

Although the involvement of mammalian cytokines and growth factors in regulating tick physiology has not yet been described, we highlight recent findings of protein homologs of macrophage migration inhibitory factor (MIF) in the hard ticks *Amblyomma americanum* (Jaworski et al., 2001) and *Haemaphysalis longicornis* (Umemiya et al., 2007) to support the argument that this phenomenon is likely to occur in ticks. In *A. americanum*, arthropod MIF is expressed in the salivary glands and in the midgut epithelium and is comparable to recombinant human MIF in its ability to inhibit migration of human macrophages (Jaworski et al., 2001). Umemiya et al. (2007) reported that the *H. longicornis* MIF (HlMIF) amino acid sequence was 77% identical to that of *A. americanum* MIF, and extended the findings of Jaworski et al. (2001) in reporting that HlMIF was upregulated in midgut epithelial cells by bloodfeeding. These findings suggest that HlMIF may interact or regulate intracellular development of tick-transmitted parasites in the genera *Babesia* and *Theileria* (Umemiya et al., 2007).

Although these findings suggest functionality of endogenous MIFs in regulating tick physiology and vector competence, they also suggest that ticks and perhaps anopheline mosquitoes could possess the signaling architecture to respond to circulating mammalian MIF ingested at feeding. In mammals, macrophage MIF appears to signal through a variety of pathways, including those that are dependent on MEK–ERK, p38 MAPK, PI-3K and nuclear factor-κB (NF-κB; Fig. 7.1; Aeberli et al., 2006; Amin et al., 2006; Yu et al., 2007). Some of these signaling proteins have been discussed above for their association with TGF-β and insulin signaling in mosquitoes, fruit flies and mammals, while various NF-κB orthologs are well known for their roles in regulating innate immunity in anopheline mosquitoes (Frolet et al., 2006; Meister et al., 2005).

In addition to suggestions of the existence of the signaling architecture in the arthropods, numerous studies have demonstrated that mammalian MIF is present in circulating blood and that levels change with relevant arthropod-borne infections. For example, peripheral blood lymphocytes of calves infected with *Theileria annulata* produced abundant MIF, suggesting that MIF plays an important role in regulating cell-mediated immunity to this tick-borne pathogen (Rehbein et al., 1981) and that mammalian MIF could be ingested by feeding ticks. Other studies have highlighted a pathological association between circulating MIF levels and infection with *P. falciparum*. Specifically, clinical malaria has been associated with altered circulating levels of TGF-β1, interleukin-12 (IL-12) and MIF (Awandare et al., 2006, 2007; Chaiyaroj et al., 2004), suggesting as observed for TGF-β1 and for insulin, that feeding mosquitoes are likely exposed to a range of concentrations of MIF that could impact physiology and malaria parasite development in the insect host.

Given the extreme conservation from insects to mammals of critical signaling pathways that transduce cellular activation by various cytokines and growth factors, it is quite reasonable to suspect that additional host blood-derived cytokines and growth factors impact the physiology of mosquitoes and other bloodfeeding arthropods, including ticks. We have carried out a preliminary screen to look for such factors. We utilized the Luminex™ platform and a multiplexing beadset specific for mouse cytokines to test the hypothesis that inflammatory cytokines in addition to TGF-β1 could be found in the *A. stephensi* midgut after a mouse blood meal. The Luminex™ platform utilizes a multiplexing beadset to detect multiple cytokines in a single, small-volume sample. Out of a panel of 25 cytokines, we detected strong signals for IL-1 and IL-10, among other cytokines (unpublished observations). Given the labile nature of most cytokines, it is quite remarkable that such cytokines would remain detectable in the midgut after feeding. This observation raises the possibility that mechanisms exist in the mosquito for selectively retaining and preserving mammalian cytokines, perhaps because such cytokines function in a beneficial way to augment the mosquito's own defenses against *Plasmodium*.

7.3   ANTIBODIES AND COMPLEMENT

The strategy of transmission-blocking immunity by targeting mammalian antibody synthesis to mosquito-stage malaria parasite epitopes is well known (reviewed by Matuschewski, 2006). An equivalent strategy to target outer surface protein A of *Borrelia burgdorferi*, the causative agent of Lyme disease, in the tick host *Ixodes scapularis* (*Ixodes dammini*) appears to block transmission of this agent to humans (Fikrig et al., 1992; Sigal et al., 1998; Steere et al., 1998). However, the actions of mammalian antibodies against mosquito and tick proteins involved in pathogen development are perhaps less well known but are at least as effective in blocking transmission as their sister anti-pathogen strategies.

In anopheline mosquitoes, several studies have highlighted the rather unexpected positive effects of digestive enzymes on malaria parasite development. For example, Lavazec et al. (2007) recently noted that *P. falciparum* induced the expression of two genes encoding carboxypeptidase B (CPB) activity in the midgut epithelium of *A. gambiae* during early parasite development in ingested blood (Lavazec et al., 2007). Furthermore, the addition of antisera against CPB to an artificial blood meal blocked parasite development in the mosquito midgut, while development of the rodent parasite *Plasmodium berghei* was significantly reduced in mosquitoes fed on infected mice that had been immunized with recombinant CPB (Lavazec et al., 2007).

In addition to blocking strategies targeting blood digestion, several studies have demonstrated that 'anti-midgut' antibodies ingested during bloodfeeding can apparently shield mosquito proteins that are used by malaria parasites for mosquito invasion. While blocking parasite invasion is not an innate immune response *per se*, it is clear that parasite invasion initiates innate responses in the mosquito host and is, hence, inextricably tied to these responses (Barillas-Mury and Kumar, 2005; Dong et al., 2006). Furthermore, like the anti-enzyme antibodies described above, it is clear that mammalian antibodies persist in the insect host, effectively bind their target immunogens and alter the mosquito response to parasite infection. In this light, Lal et al. (2001) demonstrated that both polyclonal and monoclonal antibodies produced against mosquito midgut proteins could block development of *P. falciparum* and a second human parasite *Plasmodium vivax* in numerous anopheline species. These findings extended previous studies using the mouse parasite *P. berghei* (Lal et al., 1994). In other studies, Dinglasan et al. (2003) demonstrated that a monoclonal antibody to a complex oligosaccharide epitope on mosquito midgut proteins successfully blocked development of the mouse parasite *Plasmodium yoelii* in *A. stephensi*. This work highlighted the fact that malaria parasites in mosquitoes, like many other pathogens in vector arthropods, utilize carbohydrate moieties to navigate invasion (Dinglasan et al., 2005) and perhaps even to shield themselves from damaging host responses (Osta et al., 2004).

Elements of the alternative complement cascade, which is functionally independent of antibody action in mammals, have also been found to persist and remain immunologically active against developing malaria parasites in bloodfed mosquitoes. The alternative complement pathway opsonizes and kills pathogens and is initiated by microbial surface-induced hydrolysis of the thioester bond of the protein C3, which is present in blood plasma. The cleavage product C3b, together with a factor B cleavage product Bb, catalyzes the cleavage of plasma protein C5 into C5a and C5b and the latter, together with C6, C7, C8 and C9, form the membrane attack complex (MAC). In essence, cytoplasm leakage of the inducing microbes results from the assemblage of many MACs on the microbe surface.

Two studies have highlighted the effects of complement following ingestion by mosquitoes. Tsuboi et al. (1995) revealed that mouse complement significantly reduced, but did not completely inhibit, *P. yoelii* development in *A. stephensi*. The authors found that mouse C3 was bound to the surface of parasite zygotes, which are formed in the mosquito midgut via fertilization within minutes after ingestion. This deposition apparently inhibited the transition of the immobile zygote into mobile ookinetes that penetrate the midgut epithelium during development (Tsuboi et al., 1995). In subsequent studies, Margos et al. (2001) showed that rat factor B, factor D, C3 and C5 persisted for several hours following ingestion of *P. berghei*-infected rat blood by *A. stephensi*. To discern the effects of complement on parasite development, Margos et al. (2001) used an *in vitro* culture system to produce and assay *P. berghei* ookinetes. Interestingly, parasites induced to undergo ookinete formation less than 3 h earlier were relatively resistant to complement-induced lysis; however, fewer than 4% of parasites remained alive in culture by 6 h after induction of ookinete formation (Margos et al., 2001). While it is not inherently obvious how complement activation could be enhanced to alter parasite development, it remains clear that mammalian complement could act synergistically with mosquito anti-parasite factors to reduce parasite transmission.

In ticks, numerous studies have focused on the development of anti-tick antisera to reduce the prevalence of feeding ticks by inducing detachment from the vertebrate host as a pest management strategy (reviewed by Willadsen, 2004). In addition, there has been significant interest in developing vaccines to target pathogen development in vector ticks. As mentioned previously, a successful transmission-blocking strategy has been developed based on targeting OspA of *B. burgdorferi* during its passage in the tick host (Fikrig et al., 1992; Sigal et al., 1998; Steere et al., 1998). In addition to *B. burgdorferi*, Kocan and colleagues first proposed targeting the tick-associated stages of *Anaplasma marginale* in 1996 (Kocan et al., 1996). While several anti-pathogen transmission-blocking strategies have proved promising, Kocan and colleagues have succeeded using a strategy that is analogous to the effects of the anti-mosquito antibodies on malaria parasite development described above. Specifically, de la Fuente et al. (2006b) demonstrated that immunization of mice against the tick protein subolesin, which is involved in the modulation of tick feeding and reproduction (de la Fuente et al., 2006a, b), prevented infection of feeding ticks with *Anaplasma phagocytophilum*. The authors speculated, based on the effects of subolesin gene silencing which resulted in degeneration of midgut cells, that subolesin antisera from immunized mice interferes with successful colonization of the midgut by *Anaplasma* and, therefore, successful invasion of the salivary glands of feeding ticks.

In addition to the potential role of anti-tick antibodies in mediating pathogen transmission, a fascinating hypothesis has emerged that the action of mammalian complement following ingestion by tick vectors, not physiological differences among tick species, dictates tick associations with different strains of *B. burgdorferi* (Kurtenbach et al., 1998, 2002b, 2006). In general, patterns of *B. burgdorferi* resistance or sensitivity to host complement are consistent with patterns of transmissibility (reviewed in Kurtenbach et al., 2002a). For example, spirochetes that are sensitive to lysis by the complement system of a particular host species are lysed early in the midgut of the feeding tick, and are thereby eliminated by the host (Kurtenbach et al., 2002b). These findings led to the hypothesis that spirochete strain tick host range is restricted by its repertoire of genes that encode ligands with high binding affinities for complement inhibitors (Stevenson et al., 2002; reviewed in Kurtenbach et al., 2006). This example provides one of very few mechanistic explanations for host specialization by a zoonotic pathogen (Woolhouse et al., 2001) and, as has been observed in mosquitoes, suggests that a complex interplay of arthropod and mammalian host factors regulates pathogen development in the arthropod host.

7.4 LEUKOCYTES, REACTIVE OXYGEN SPECIES AND NITRIC OXIDE

Within the bolus of the blood meal in anopheline mosquitoes, malaria parasites transition from an intracellular to an extracellular existence. As such, gametes, zygotes and ookinetes are not only directly exposed to active humoral components of the host blood, as we have already discussed, but they are exposed to active cellular components as well. Based on ingestion of 1–2 μl of blood by anophelines (Clements, 1992), the midgut can contain up to 30,000 leukocytes, the majority of which are neutrophils (Lensen et al., 1997). These cells can survive for several hours in the midgut lumen (Lensen et al., 1997), a feat made possible by the fact that blood digestion proceeds from the outside of the blood bolus toward the center of the midgut lumen. This digestion pattern allows cells in the center of the blood mass to remain physically intact and viable for an extended period.

Leukocyte-dependent killing of asexual malaria parasites has been proposed as a potentially important mechanism of protective immunity to malaria within the mammalian host (Kharazmi and Jepsen, 1984; Khusmith and Druilhe, 1983), but the extent to which killing is mediated by phagocytosis rather than by other antiparasitic mechanisms is not clear (Bouharoun-Tayoun et al., 1995). Further, in contrast to the large body of work on cell-mediated killing of asexual parasites, relatively little is known about cell-mediated killing of sexual stage gametocytes in mammals. As such, several studies have attempted to examine this phenomenon and its impact on parasite transmission to mosquitoes. Lensen et al. (1997, 1998) reported that the presence of immune serum and leukocytes significantly reduced infectivity of *P. falciparum* gametocytes to *A. gambiae* mosquitoes and suggested that phagocytosis was the likely explanation for this phenomenon. However, a direct effect of phagocytosis in the mosquito was not demonstrated. Further, Healer et al. (1999) demonstrated that, indeed, phagocytosis does not occur to a significant degree in the mosquito midgut and that neither addition of leukocytes alone nor in combination with immune serum had any effect on transmission of *P. falciparum* to *A. stephensi*. Although the findings of Healer et al. (1999) contradicted those of Lensen et al. (1997, 1998), the fact remains that leukocyte-dependent processes other than phagocytosis could influence parasite transmission to anopheline mosquitoes and synergize with mosquito responses to regulate parasite development. For example, the role of NO in cell-dependent parasite killing by mammalian cells has been discussed above and has also been implicated in the action of activated leukocytes against gametocytes prior to and perhaps following mosquito infection (Motard et al., 1993; Naotunne et al., 1993; reviewed by Hurd and Carter, 2004; Hurd et al., 2006).

While the effects of mammalian cell-dependent phagocytosis and toxic oxidants on malaria parasite development remain to be resolved, our recent findings suggest that leukocyte-derived reactive oxygen species (ROS) and nitrogen species could play additional roles as secondary messengers in regulating malaria parasite development in the insect host. As opposed to mammalian proteins, which require high levels of conservation to activate mosquito cells and signaling pathways, mosquito cells are likely to respond in a similar fashion to distinct ROS and NO species, whether these species are produced endogenously or in the mammalian host. Recent findings from our laboratory indicate that NO, and perhaps ROS as well, function as second messengers as well as killing molecules in the mosquito midgut epithelium. Viable ingested leukocytes, therefore, could produce ROS and NO species that would be perceived as signals by the mosquito midgut epithelium to regulate innate responses to malaria parasite development.

In support of this hypothesis, we recently demonstrated that induction of *NOS* expression in the mosquito midgut by mammalian TGF-β1 is negatively regulated by NO synthesis at the highest levels of *NOS* induction (Luckhart et al., in press). That is, high doses of TGF-β1 induce relatively higher levels of *NOS* expression and negative feedback of NO occurs under these conditions. These findings are consistent with numerous observations from mammalian cells that TGF-β1 regulates NOS activity at transcriptional, post-transcriptional and translational levels (Vodovotz, 1997) and, further, that TGF-β1 participates in the negative feedback regulation of NOS by NO (Vodovotz et al., 1996). Our observations of NO feedback are intriguing in light of the fact that provision of TGF-β1 at lower doses by artificial blood meal can also reduce the burden of *P. falciparum* in *A. stephensi* (Luckhart et al., 2003). These findings suggest that NO functions simultaneously as a killing molecule, presumably at high local concentrations and via the synthesis of toxic reactive nitrogen oxides in the blood mass, and as a signaling molecule at the surface of the epithelial cell. The latter would function to reduce the synthesis of NO at high levels of *NOS* induction and minimize damage to mosquito host tissues.

Identifying the individual effects of blood cell-derived NO and mosquito cell-derived NO in the midgut compartment is likely to be difficult, given the dual complexities of the chemistry of NO (Wink et al., 1999) along with the myriad processes that NO modulates in both mammals and invertebrates (Bicker, 2007; Bogdan et al., 2000; Rivero, 2006). However, it is intriguing to speculate that high levels of NO produced in the mammalian host would be consistent with a hostile environment for the parasite, thus 'escape' from the host would be necessary for parasite survival. The carry-over of activated leukocytes, NO and reactive nitrogen oxides into the mosquito midgut may serve some function in killing, but could also act to negatively regulate innate responses of the mosquito as well. As such, parasite development in the mosquito could be in fact favored following feeding on a 'hostile' host, thus completing the circle of transmission predicted by Ewald (1987) in his models of host–pathogen virulence.

7.5 ENZYMES

Mosquito chitinase is involved in the transient formation and breakdown of the midgut peritrophic matrix (PM), a physical barrier between the ingested blood mass and the midgut epithelium, and is orthologous to malaria parasite chitinase which is involved in penetration of the PM (reviewed by Langer and Vinetz, 2001). Studies by Bhatnagar et al. (2003) suggested that mosquito chitinase had positive effects on parasite development. Specifically, Bhatnagar et al. (2003) showed that a propeptide-derived inhibitor of mosquito chitinase blocked *P. falciparum* development in the mosquitoes *Anopheles freeborni* and *A. gambiae* and also blocked development of the chicken parasite *Plasmodium gallinaceum* in *A. aegypti*. Although it is not clear that the inhibitor used by Bhatnagar et al. (2003) was specific to the mosquito chitinase, their studies and those of Lavazec et al. (2007) suggest that blood digestion, which was previously only thought to create a hostile environment for parasite development, can be manipulated to enhance anti-parasite immunity in the mosquito host.

Production of chitotriosidase in the peripheral blood of malaria patients may be another example of an intricate interplay between arthropods and mammalian host factors that facilitate successful pathogen development in the arthropod host. The enzyme chitotriosidase is the human analog of invertebrate chitinase and has the capability to hydrolyze chitin (reviewed by Malaguarnera, 2006). Chitotriosidase is primarily secreted by macrophages upon activation and Barone et al. (2003) have shown that plasma chitotriosidase activity is elevated in patients infected with *P. falciparum*. Recently, Di Rosa et al. (2005) demonstrated that interferon-γ (IFN-γ), tumor necrosis factor-α (TNF-α) and lipopolysaccharide (LPS) can upregulate chitotriosidase gene expression in human macrophages. Thus, chitotriosidase likely plays a role in anti-malarial immunity in human hosts.

It has been suggested that accumulation of erythrocyte membrane degradation products in macrophages might trigger the overproduction of chitotriosidase in malaria patients (Barone et al., 2003). This is an intriguing hypothesis since it implies that *P. falciparum* manipulation of blood factors can ultimately help parasites invade the midgut epithelium via human chitotriosidase-facilitated PM degradation. In support of this hypothesis, Di Luca et al. (2006) have recently shown that human chitotriosidase can alter the structure of *A. stephensi* PM. Specifically, PM thickness was reduced in midguts from *A. stephensi* fed on blood supplemented with malaria patient plasma containing 0.68 and 0.95 mU/ml chitotriosidase compared to controls. Although it is still unknown whether human chitotriosidase in addition to the parasite chitinase can facilitate parasite invasion of PM, it is clear that human chitotriosidase can alter the integrity of PM barrier in the mosquito midgut and thus might play a role in malaria parasite development in *Anopheles* mosquitoes.

7.6   HEMOGLOBIN FRAGMENTS

Antimicrobial peptides play a major role in innate immunity and have been studied extensively in insects and mammals. Ticks transmit a wide spectrum of pathogens, including viruses, bacteria, rickettsia, protozoa and fungi, which is indicative of remarkably evadable innate immunity. Nevertheless, little is known about tick defense mechanisms against ingested microorganisms and presence of inducible antimicrobial peptides has only recently been discovered in tick midguts (Nakajima et al., 2001, 2002a, b, 2003).

While studying antimicrobial compounds in gut contents of the cattle tick, *Boophilus microplus*, Fogaca et al. (1999) discovered a small peptide with antibacterial and antifungal activities and showed that this peptide was, surprisingly, identical to the residues 33–61 of bovine α-hemoglobin. A synthetic peptide corresponding to this α-hemoglobin fragment was found to be active against Gram-positive bacteria as well as yeast and fungi at micromolar concentrations. In contrast, neither the intact bovine hemoglobin nor lysed erythrocytes which likely contain hemoglobin degradation products had any activity against *Micrococcus luteus*, a bacterial strain commonly used for detection of antibiotic activity (Fogaca et al., 1999). Thus, the authors suggested that the active hemoglobin fragment originated inside the tick gut, possibly through enzymatic cleavage of bovine hemoglobin, and that this fragment functions to protect ticks against gut microorganisms (Fogaca et al., 1999). Since the first discovery of an antimicrobial hemoglobin fragment in cattle ticks by Fogaca et al., several studies have affirmed the presence of hemoglobin fragments with antimicrobial activities in other hard ticks, including *Dermacentor variabilis* (Sonenshine et al., 2005), as well as in the soft tick *Ornithodoros moubata* (Nakajima et al., 2003a). Thus, ticks appear to generally utilize host hemoglobin fragments as immune effectors against ingested microorganisms. In contrast to inducible antimicrobial peptides in other animals, however, the activity of hemoglobin fragments in ticks is independent of pathogen exposure and directly dependent on blood digestion.

The involvement of hemoglobin fragments in modulation of host–pathogen interactions has been identified in other arthropods as well. Specifically, a peptide corresponding to residues 1–40 of the α-D-globin chain of chicken hemoglobin purified from the hindguts of the reduviid *Triatoma infestans* was found to bind to membranes of epimastigote stages of *Trypanosoma cruzi*, the causative agent of Chagas disease, and to induce differentiation to metacyclic trypomastigotes *in vitro* through activation of parasite adenylyl cyclase (Fraidenraich et al., 1993). Similarly, Garcia et al. (1995) showed that hemoglobin and synthetic peptides corresponding to 30–49 and 35–73 of the α-D-globin chain induced *T. cruzi* metacyclogenesis in the gut of the reduviid *Rhodnius prolixus*. Thus, α-D-globin fragments appear to play a role in mediating transmission of *T. cruzi*.

Following the discovery of antimicrobial hemoglobin fragments in ticks, Parish et al. (2001) found that intact hemoglobin tetramers, including those from humans, have considerable activity against bacteria and fungi and that different hemoglobin-derived peptides exhibit distinct antimicrobial activities. As such, the generation of antimicrobial hemoglobin fragments appears to be another defense mechanism that is shared between arthropods and mammals. In support of this hypothesis, Liepke et al. (2003) reported that naturally occurring human hemoglobin fragments were antibacterial and that bacterial growth inhibition was detectable in erythrocyte lysates. These findings conflicted with the earlier findings of Fogaca et al. (1999) in that the authors of the tick study found no direct antimicrobial activity of intact hemoglobin or erythrocyte lysates. The discrepancy between the findings of Fogaca et al. (1999) and Liepke et al. (2003) could result from the use of different hemoglobins (e.g. bovine versus human) and the use of different indicator bacterial strains. Despite these differences, these studies suggest that hemoglobin fragments are clearly important in innate immunity in a wide range of invertebrate and vertebrate hosts.

7.7   *PLASMODIUM*-ASSOCIATED MOLECULES

Innate immune responses against microorganisms are initiated upon recognition of pathogen-associated molecular patterns (PAMPs) through pattern recognition receptors (PRR) (reviewed in Akira et al., 2006; Lee and Kim, 2007; Strand, this volume). When mosquitoes feed on infected hosts, the ingested blood may contain viruses, bacteria, protozoa as well as molecules associated with these pathogens. *Anopheles gambiae* has been shown to respond to infection via Toll/immunodeficiency (Imd) (Luna et al., 2002; Meister et al., 2005) and signal transducers and activators of transcription (STAT) pathways (Barillas-Mury et al., 1999) and to upregulate the expression of genes that encode peptidoglycan recognition proteins (PGRPs) (Christophides et al., 2002), thioester-containing proteins (TEPs) (Oduol et al., 2000) and galectins (GALE) (Dimopoulos et al., 1998) among others. Although a wide range of microorganisms and pathogen-associated molecules such as LPS (Oduol et al., 2000) can induce innate immune responses in *Anopheles* mosquitoes, here we focus on pathogen-associated inducing molecules derived from *P. falciparum*.

Richman et al. (1997) and Dimopoulos et al. (1997) were the first to demonstrate that *A. gambiae* can respond to *Plasmodium* parasites within the ingested blood meal by mounting an immune response both locally in the midgut epithelium and systemically in the rest of the body. These studies identified some of the *A. gambiae* genes that are responsive to *P. berghei* infection, including those encoding defensin, galectin (GALE) and Gram-negative bacteria binding protein (GNBP). Studies by Luckhart et al. (1998) demonstrated that early induction of *NOS* in the midgut and carcass by *P. berghei* was critical to limiting parasite development in *A. stephensi*. Mosquito immune responses against *Plasmodium* spp. have been studied and reviewed extensively by Osta et al. (2004), Christophides et al. (2004), Meister et al. (2004), Hurd and Carter (2004), Michel and Kafatos (2005) and Whitten et al. (2006), and are discussed in chapter 8.

In addition to the studies mentioned above, studies by Tahar et al. (2002) revealed that important differences exist in mosquito immune responses against different *Plasmodium* species. Clearly, it is also important to distinguish between immune responses against different parasite-associated molecules in order to better understand anti-*Plasmodium* immunity in general. However, the exact identity of the parasite component(s) initiating immune signaling in mosquitoes was unknown until the work of Lim et al. (2005). We describe those findings here below in greater detail, but first highlight the studies in mammals that reveal the conservation of action of these parasite signaling factors.

In mammals, two *Plasmodium*-associated proinflammatory molecules, glycosylphosphatidylinositol (GPI) and hemozoin (Hz), can modulate immune responses and upregulate *NOS* expression (reviewed in Riley et al., 2006). GPIs are glycolipids that are ubiquitous in eukaryotic cells and function to anchor proteins to cell surfaces. Like mammalian GPIs (reviewed by Field, 1997; Muller, 2002; Saltiel, 1991), *P. falciparum* GPIs (*Pf*GPIs) have been shown to be insulin mimetic (Caro et al., 1996; Schofield and Hackett, 1993) and most importantly, they are considered to be the major parasite toxin contributing to malaria pathogenesis via induction of proinflammatory responses (reviewed by Gowda, 2002; Nebl et al., 2005). *Plasmodium falciparum* GPIs (*Pf*GPIs), free or associated with proteins, can induce TNF-α and IL-1 secretion in murine macrophages and increase the expression of cell adhesion molecules in human endothelial cells (Schofield and Hackett, 1993). Tachado et al. (1996) extended these findings by showing that *Pf*GPIs can also induce NO production in macrophages and endothelial cells and this induction can be blocked by the protein kinase C (PKC) inhibitor calphostin C and by pyrrolidine dithiocarbamate, which implicates the NF-κB/c-rel family of transcription factors in downstream signaling in response to *Pf*GPIs. Signaling in response to *Pf*GPIs has recently been shown to be mediated mainly through recognition by Toll-like receptor 2 (TLR-2) and to a lesser extent by TLR-4, and to involve myeloid differentiation factor 88 (MyD88)-dependent activation of ERK, JNK, p38 MAPK and NF-κB signaling pathways (Fig. 7.1; Krishnegowda et al., 2005). Zhu et al. (2005) showed TNF-α, IL-12, IL-6 and NO production by *Pf*GPI-stimulated macrophages to be dependent on NF-κB and JNK pathways. Interestingly, in these studies the ERK-dependent signaling pathway was found to negatively regulate the expression of IL-6 and IL-12 and not to be involved in TNF-α and NO production, whereas p38 MAPK was found to be critical for the production of IL-6 and IL-12 and only marginally required for TNF-α and NO production (Zhu et al., 2005). Using peritoneal macrophages from CD36 deficient mice, Patel et al. (2007) provided direct evidence that *Pf*GPI-induced TNF-α secretion is CD36 scavenger receptor dependent and that *Pf*GPI-induced phosphorylation of JNK, ERK and the transcription factor c-Jun is impaired in CD36$^{-/-}$ macrophages. These studies demonstrated, for the first time, that CD36 contributes to the induction of innate inflammatory responses to *P. falciparum* infection and that *Pf*GPI signaling involves multiple PRR.

In parallel studies, Lim et al. (2005) showed that *Pf*GPIs can also induce *NOS* expression in *A. stephensi* cells *in vitro* and in the midgut epithelium *in vivo* and thus demonstrated for the first time that *Pf*GPIs can be recognized by mosquito cells as well. Specifically, 2.5 μM *Pf*GPIs, corresponding to ingestion of approximately 312,000 parasites (~8% blood parasitemia level), were provided to cohorts of *A. stephensi* in artificial blood meals. Total RNA isolated from midguts dissected at various time points were analyzed for *NOS* expression by quantitative reverse transcription PCR (qRT-PCR). *Plasmodium falciparum* GPIs significantly induced *NOS* expression 1.4-fold immediately after bloodfeeding and >2-fold at 24 and 48 h after feeding, relative to controls. Furthermore, in *A. stephensi*, signaling by *Pf*GPIs was found to be mediated through Akt/PKB and ERK (Fig. 7.1; Lim et al., 2005). Surprisingly, although these kinases are associated with insulin signaling in *Anopheles* (reviewed by Luckhart and Riehle, 2007), neither *P. falciparum* merozoites nor *Pf*GPIs were found to be insulin mimetic to *A. stephensi* cells. Therefore, insulin mimicry by *Pf*GPIs seems to be unique to the mammalian host of *P. falciparum*. However, apparent similarities in GPI signaling between human and mosquito cells indicate that *Pf*GPIs are common inflammatory mediators in both the mammalian and the invertebrate host. These studies are expected to help identify additional molecules that can be targeted to increase mosquito resistance to parasite development.

In addition to identifying *Pf*GPIs as proinflammatory signals to mosquitoes, Lim et al. (2005) suggested that other *P. falciparum*-associated factors in addition to GPIs could induce mosquito *NOS* expression. Specifically, 15.6 merozoites/cell (corresponding to 0.25 μM *Pf*GPIs) induced *NOS* expression 3-fold, whereas 0.25 μM *Pf*GPIs induced *NOS* expression only 1.7-fold relative to controls. These findings pointed to a 'gap' in *NOS* induction between assays with *Pf*GPIs and assays with *P. falciparum* merozoites. Based on studies with mammalian cells described here below, we hypothesized that malaria parasite hemozoin could function as a second parasite-derived factor to induce mosquito *NOS* expression.

Hemozoin is an insoluble brown pigment produced in the parasite digestive vacuole during malaria parasite hemoglobin catabolism in host RBCs (reviewed in Francis et al., 1997). As RBCs burst, hemozoin is released along with merozoites and subsequently engulfed by monocytes, neutrophils and macrophages (reviewed by Arese and Schwarzer, 1997). Thus, mosquitoes are likely to be exposed to the malaria pigment in a *Plasmodium*-infected blood meal, either through ingestion of free hemozoin or cells that contain hemozoin (e.g. parasitized RBCs and leukocytes).

In mammals, hemozoin has been shown to be both pro- and anti-inflammatory (Millington et al., 2006; Pichyangkul et al., 1994; Schwarzer et al., 1993; Skorokhod et al., 2004; reviewed by Urban and Todryk, 2006). For example, crude *P. falciparum* (*Pf*Hz) preparations have been reported to be immunosuppressive (Schwarzer and Arese, 1996; Schwarzer et al., 1993, 1998, 2003; Taramelli et al., 1995, 1998), but the nature of the factors in addition to hemozoin that could account for this activity were not identified. In contrast, studies by Coban et al. (2002) demonstrated that purified *Pf*Hz was pro-inflammatory based on *Pf*Hz induction of dendritic cell maturation and IL-12 production. Purified *Pf*Hz has also been shown to induce secretion of pro-inflammatory mediators, including TNF-α, IL-1β (Pichyangkul et al., 1994) and macrophage inflammatory protein (MIP)-1α and MIP-1β (Sherry et al., 1995) in monocytes and macrophages. Jaramillo et al. (2003, 2005) showed that *Pf*Hz increased IFN-γ-mediated NO generation and this induction was mediated through ERK- and NF-κB-dependent signaling pathways in murine macrophages. These authors also demonstrated that *Pf*Hz can induce leukocyte recruitment *in vivo* as well as stimulate the expression of chemokines (MIP-1α, MIP-1β and MIP-2) and chemokine receptors (CCR1, CCR2, CCR5, CXCR2 and CXCR4) and cytokines (IL-1β and IL-6) in leukocytes from the BALB/c mice air pouch exudates (Jaramillo et al., 2004). *Plasmodium falciparum* hemozoin has also been shown to be a novel non-DNA ligand for TLR-9 (Coban et al., 2005), although Parroche et al. (2007) have argued that contaminating parasite DNA rather than *Pf*Hz itself accounted for TLR-9 activation. Despite some contradictory findings, the overwhelming body of evidence suggests that *Pf*Hz, together with PfGPIs, activates innate responses that ultimately influence parasite development.

In accord with observations in mammalian cells, we have recently demonstrated that *Pf*Hz can induce *NOS* expression in *A. stephensi* and *A. gambiae* cells *in vitro* and in *A. stephensi* midgut tissue *in vivo* (Akman-Anderson et al., 2007). As in mammalian cells, *Pf*Hz signaling in *Anopheles* cells appears to involve multiple pathways and is mediated through activation of TGF-β-associated kinase 1 (TAK1), Akt/PKB, ERK and atypical PKCζ/λ (Fig. 7.1; Akman-Anderson et al., 2007). Although *Pf*GPIs and *Pf*Hz are recognized through TLRs in mammals, the identity of the midgut PRRs that might be interacting specifically with these PAMPs is still unknown. Further studies of the signaling molecules and potential receptors that initiate pro-inflammatory responses in *Anopheles* will undoubtedly reveal novel targets for genetic enhancement of mosquito resistance to *Plasmodium*. Whether there are additional *Plasmodium*-associated molecules (and perhaps some that are unique to the mosquito stages) that can induce immune responses in *Anopheles* is an intriguing question that awaits further investigation. Taken together, studies to date indicate that the same parasite factors activate conserved innate signaling pathways to regulate conserved anti-parasite responses in the mosquito and the mammalian hosts of malaria parasites.

7.8   SUMMARY: THE NEED FOR A SYSTEMS BIOLOGY APPROACH

In this chapter, we have highlighted a large number of mammalian proteins that remain active or become immunologically active in the vector arthropod. These proteins alter the host arthropod physically and immunologically and these changes can result in direct effects on pathogen development. Many of these factors act or would be predicted to act on conserved signaling pathways in the arthropod vectors, indicating that a surprising number of innate responses are conserved between arthropod and mammalian hosts of these pathogens. In addition, some parasite-derived factors delivered in the blood meal can activate analogous responses in the mammalian host and in the arthropod vector. These interactions are summarized graphically in Fig. 7.2. Given the level of conservation of both molecules and pathways involved in innate immune responses, it is likely that the list of pathogen-derived factors that are functional in the mammal and in the arthropod will grow with future studies. With multiple interactions occurring simultaneously, how can the interface of bloodfeeding, which brings together the arthropod vector, the mammalian host and the pathogen, be better studied to understand the arthropod host response to pathogen infection?

We propose that systems biology approaches may help to unravel this daunting complexity. Systems biology is usually seen as a holistic approach to the mechanisms underlying complex biological processes, in which processes are viewed as integrated systems of many diverse, interacting components. Systems biology involves the use of computational approaches to extract patterns from '-omic' datasets, and is becoming highly mathematical through inclusion of prior knowledge embodied in multi-scale computational models (Kitano, 2002). Systems biology involves: (a) collection of large sets of experimental data (by high-throughput technologies and/or by mining the literature of reductionist molecular biology and biochemistry); (b) the proposal of conceptual frameworks of integration based from existing biological knowledge; (c) the embodiment of this conceptual framework in mathematical formalisms or models from which are derived, quantitative and/or qualitative predictions; (d) assessment of the validity of the models by comparing these predictions with the experimental data and (e) iterative revision of the conceptual framework and guiding of punctual experiments on

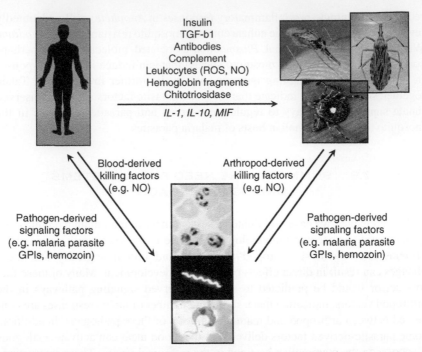

FIGURE 7.2 The act of bloodfeeding creates an interface of mammalian and arthropod immunity. Several mammalian blood-derived factors are transferred into the midguts of feeding arthropods and remain immunologically active for extended periods. These factors can activate conserved arthropod signaling pathways or interact with arthropod proteins to regulate the vector response to ingested pathogens. Examples discussed in the text are highlighted; italicized proteins may be involved based on unpublished observations. Some pathogen-derived signaling factors activate analogous defense responses (e.g. NO synthesis) in mammalian and arthropod hosts. Photo credits: *Anopheles gambiae* (Centers for Disease Control/Jim Gathany), *Amblyomma americanum* (Centers for Disease Control, Division of Vector-borne Infectious Diseases), *Rhodnius pallescens* (Pan-American Health Organization), *Plasmodium* merozoites (Tropical Disease Research, World Health Organization), *Borrelia burgdorferi* (Centers for Disease Control, Dr. Robert D. Gilmore), *Trypanosoma cruzi* (Centers for Disease Control, Division of Parasitic Diseases).

the basis of the quality of the predictions (Kitano, 2002). Such an approach may typically, yet not necessarily, provide a framework of interpretation of emergent phenomena and high-level disease manifestations or phenotypes that could not have been inferred from isolated pieces of knowledge or associations. Systems biology approaches have been applied in analyses of complex biological processes such as sepsis, trauma and wound healing (Aderem and Smith, 2004; Vodovotz, 2006; Vodovotz et al., 2004a), as well as host–pathogen interactions (Forst, 2006). We suggest that through a systems biology approach that involves examining the transfer of agents from the mammalian to the mosquito host, and the aggregate effect on the parasite as well as the two hosts, we will gain a true understanding of the complex inflammatory biology described herein.

ACKNOWLEDGMENTS

This work was supported by grants from the National Institutes of Health (NIH) NIAID AI50663, AI60664, and an NIH NCRR Research Facilities Improvement Grant C06 RR-12088-01.

REFERENCES

Aderem, A., and Smith, K. D. (2004). A systems approach to dissecting immunity and inflammation. *Semin. Immunol.* **16**, 55–67.

Aeberli, D., Yang, Y., Mansell, A., Santos, L., Leech, M., and Morand, E. F. (2006). Endogenous macrophage migration inhibitory factor modulates glucocorticoid sensitivity in macrophages via effects on MAP kinase phosphatase-1 and p38 MAP kinase. *FEBS Lett.* **580**, 974–981.

Akira, S., Uematsu, S., and Takeuchi, O. (2006). Pathogen recognition and innate immunity. *Cell* **124**, 783–801.

Akman-Anderson, L., Olivier, M., and Luckhart, S. (2007). Induction of *nitric oxide synthase* and activation of signaling proteins in *Anopheles* mosquitoes by malaria pigment hemozoin. *Infect. Immun.* **75**, 4012–4019.

Amin, M. A., Haas, C. S., Zhu, K., Mansfield, P. J., Kim, M. J., Lackowski, N. P., and Koch, A. E. (2006). Migration inhibitory factor up-regulates vascular cell adhesion molecule-1 and intercellular adhesion molecule-1 via Src, PI3 kinase, and NFkappaB. *Blood* **107**, 2252–2261.

Arese, P., and Schwarzer, E. (1997). Malarial pigment (haemozoin): A very active 'inert' substance. *Ann. Trop. Med. Parasitol.* **91**, 501–516.

Awandare, G. A., Hittner, J. B., Kremsner, P. G., Ochiel, D. O., Keller, C. C., Weinberg, J. B., Clark, I. A., and Perkins, D. J. (2006). Decreased circulating macrophage migration inhibitory factor (MIF) protein and blood mononuclear cell MIF transcripts in children with *Plasmodium falciparum* malaria. *Clin. Immunol.* **119**, 219–225.

Awandare, G. A., Ouma, Y., Ouma, C., Were, T., Otieno, R., Keller, C. C., Davenport, G. C., Hittner, J. B., Vulule, J., Ferrell, R., Ong'echa, J. M., and Perkins, D. J. (2007). Role of monocyte-acquired hemozoin in suppression of macrophage migration inhibitory factor in children with severe malarial anemia. *Infect. Immun.* **75**, 201–210.

Barillas-Mury, C., and Kumar, S. (2005). *Plasmodium*–mosquito interactions: A tale of dangerous liaisons. *Cell Microbiol.* **7**, 1539–1545.

Barillas-Mury, C., Han, Y. S., Seeley, D., and Kafatos, F. C. (1999). *Anopheles gambiae* Ag-STAT, a new insect member of the STAT family, is activated in response to bacterial infection. *EMBO J.* **18**, 959–967.

Barone, R., Simpore, J., Malaguarnera, L., Pignatelli, S., and Musumeci, S. (2003). Plasma chitotriosidase activity in acute *Plasmodium falciparum* malaria. *J. Trop. Pediatr.* **49**, 63–64.

Beier, M. S., Pumpuni, C. B., Beier, J. C., and Davis, J. R. (1994). Effects of para-aminobenzoic acid, insulin, and gentamicin on *Plasmodium falciparum* development in anopheline mosquitoes (Diptera: Culicidae). *J. Med. Entomol.* **31**, 561–565.

Bhatnagar, R. K., Arora, N., Sachidanand, S., Shahabuddin, M., Keister, D., and Chauhan, V. S. (2003). Synthetic propeptide inhibits mosquito midgut chitinase and blocks sporogonic development of malaria parasite. *Biochem. Biophys. Res. Commun.* **304**, 783–787.

Bicker, G. (2007). Pharmacological approaches to nitric oxide signalling during neural development of locusts and other model insects. *Arch. Insect Biochem. Physiol.* **64**, 43–58.

Billingsley, P. F. (1990). Blood digestion in the mosquito, *Anopheles stephensi* Liston (Diptera: Culicidae): Partial characterization and post-feeding activity of midgut aminopeptidases. *Arch. Insect. Biochem. Physiol.* **15**, 149–163.

Billingsley, P. F., and Hecker, H. (1991). Blood digestion in the mosquito, *Anopheles stephensi* Liston (Diptera: Culicidae): Activity and distribution of trypsin, aminopeptidase, and alpha-glucosidase in the midgut. *J. Med. Entomol.* **28**, 865–871.

Billingsley, P. F., Baird, J., Mitchell, J. A., and Drakeley, C. (2006). Immune interactions between mosquitoes and their hosts. *Parasite Immunol.* **28**, 143–153.

Bogdan, C., Rollinghoff, M., and Diefenbach, A. (2000). Reactive oxygen and reactive nitrogen intermediates in innate and specific immunity. *Curr. Opin. Immunol.* **12**, 64–76.

Bouharoun-Tayoun, H., Oeuvray, C., Lunel, F., and Druilhe, P. (1995). Mechanisms underlying the monocyte-mediated antibody-dependent killing of *Plasmodium falciparum* asexual blood stages. *J. Exp. Med.* **182**, 409–418.

Caro, H. N., Sheikh, N. A., Taverne, J., Playfair, J. H., and Rademacher, T. W. (1996). Structural similarities among malaria toxins insulin second messengers, and bacterial endotoxin. *Infect. Immun.* **64**, 3438–3441.

Chaiyaroj, S. C., Rutta, A. S., Muenthaisong, K., Watkins, P., Na Ubol, M., and Looareesuwan, S. (2004). Reduced levels of transforming growth factor-beta1, interleukin-12 and increased migration inhibitory factor are associated with severe malaria. *Acta Trop.* **89**, 319–327.

Christophides, G. K., Zdobnov, E., Barillas-Mury, C., Birney, E., Blandin, S., Blass, C., Brey, P. T., Collins, F. H., Danielli, A., Dimopoulos, G., Hetru, C., Hoa, N. T., Hoffmann, J. A., Kanzok, S. M., Letunic, I., Levashina, E. A., Loukeris, T. G., Lycett, G., Meister, S., Michel, K., Moita, L. F., Muller, H. M., Osta, M. A., Paskewitz, S. M., Reichhart, J. M., Rzhetsky, A., Troxler, L., Vernick, K. D., Vlachou, D., Volz, J., von Mering, C., Xu, J., Zheng, L., Bork, P., and Kafatos, F. C. (2002). Immunity-related genes and gene families in *Anopheles gambiae. Science* **298**, 159–165.

Christophides, G. K., Vlachou, D., and Kafatos, F. C. (2004). Comparative and functional genomics of the innate immune system in the malaria vector *Anopheles gambiae. Immunol. Rev.* **198**, 127–148.

Clark, I. A., and Cowden, W. B. (2003). The pathophysiology of *falciparum* malaria. *Pharmacol. Ther.* **99**, 221–260.

Clements, A. N. (1992). Adult food and feeding mechanisms. In *The Biology of Mosquitoes: Development, Nutrition and Reproduction*. Chapman & Hall, London.

Coban, C., Ishii, K. J., Sullivan, D. J., and Kumar, N. (2002). Purified malaria pigment (hemozoin) enhances dendritic cell maturation and modulates the isotype of antibodies induced by a DNA vaccine. *Infect. Immun.* **70**, 3939–3943.

Coban, C., Ishii, K. J., Kawai, T., Hemmi, H., Sato, S., Uematsu, S., Yamamoto, M., Takeuchi, O., Itagaki, S., Kumar, N., Horii, T., and Akira, S. (2005). Toll-like receptor 9 mediates innate immune activation by the malaria pigment hemozoin. *J. Exp. Med.* **201**, 19–25.

Coons, L. B., Rosell-Davis, R., and Tarnowski, B. I. (1986). *Bloodmeal Digestion in Ticks*. Ellis Horwood, Chicester, UK.

Danielpour, D., and Song, K. (2006). Cross-talk between IGF-I and TGF-beta signaling pathways. *Cytokine Growth Factor Rev.* **17**, 59–74.

Darby, S. M., Miller, M. L., Allen, R. O., and LeBeau, M. (2001). A mass spectrometric method for quantitation of intact insulin in blood samples. *J. Anal. Toxicol.* **25**, 8–14.

de la Fuente, J., Almazan, C., Blas-Machado, U., Naranjo, V., Mangold, A. J., Blouin, E. F., Gortazar, C., and Kocan, K. M. (2006a). The tick protective antigen, 4D8, is a conserved protein involved in modulation of tick blood ingestion and reproduction. *Vaccine* **24**, 4082–4095.

de la Fuente, J., Almazan, C., Blouin, E. F., Naranjo, V., and Kocan, K. M. (2006b). Reduction of tick infections with *Anaplasma marginale* and *A. phagocytophilum* by targeting the tick protective antigen subolesin. *Parasitol. Res.* **100**, 85–91.

Devenport, M., Fujioka, H., and Jacobs-Lorena, M. (2004). Storage and secretion of the peritrophic matrix protein Ag-Aper1 and trypsin in the midgut of *Anopheles gambiae. Insect Mol. Biol.* **13**, 349–358.

Di Luca, M., Romi, R., Severini, F., Toma, L., Musumeci, M., Fausto, A. M., Mazzini, M., Gambellini, G., and Musumeci, S. (2006). Human chitotriosidase helps *Plasmodium falciparum* in the *Anopheles* midgut. *J. Vector Borne Dis.* **43**, 144–146.

Di Rosa, M., Musumeci, M., Scuto, A., Musumeci, S., and Malaguarnera, L. (2005). Effect of interferon-gamma, interleukin-10, lipopolysaccharide and tumor necrosis factor-alpha on chitotriosidase synthesis in human macrophages. *Clin. Chem. Lab. Med.* **43**, 499–502.

Dimopoulos, G., Richman, A., Muller, H. M., and Kafatos, F. C. (1997). Molecular immune responses of the mosquito *Anopheles gambiae* to bacteria and malaria parasites. *Proc. Natl. Acad. Sci. USA* **94**, 11508–11513.

Dimopoulos, G., Seeley, D., Wolf, A., and Kafatos, F. C. (1998). Malaria infection of the mosquito *Anopheles gambiae* activates immune-responsive genes during critical transition stages of the parasite life cycle. *EMBO J.* **17**, 6115–6123.

Dinglasan, R. R., Fields, I., Shahabuddin, M., Azad, A. F., and Sacci Jr., J. B. (2003). Monoclonal antibody MG96 completely blocks *Plasmodium yoelii* development in *Anopheles stephensi*. *Infect. Immun.* **71**, 6995–7001.

Dinglasan, R. R., Valenzuela, J. G., and Azad, A. F. (2005). Sugar epitopes as potential universal disease transmission blocking targets. *Insect Biochem. Mol. Biol.* **35**, 1–10.

Dong, Y., Aguilar, R., Xi, Z., Warr, E., Mongin, E., and Dimopoulos, G. (2006). *Anopheles gambiae* immune responses to human and rodent *Plasmodium* parasite species. *PLoS Pathog.* **2**, e52.

Elased, K. M., Gumaa, K. A., de Souza, J. B., Playfair, J. H., and Rademacher, T. W. (2004). Improvement of glucose homeostasis in obese diabetic db/db mice given *Plasmodium yoelii* glycosylphosphatidylinositols. *Metabolism* **53**, 1048–1053.

Ewald, P. W. (1987). Transmission modes and evolution of the parasitism–mutualism continuum. *Ann. NY Acad. Sci.* **503**, 295–306.

Field, M. C. (1997). Is there evidence for phospho-oligosaccharides as insulin mediators? *Glycobiology* **7**, 161–168.

Fikrig, E., Telford III, S. R., Barthold, S. W., Kantor, F. S., Spielman, A., and Flavell, R. A. (1992). Elimination of *Borrelia burgdorferi* from vector ticks feeding on OspA-immunized mice. *Proc. Natl. Acad. Sci. USA* **89**, 5418–5421.

Fogaca, A. C., da Silva Jr., P. I., Miranda, M. T., Bianchi, A. G., Miranda, A., Ribolla, P. E., and Daffre, S. (1999). Antimicrobial activity of a bovine hemoglobin fragment in the tick *Boophilus microplus*. *J. Biol. Chem.* **274**, 25330–25334.

Forst, C. V. (2006). Host–pathogen systems biology. *Drug Discov. Today* **11**, 220–227.

Fraidenraich, D., Pena, C., Isola, E. L., Lammel, E. M., Coso, O., Anel, A. D., Pongor, S., Baralle, F., Torres, H. N., and Flawia, M. M. (1993). Stimulation of *Trypanosoma cruzi* adenylyl cyclase by an alpha D-globin fragment from *Triatoma* hindgut: Effect on differentiation of epimastigote to trypomastigote forms. *Proc. Natl. Acad. Sci. USA* **90**, 10140–10144.

Francis, S. E., Sullivan Jr., D. J., and Goldberg, D. E. (1997). Hemoglobin metabolism in the malaria parasite *Plasmodium falciparum*. *Annu. Rev. Microbiol.* **51**, 97–123.

Frolet, C., Thoma, M., Blandin, S., Hoffmann, J. A., and Levashina, E. A. (2006). Boosting NF-kappaB-dependent basal immunity of *Anopheles gambiae* aborts development of *Plasmodium berghei*. *Immunity* **25**, 677–685.

Garcia, E. S., Gonzalez, M. S., de Azambuja, P., Baralle, F. E., Fraidenraich, D., Torres, H. N., and Flawia, M. M. (1995). Induction of *Trypanosoma cruzi* metacyclogenesis in the gut of the hematophagous insect vector, *Rhodnius prolixus*, by hemoglobin and peptides carrying alpha D-globin sequences. *Exp. Parasitol.* **81**, 255–261.

Gowda, C. D. (2002). Structure and activity of glycosylphosphatidylinositol anchors of *Plasmodium falciparum*. *Microb. Infect.* **4**, 983–990.

Healer, J., Graszynski, A., and Riley, E. (1999). Phagocytosis does not play a major role in naturally acquired transmission-blocking immunity to *Plasmodium falciparum* malaria. *Infect. Immun.* **67**, 2334–2339.

Heemskerk, V. H., Daemen, M. A., and Buurman, W. A. (1999). Insulin-like growth factor-1 (IGF-1) and growth hormone (GH) in immunity and inflammation. *Cytokine Growth Factor Rev.* **10**, 5–14.

Hurd, H., and Carter, V. (2004). The role of programmed cell death in *Plasmodium*–mosquito interactions. *Int. J. Parasitol.* **34**, 1459–1472.

Hurd, H., Grant, K. M., and Arambage, S. C. (2006). Apoptosis-like death as a feature of malaria infection in mosquitoes. *Parasitology* **132**(Suppl), S33–S47.

Jahan, N., Docherty, P. T., Billingsley, P. F., and Hurd, H. (1999). Blood digestion in the mosquito, *Anopheles stephensi*: The effects of *Plasmodium yoelii nigeriensis* on midgut enzyme activities. *Parasitology* **119**(Pt 6), 535–541.

Jaramillo, M., Gowda, D. C., Radzioch, D., and Olivier, M. (2003). Hemozoin increases IFN-gamma-inducible macrophage nitric oxide generation through extracellular signal-regulated kinase- and NF-kappa B-dependent pathways. *J. Immunol.* **171**, 4243–4253.

Jaramillo, M., Plante, I., Ouellet, N., Vandal, K., Tessier, P. A., and Olivier, M. (2004). Hemozoin-inducible proinflammatory events *in vivo*: Potential role in malaria infection. *J. Immunol.* **172**, 3101–3110.

Jaramillo, M., Godbout, M., and Olivier, M. (2005). Hemozoin induces macrophage chemokine expression through oxidative stress-dependent and -independent mechanisms. *J. Immunol.* **174**, 475–484.

Jaworski, D. C., Jasinskas, A., Metz, C. N., Bucala, R., and Barbour, A. G. (2001). Identification and characterization of a homologue of the pro-inflammatory cytokine macrophage migration inhibitory factor in the tick, *Amblyomma americanum*. *Insect Mol. Biol.* **10**, 323–331.

Kharazmi, A., and Jepsen, S. (1984). Enhanced inhibition of *in vitro* multiplication of *Plasmodium falciparum* by stimulated human polymorphonuclear leucocytes. *Clin. Exp. Immunol.* **57**, 287–292.

Khusmith, S., and Druilhe, P. (1983). Cooperation between antibodies and monocytes that inhibit *in vitro* proliferation of *Plasmodium falciparum*. *Infect. Immun.* **41**, 219–223.

Kitano, H. (2002). Systems biology: A brief overview. *Science* **295**, 1662–1664.

Kocan, K. M., Blouin, E. F., Palmer, G. H., Eriks, I. S., and Edwards, W. L. (1996). Strategies to interrupt the development of *Anaplasma marginale* in its tick vector. The effect of bovine-derived antibodies. *Ann. NY Acad. Sci.* **791**, 157–165.

Krishnegowda, G., Hajjar, A. M., Zhu, J., Douglass, E. J., Uematsu, S., Akira, S., Woods, A. S., and Gowda, D. C. (2005). Induction of proinflammatory responses in macrophages by the glycosylphosphatidylinositols of *Plasmodium falciparum*: Cell signaling receptors, glycosylphosphatidylinositol (GPI) structural requirement, and regulation of GPI activity. *J. Biol. Chem.* **280**, 8606–8616.

Kurtenbach, K., Sewell, H. S., Ogden, N. H., Randolph, S. E., and Nuttall, P. A. (1998). Serum complement sensitivity as a key factor in Lyme disease ecology. *Infect. Immun.* **66**, 1248–1251.

Kurtenbach, K., De Michelis, S., Etti, S., Schafer, S. M., Sewell, H. S., Brade, V., and Kraiczy, P. (2002a). Host association of *Borrelia burgdorferi* sensu lato – The key role of host complement. *Trends Microbiol.* **10**, 74–79.

Kurtenbach, K., Schafer, S. M., Sewell, H. S., Peacey, M., Hoodless, A., Nuttall, P. A., and Randolph, S. E. (2002b). Differential survival of Lyme borreliosis spirochetes in ticks that feed on birds. *Infect. Immun.* **70**, 5893–5895.

Kurtenbach, K., Hanincova, K., Tsao, J. I., Margos, G., Fish, D., and Ogden, N. H. (2006). Fundamental processes in the evolutionary ecology of Lyme borreliosis. *Nat. Rev. Microbiol.* **4**, 660–669.

Lal, A. A., Schriefer, M. E., Sacci, J. B., Goldman, I. F., Louis-Wileman, V., Collins, W. E., and Azad, A. F. (1994). Inhibition of malaria parasite development in mosquitoes by anti-mosquito-midgut antibodies. *Infect. Immun.* **62**, 316–318.

Lal, A. A., Patterson, P. S., Sacci, J. B., Vaughan, J. A., Paul, C., Collins, W. E., Wirtz, R. A., and Azad, A. F. (2001). Anti-mosquito midgut antibodies block development of *Plasmodium falciparum* and *Plasmodium vivax* in multiple species of *Anopheles* mosquitoes and reduce vector fecundity and survivorship. *Proc. Natl. Acad. Sci. USA* **98**, 5228–5233.

Langer, R. C., and Vinetz, J. M. (2001). *Plasmodium* ookinete-secreted chitinase and parasite penetration of the mosquito peritrophic matrix. *Trends Parasitol.* **17**, 269–272.

Lavazec, C., Boudin, C., Lacroix, R., Bonnet, S., Diop, A., Thiberge, S., Boisson, B., Tahar, R., and Bourgouin, C. (2007). Carboxypeptidases B of *Anopheles gambiae* as targets for a *Plasmodium falciparum* transmission-blocking vaccine. *Infect. Immun.* **75**, 1635–1642.

Lee, M. S., and Kim, Y. J. (2007). Signaling pathways downstream of pattern-recognition receptors and their cross talk. *Annu. Rev. Biochem.* **76**, 447–480.

Lensen, A. H., Bolmer-Van de Vegte, M., van Gemert, G. J., Eling, W. M., and Sauerwein, R. W. (1997). Leukocytes in a *Plasmodium falciparum*-infected blood meal reduce transmission of malaria to *Anopheles* mosquitoes. *Infect. Immun.* **65**, 3834–3837.

Lensen, A., Mulder, L., Tchuinkam, T., Willemsen, L., Eling, W., and Sauerwein, R. (1998). Mechanisms that reduce transmission of *Plasmodium falciparum* malaria in semiimmune and non-immune persons. *J. Infect. Dis.* **177**, 1358–1363.

Lieber, M. J., and Luckhart, S. (2004). Transforming growth factor-betas and related gene products in mosquito vectors of human malaria parasites: Signaling architecture for immunological crosstalk. *Mol. Immunol.* **41**, 965–977.

Liepke, C., Baxmann, S., Heine, C., Breithaupt, N., Standker, L., and Forssmann, W. G. (2003). Human hemoglobin-derived peptides exhibit antimicrobial activity: A class of host defense peptides. *J. Chromatogr. B. Analyt. Technol. Biomed. Life Sci.* **791**, 345–356.

Lim, J., Gowda, D. C., Krishnegowda, G., and Luckhart, S. (2005). Induction of nitric oxide synthase in *Anopheles stephensi* by *Plasmodium falciparum*: Mechanism of signaling and the role of parasite glycosylphosphatidylinositols. *Infect. Immun.* **73**, 2778–2789.

Luckhart, S., and Riehle, M. A. (2007). The insulin signaling cascade from nematodes to mammals: Insights into innate immunity of *Anopheles* mosquitoes to malaria parasite infection. *Dev. Comp. Immunol.* **31**, 647–656.

Luckhart, S., Vodovotz, Y., Cui, L., and Rosenberg, R. (1998). The mosquito *Anopheles stephensi* limits malaria parasite development with inducible synthesis of nitric oxide. *Proc. Natl. Acad. Sci. USA* **95**, 5700–5705.

Luckhart, S., Lieber, M. J., Singh, N., Zamora, R., Vodovotz, Y. Low levels of mammalian TGF-β1 are protective against malaria parasite infection, a paradox clarified in the mosquito host. *Exp. Parasitol.* (in press).

Luckhart, S., Crampton, A. L., Zamora, R., Lieber, M. J., Dos Santos, P. C., Peterson, T. M., Emmith, N., Lim, J., Wink, D. A., and Vodovotz, Y. (2003). Mammalian transforming growth factor beta1 activated after ingestion by *Anopheles stephensi* modulates mosquito immunity. *Infect. Immun.* **71**, 3000–3009.

Luna, C., Wang, X., Huang, Y., Zhang, J., and Zheng, L. (2002). Characterization of four Toll related genes during development and immune responses in *Anopheles gambiae*. *Insect Biochem. Mol. Biol.* **32**, 1171–1179.

Malaguarnera, L. (2006). Chitotriosidase: The yin and yang. *Cell Mol. Life Sci.* **63**, 3018–3029.

Margos, G., Navarette, S., Butcher, G., Davies, A., Willers, C., Sinden, R. E., and Lachmann, P. J. (2001). Interaction between host complement and mosquito-midgut-stage *Plasmodium berghei*. *Infect. Immun.* **69**, 5064–5071.

Matsuo, T., Cerruto Noya, C. A., Taylor, D., and Fujisaki, K. (2007). Immunohistochemical examination of PDGF-AB, TGF-beta and their receptors in the hemocytes of a tick, *Ornithodoros moubata* (Acari: Argasidae). *J. Vet. Med. Sci.* **69**, 317–320.

Matuschewski, K. (2006). Vaccine development against malaria. *Curr. Opin. Immunol.* **18**, 449–457.

Meister, S., Koutsos, A. C., and Christophides, G. K. (2004). The *Plasmodium* parasite – A 'new' challenge for insect innate immunity. *Int. J. Parasitol.* **34**, 1473–1482.

Meister, S., Kanzok, S. M., Zheng, X. L., Luna, C., Li, T. R., Hoa, N. T., Clayton, J. R., White, K. P., Kafatos, F. C., Christophides, G. K., and Zheng, L. (2005). Immune signaling pathways regulating bacterial and malaria parasite infection of the mosquito *Anopheles gambiae*. *Proc. Natl. Acad. Sci. USA* **102**, 11420–11425.

Mellouk, S., Hoffman, S. L., Liu, Z. Z., de la Vega, P., Billiar, T. R., and Nussler, A. K. (1994). Nitric oxide-mediated antiplasmodial activity in human and murine hepatocytes induced by gamma interferon and the parasite itself: Enhancement by exogenous tetrahydrobiopterin. *Infect. Immun.* **62**, 4043–4046.

Michel, K., and Kafatos, F. C. (2005). Mosquito immunity against *Plasmodium*. *Insect Biochem. Mol. Biol.* **35**, 677–689.

Millington, O. R., Di Lorenzo, C., Phillips, R. S., Garside, P., and Brewer, J. M. (2006). Suppression of adaptive immunity to heterologous antigens during *Plasmodium* infection through hemozoin-induced failure of dendritic cell function. *J. Biol.* **5**, 5.

Mizutani, T., Kobayashi, M., Eshita, Y., Shirato, K., Kimura, T., Ako, Y., Miyoshi, H., Takasaki, T., Kurane, I., Kariwa, H., Umemura, T., and Takashima, I. (2003). Involvement of the JNK-like protein of the *Aedes albopictus* mosquito cell line, C6/36, in phagocytosis, endocytosis and infection of West Nile virus. *Insect Mol. Biol.* **12**, 491–499.

Motard, A., Landau, I., Nussler, A., Grau, G., Baccam, D., Mazier, D., and Targett, G. A. (1993). The role of reactive nitrogen intermediates in modulation of gametocyte infectivity of rodent malaria parasites. *Parasite Immunol.* **15**, 21–26.

Moustakas, A., and Heldin, C. H. (2005). Non-Smad TGF-beta signals. *J. Cell Sci.* **118**, 3573–3584.

Muller, G. (2002). Dynamics of plasma membrane microdomains and cross-talk to the insulin signalling cascade. *FEBS Lett.* **531**, 81–87.

Muller, H. M., Crampton, J. M., della Torre, A., Sinden, R., and Crisanti, A. (1993a). Members of a trypsin gene family in *Anopheles gambiae* are induced in the gut by blood meal. *EMBO J.* **12**, 2891–2900.

Muller, H. M., Vizioli, I., della Torre, A., and Crisanti, A. (1993b). Temporal and spatial expression of serine protease genes in *Anopheles gambiae*. *Parassitologia* **35**(Suppl), 73–76.

Muller, H. M., Catteruccia, F., Vizioli, J., della Torre, A., and Crisanti, A. (1995). Constitutive and blood meal-induced trypsin genes in *Anopheles gambiae*. *Exp. Parasitol.* **81**, 371–385.

Nakajima, Y., van der Goes van Naters-Yasui, A., Taylor, D., and Yamakawa, M. (2001). Two isoforms of a member of the arthropod defensin family from the soft tick, *Ornithodoros moubata* (Acari: Argasidae). *Insect Biochem. Mol. Biol.* **31**, 747–751.

Nakajima, Y., Taylor, D., and Yamakawa, M. (2002a). Involvement of antibacterial peptide defensin in tick midgut defense. *Exp. Appl. Acarol.* **28**, 135–140.

Nakajima, Y., van der Goes van Naters-Yasui, A., Taylor, D., and Yamakawa, M. (2002b). Antibacterial peptide defensin is involved in midgut immunity of the soft tick, *Ornithodoros moubata*. *Insect Mol. Biol.* **11**, 611–618.

Nakajima, Y., Ogihara, K., Taylor, D., and Yamakawa, M. (2003a). Antibacterial hemoglobin fragments from the midgut of the soft tick, *Ornithodoros moubata* (Acari: Argasidae). *J. Med. Entomol.* **40**, 78–81.

Nakajima, Y., Saido-Sakanaka, H., Taylor, D., and Yamakawa, M. (2003b). Up-regulated humoral immune response in the soft tick, *Ornithodoros moubata* (Acari: Argasidae). *Parasitol. Res.* **91**, 476–481.

Naotunne, T. S., Karunaweera, N. D., Mendis, K. N., and Carter, R. (1993). Cytokine-mediated inactivation of malarial gametocytes is dependent on the presence of white blood cells and involves reactive nitrogen intermediates. *Immunology* **78**, 555–562.

Nebl, T., De Veer, M. J., and Schofield, L. (2005). Stimulation of innate immune responses by malarial glycosylphosphatidylinositol via pattern recognition receptors. *Parasitology* **130**(Suppl), S45–S62.

Oduol, F., Xu, J., Niare, O., Natarajan, R., and Vernick, K. D. (2000). Genes identified by an expression screen of the vector mosquito *Anopheles gambiae* display differential molecular immune response to malaria parasites and bacteria. *Proc. Natl. Acad. Sci. USA* **97**, 11397–11402.

Omer, F. M., and Riley, E. M. (1998). Transforming growth factor beta production is inversely correlated with severity of murine malaria infection. *J. Exp. Med.* **188**, 39–48.

Omer, F. M., Kurtzhals, J. A., and Riley, E. M. (2000). Maintaining the immunological balance in parasitic infections: A role for TGF-beta? *Parasitol. Today* **16**, 18–23.

Osta, M. A., Christophides, G. K., and Kafatos, F. C. (2004). Effects of mosquito genes on *Plasmodium* development. *Science* **303**, 2030–2032.

Parish, C. A., Jiang, H., Tokiwa, Y., Berova, N., Nakanishi, K., McCabe, D., Zuckerman, W., Xia, M. M., and Gabay, J. E. (2001). Broad-spectrum antimicrobial activity of hemoglobin. *Bioorg. Med. Chem.* **9**, 377–382.

Parroche, P., Lauw, F. N., Goutagny, N., Latz, E., Monks, B. G., Visintin, A., Halmen, K. A., Lamphier, M., Olivier, M., Bartholomeu, D. C., Gazzinelli, R. T., and Golenbock, D. T. (2007).

From the cover: Malaria hemozoin is immunologically inert but radically enhances innate responses by presenting malaria DNA to Toll-like receptor 9. *Proc. Natl. Acad. Sci. USA* **104**, 1919–1924.

Patel, S. N., Lu, Z., Ayi, K., Serghides, L., Gowda, D. C., and Kain, K. C. (2007). Disruption of CD36 impairs cytokine response to *Plasmodium falciparum* glycosylphosphatidylinositol and confers susceptibility to severe and fatal malaria *in vivo. J. Immunol.* **178**, 3954–3961.

Perkins, D. J., Weinberg, J. B., and Kremsner, P. G. (2000). Reduced interleukin-12 and transforming growth factor-beta1 in severe childhood malaria: Relationship of cytokine balance with disease severity. *J. Infect. Dis.* **182**, 988–992.

Peterson, T. M., Gow, A. J., and Luckhart, S. (2007). Nitric oxide metabolites induced in *Anopheles stephensi* control malaria parasite infection. *Free Radic. Biol. Med.* **42**, 132–142.

Pichyangkul, S., Saengkrai, P., and Webster, H. K. (1994). *Plasmodium falciparum* pigment induces monocytes to release high levels of tumor necrosis factor-alpha and interleukin-1 beta. *Am. J. Trop. Med. Hyg.* **51**, 430–435.

Planche, T., Dzeing, A., Ngou-Milama, E., Kombila, M., and Stacpoole, P. W. (2005). Metabolic complications of severe malaria. *Curr. Top. Microbiol. Immunol.* **295**, 105–136.

Prevot, G. I., Laurent-Winter, C., Rodhain, F., and Bourgouin, C. (2003). Sex-specific and blood meal-induced proteins of *Anopheles gambiae* midguts: Analysis by two-dimensional gel electrophoresis. *Malar. J.* **2**, 1.

Raftery, L. A., and Sutherland, D. J. (1999). TGF-beta family signal transduction in *Drosophila* development: From Mad to Smads. *Dev. Biol.* **210**, 251–268.

Rehbein, G., Ahmed, J. S., Schein, E., Horchner, F., and Zweygarth, E. (1981). Immunological aspects of *Theileria annulata* infection calves. 2. Production of macrophage migration inhibition factor (MIF) by sensitized lymphocytes from *Theileria annulata*-infected calves. *Tropenmed. Parasitol.* **32**, 154–156.

Richman, A. M., Dimopoulos, G., Seeley, D., and Kafatos, F. C. (1997). *Plasmodium* activates the innate immune response of *Anopheles gambiae* mosquitoes. *EMBO J.* **16**, 6114–6119.

Riehle, M. A., and Brown, M. R. (1999). Insulin stimulates ecdysteroid production through a conserved signaling cascade in the mosquito *Aedes aegypti. Insect Biochem. Mol. Biol.* **29**, 855–860.

Riehle, M. A., and Brown, M. R. (2002). Insulin receptor expression during development and a reproductive cycle in the ovary of the mosquito *Aedes aegypti. Cell Tissue Res.* **308**, 409–420.

Riehle, M. A., and Brown, M. R. (2003). Molecular analysis of the serine/threonine kinase Akt and its expression in the mosquito *Aedes aegypti. Insect Mol. Biol.* **12**, 225–232.

Riehle, M. A., Garczynski, S. F., Crim, J. W., Hill, C. A., and Brown, M. R. (2002). Neuropeptides and peptide hormones in *Anopheles gambiae. Science* **298**, 172–175.

Riley, E. M., Wahl, S., Perkins, D. J., and Schofield, L. (2006). Regulating immunity to malaria. *Parasite Immunol.* **28**, 35–49.

Rivero, A. (2006). Nitric oxide: An antiparasitic molecule of invertebrates. *Trends Parasitol.* **22**, 219–225.

Saltiel, A. R. (1991). The role of glycosyl-phosphoinositides in hormone action. *J. Bioenerg. Biomembr.* **23**, 29–41.

Schoeler, G. B., and Wikel, S. K. (2001). Modulation of host immunity by haematophagous arthropods. *Ann. Trop. Med. Parasitol.* **95**, 755–771.

Schofield, L., and Hackett, F. (1993). Signal transduction in host cells by a glycosylphosphatidylinositol toxin of malaria parasites. *J. Exp. Med.* **177**, 145–153.

Schwarzer, E., and Arese, P. (1996). Phagocytosis of malarial pigment hemozoin inhibits NADPH-oxidase activity in human monocyte-derived macrophages. *Biochim. Biophys. Acta* **1316**, 169–175.

Schwarzer, E., Turrini, F., Giribaldi, G., Cappadoro, M., and Arese, P. (1993). Phagocytosis of *P. falciparum* malarial pigment hemozoin by human monocytes inactivates monocyte protein kinase C. *Biochim. Biophys. Acta* **1181**, 51–54.

Schwarzer, E., Alessio, M., Ulliers, D., and Arese, P. (1998). Phagocytosis of the malarial pigment, hemozoin, impairs expression of major histocompatibility complex class II antigen, CD54, and CD11c in human monocytes. *Infect. Immun.* **66**, 1601–1606.

Schwarzer, E., Kuhn, H., Valente, E., and Arese, P. (2003). Malaria-parasitized erythrocytes and hemozoin nonenzymatically generate large amounts of hydroxy fatty acids that inhibit monocyte functions. *Blood* **101**, 722–728.

Sherry, B. A., Alava, G., Tracey, K. J., Martiney, J., Cerami, A., and Slater, A. F. (1995). Malaria-specific metabolite hemozoin mediates the release of several potent endogenous pyrogens (TNF, MIP-1 alpha, and MIP-1 beta) *in vitro*, and altered thermoregulation *in vivo*. *J. Inflamm.* **45**, 85–96.

Sigal, L. H., Zahradnik, J. M., Lavin, P., Patella, S. J., Bryant, G., Haselby, R., Hilton, E., Kunkel, M., Adler-Klein, D., Doherty, T., Evans, J., Molloy, P. J., Seidner, A. L., Sabetta, J. R., Simon, H. J., Klempner, M. S., Mays, J., Marks, D., and Malawista, S. E. (1998). A vaccine consisting of recombinant *Borrelia burgdorferi* outer-surface protein A to prevent Lyme disease. Recombinant Outer-Surface Protein A Lyme Disease Vaccine Study Consortium. *New Engl. J. Med.* **339**, 216–222.

Skorokhod, O. A., Alessio, M., Mordmuller, B., Arese, P., and Schwarzer, E. (2004). Hemozoin (malarial pigment) inhibits differentiation and maturation of human monocyte-derived dendritic cells: A peroxisome proliferator-activated receptor-gamma-mediated effect. *J. Immunol.* **173**, 4066–4074.

Sonenshine, D. E., Hynes, W. L., Ceraul, S. M., Mitchell, R., and Benzine, T. (2005). Host blood proteins and peptides in the midgut of the tick *Dermacentor variabilis* contribute to bacterial control. *Exp. Appl. Acarol.* **36**, 207–223.

Steere, A. C., Sikand, V. K., Meurice, F., Parenti, D. L., Fikrig, E., Schoen, R. T., Nowakowski, J., Schmid, C. H., Laukamp, S., Buscarino, C., and Krause, D. S. (1998). Vaccination against Lyme disease with recombinant *Borrelia burgdorferi* outer-surface lipoprotein A with adjuvant. Lyme Disease Vaccine Study Group. *New Engl. J. Med.* **339**, 209–215.

Stevenson, B., El-Hage, N., Hines, M. A., Miller, J. C., and Babb, K. (2002). Differential binding of host complement inhibitor factor H by *Borrelia burgdorferi* Erp surface proteins: A possible mechanism underlying the expansive host range of Lyme disease spirochetes. *Infect. Immun.* **70**, 491–497.

Tachado, S. D., Gerold, P., McConville, M. J., Baldwin, T., Quilici, D., Schwarz, R. T., and Schofield, L. (1996). Glycosylphosphatidylinositol toxin of *Plasmodium* induces nitric oxide synthase expression in macrophages and vascular endothelial cells by a protein tyrosine kinase-dependent and protein kinase C-dependent signaling pathway. *J. Immunol.* **156**, 1897–1907.

Tahar, R., Boudin, C., Thiery, I., and Bourgouin, C. (2002). Immune response of *Anopheles gambiae* to the early sporogonic stages of the human malaria parasite *Plasmodium falciparum*. *EMBO J.* **21**, 6673–6680.

Taramelli, D., Basilico, N., Pagani, E., Grande, R., Monti, D., Ghione, M., and Olliaro, P. (1995). The heme moiety of malaria pigment (beta-hematin) mediates the inhibition of nitric oxide and tumor necrosis factor-alpha production by lipopolysaccharide-stimulated macrophages. *Exp. Parasitol.* **81**, 501–511.

Taramelli, D., Basilico, N., De Palma, A. M., Saresella, M., Ferrante, P., Mussoni, L., and Olliaro, P. (1998). The effect of synthetic malaria pigment (beta-haematin) on adhesion molecule expression and interleukin-6 production by human endothelial cells. *Trans. R. Soc. Trop. Med. Hyg.* **92**, 57–62.

Tsuboi, T., Cao, Y. M., Torii, M., Hitsumoto, Y., and Kanbara, H. (1995). Murine complement reduces infectivity of *Plasmodium yoelii* to mosquitoes. *Infect. Immun.* **63**, 3702–3704.

Tsutsui, N., and Kamiyama, T. (1999). Transforming growth factor beta-induced failure of resistance to infection with blood-stage *Plasmodium chabaudi* in mice. *Infect. Immun.* **67**, 2306–2311.

Umemiya, R., Hatta, T., Liao, M., Tanaka, M., Zhou, J., Inoue, N., and Fujisaki, K. (2007). *Haemaphysalis longicornis*: Molecular characterization of a homologue of the macrophage migration inhibitory factor from the partially fed ticks. *Exp. Parasitol.* **115**, 135–142.

Urban, B. C., and Todryk, S. (2006). Malaria pigment paralyzes dendritic cells. *J. Biol.* **5**, 4.

Vizioli, J., Catteruccia, F., della Torre, A., Reckmann, I., and Muller, H. M. (2001). Blood digestion in the malaria mosquito *Anopheles gambiae*: Molecular cloning and biochemical characterization of two inducible chymotrypsins. *Eur. J. Biochem.* **268**, 4027–4035.

Vodovotz, Y. (1997). Control of nitric oxide production by transforming growth factor-beta1: Mechanistic insights and potential relevance to human disease. *Nitric Oxide* **1**, 3–17.

Vodovotz, Y. (2006). Deciphering the complexity of acute inflammation using mathematical models. *Immunol. Res.* **36**, 237–246.

Vodovotz, Y., Bogdan, C., Paik, J., Xie, Q. W., and Nathan, C. (1993). Mechanisms of suppression of macrophage nitric oxide release by transforming growth factor beta. *J. Exp. Med.* **178**, 605–613.

Vodovotz, Y., Geiser, A. G., Chesler, L., Letterio, J. J., Campbell, A., Lucia, M. S., Sporn, M. B., and Roberts, A. B. (1996). Spontaneously increased production of nitric oxide and aberrant expression of the inducible nitric oxide synthase *in vivo* in the transforming growth factor beta 1 null mouse. *J. Exp. Med.* **183**, 2337–2342.

Vodovotz, Y., Clermont, G., Chow, C., and An, G. (2004a). Mathematical models of the acute inflammatory response. *Curr. Opin. Crit. Care* **10**, 383–390.

Vodovotz, Y., Zamora, R., Lieber, M. J., and Luckhart, S. (2004b). Cross-talk between nitric oxide and transforming growth factor-beta1 in malaria. *Curr. Mol. Med.* **4**, 787–797.

Wenisch, C., Parschalk, B., Burgmann, H., Looareesuwan, S., and Graninger, W. (1995). Decreased serum levels of TGF-beta in patients with acute *Plasmodium falciparum* malaria. *J. Clin. Immunol.* **15**, 69–73.

White, N. J., Warrell, D. A., Chanthavanich, P., Looareesuwan, S., Warrell, M. J., Krishna, S., Williamson, D. H., and Turner, R. C. (1983). Severe hypoglycemia and hyperinsulinemia in *falciparum* malaria. *New Engl. J. Med.* **309**, 61–66.

White, N. J., Miller, K. D., Marsh, K., Berry, C. D., Turner, R. C., Williamson, D. H., and Brown, J. (1987). Hypoglycaemia in African children with severe malaria. *Lancet* **1**, 708–711.

Whitten, M. M., Shiao, S. H., and Levashina, E. A. (2006). Mosquito midguts and malaria: Cell biology, compartmentalization and immunology. *Parasite Immunol.* **28**, 121–130.

Willadsen, P. (2004). Anti-tick vaccines. *Parasitology* **129**(Suppl), S367–S387.

Wink, D. A., Feelisch, M., Vodovotz, Y., Fukuto, J., and Grisham, M. B. (1999). The chemical biology of nitric oxide. In *Reactive Oxygen Species in Biological Systems: An Interdisciplinary Approach* (C. A. Colton, and D. L. Gilbert, Eds.), pp. 245–291. Kluwer Academic/Plenum Publishing, New York.

Woolhouse, M. E., Taylor, L. H., and Haydon, D. T. (2001). Population biology of multihost pathogens. *Science* **292**, 1109–1112.

Yu, X., Lin, S. G., Huang, X. R., Bacher, M., Leng, L., Bucala, R., and Lan, H. Y. (2007). Macrophage migration inhibitory factor induces MMP-9 expression in macrophages via the MEK–ERK MAP kinase pathway. *J. Interferon Cytokine Res.* **27**, 103–109.

Yue, J., and Mulder, K. M. (2000). Activation of the mitogen-activated protein kinase pathway by transforming growth factor-beta. *Meth. Mol. Biol.* **142**, 125–131.

Zhu, J., Krishnegowda, G., and Gowda, D. C. (2005). Induction of proinflammatory responses in macrophages by the glycosylphosphatidylinositols of *Plasmodium falciparum*: The requirement of extracellular signal-regulated kinase, p38, c-Jun *N*-terminal kinase and NF-kappaB pathways for the expression of proinflammatory cytokines and nitric oxide. *J. Biol. Chem.* **280**, 8617–8627.

8

MOSQUITO IMMUNITY TO THE MALARIA PARASITE

LINDSEY S. GARVER, LUKE BATON AND
GEORGE DIMOPOULOS

W. Harry Feinstone Department of Molecular Microbiology and Immunology, Johns Hopkins Bloomberg School of Public Health, Baltimore, MD 21205, USA

ABSTRACT: *Anopheles gambiae*, the insect vector of human malaria, has evolved into a unique model system for the study of innate immunity, mainly because of advances in the functional genomics of this organism. The purpose of this chapter is to provide a detailed outline of the cellular and molecular components and mechanisms that are involved in the mosquito's defense against malaria parasites. In general, immune responses are initiated when the pathogen is recognized by pattern recognition receptors (PRRs) (Section 8.3), the signal is propagated by amplification (Section 8.4) and transduction cascades (Section 8.5), and carried out by effector proteins (Section 8.6) and cellular defense mechanisms (Sections 8.7 and 8.8). Section 8.10 outlines transcriptional profiling studies that have been widely used to identify anti-*Plasmodium* factors and show how the mosquito's physiology responds to infection. *Plasmodium* parasites undergo several developmental transitions within the mosquito and immune responses can be specific toward these stages (Section 8.11 and throughout). Finally, mosquito immunity to *Plasmodium* has shown specificity with regard to the species of both the vector and the parasite; these differences are described in Section 8.12. The material presented here will provide a solid understanding of mosquito immunity toward the malaria parasite and address some of the intricacies of this complex vector–parasite interaction.

Abbreviations:

PRR   = pattern recognition receptor
AMP   = antimicrobial peptide
RNAi = RNA interference
QTI   = quantitative trait locus/loci.

8.1   INTRODUCTION

The evolutionary relationship of *A. gambiae* to the model organism *Drosophila melanogaster* builds a solid foundation on which to begin immune system analyses, and its essential role in disease transmission makes its immune system a potential target for public health interventions. This public health relevancy has garnered enthusiasm for research in *Anopheles* immunity, while the availability of the requisite molecular tools and our advanced knowledge of a closely related organism have facilitated such research.

The mosquito's defense against pathogens involves not only structural barriers, such as the exoskeleton and the peritrophic matrix that is secreted to line the midgut after feeding, but also a variety of cellular and humoral mechanisms, which will be discussed in subsequent sections. These cellular and humoral defenses are mounted by a variety of mosquito organs and cell types and are particularly critical to the mosquito's defense against the *Plasmodium* parasites that are the etiological agents of malaria.

Transmission of *Plasmodium* requires the successful completion of a complex life cycle in the *Anopheles* mosquito that involves several developmental and spatial transitions and interactions with multiple mosquito cell types and tissues.

When the female mosquito takes a blood meal from an infected vertebrate, game-tocytes, the sexual blood stage of *Plasmodium*, are taken into the midgut lumen, where they form male microgametes and female macrogametes. The microga-metes are stimulated to exflagellate and fertilize the macrogametes, forming a zygote (within 2 days post-ingestion). The zygote then undergoes a stage shift to become the elongated ookinete, which is motile and penetrates through the peritrophic matrix and epithelial cells into the wall of the midgut. Peak ookinete densities occur at 16–32 h post-ingestion. The ookinete then transitions to an oocyst, which docks between the basal lamina and the basal membrane of the midgut epithelial cells (7–9 days post-ingestion). Asexual reproduction occurs within the oocyst, which then bursts, releasing thousands of sporozoites into the mos-quito's hemolymph. Between days 10 and 14, the sporozoites migrate to the salivary glands and are only then able to be transmitted to a vertebrate when the mosquito bites. The parasite population experiences a bottleneck at each stage transition resulting in over 90% total average loss (reviewed by Vaughan, 2007), some of which can be attributed to the vector's active defense mechanisms.

This complex life cycle is a challenge to the parasite, since it must encounter such a variety of environments within the mosquito before being transmitted to the next human host. However, it is also a challenge to the mosquito's immune system, because it requires immune competency from many cells and tissues, as well as the ability to recognize and respond to *Plasmodium*'s changing life stages.

This chapter will address the *A. gambiae* anti-*Plasmodium* defense system with regard to its molecular constituents, mechanistic and spatial specificities, and relationship to the defense against other pathogens.

8.2 THE MOSQUITO IMMUNE RESPONSE: AN INTEGRATED OVERVIEW

In broad terms, the *Anopheles* immune system has four functional arms that involve both cellular and molecular immune responses: (1) pathogen recognition, (2) signaling amplification cascades, (3) immune signaling pathways, and (4) down-stream effector molecules and mechanisms (Fig. 8.1). PRRs are responsible for detecting the *Plasmodium* parasite and identifying it as foreign, then triggering the signaling pathways that ultimately lead to killing or containing the pathogen. Sig-naling amplification cascades generally lead to the encapsulation of *Plasmodium*, while immune signaling pathways allow transcription factors to translocate to the nucleus, where they initiate the transcription of immune genes. Effector genes gen-erally encode products such as anti-microbial peptides (AMPs) that act directly on the pathogen, while effector mechanisms, such as melanotic encapsulation, are concerted events that kill or contain the parasite. The recognition, signaling, effec-tor function, and regulation of these processes must be coordinated and efficient if they are to limit *Plasmodium* without damaging or exhausting the mosquito. Here we outline each component of mosquito immunity separately, but it is important to note that, like other pathogens, *Plasmodium* is bound by PRRs upon infection

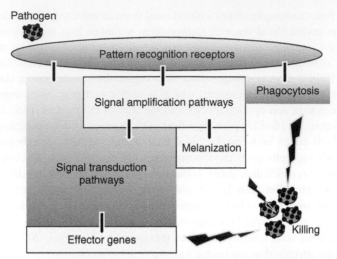

FIGURE 8.1   A simplified model of the insect innate immune system that comprises a pattern recognition repertoire that identifies pathogens, serine protease cascades that amplify the recognition signal and intracellular signal transduction pathways that lead to the transcription of effector genes. Some defense reactions, such as phagocytosis and melanization, can become activated directly upon pathogen recognition and signal amplification.

which triggers activation of immune signaling and signal amplification pathways that lead to transcription of effector molecules and initiation of effector mechanism.

The study of mosquito–*Plasmodium* interactions employs a variety of different experimental techniques, but much of what we know today stems from analysis of differential gene expression (many anti-*Plasmodium* genes display increased transcription after an infectious blood meal) and targeted gene silencing using RNA interference (RNAi; see chapter 12). It is noteworthy that although there is redundancy in the mosquito's defense system, disruption of one or more components of that system can compromise the mosquito's defense. Furthermore, this situation has been exploited by researchers as a means of determining the relative contribution of individual molecules to the overall anti-*Plasmodium* defense. Details concerning *A. gambiae*'s anti-malarial immune molecules, identified by these and other methods, will be discussed in the subsequent sections. The current view of the total network of PRRs, pathways, and effectors is far from complete, but many critical molecules and events have been established, and it is clear that the mosquito actively recognizes the parasite as foreign and tries to eliminate it.

8.3   PATTERN RECOGNITION RECEPTORS

The *A. gambiae* genome harbors approximately 150 germ line-encoded PRR genes. The majority of mosquito PRRs are secreted proteins with adhesive domains that can interact with pathogen-associated molecular patterns (PAMPs), such as lipopolysaccharide and peptidoglycan. Some PRRs comprise a single pattern recognition domain, while others display a more complex gene organization pattern, with multiple domains that may have catalytic or signal transduction roles. A common feature of mosquito PRRs is that they belong to larger gene families, most of which show expansion when compared to *D. melanogaster*. A variety of *A. gambiae* PRRs have been linked with anti-*Plasmodium* defense, and some have also been shown to associate with the parasite.

One of the first-studied PRRs that recognizes the malaria parasite is the thioester-containing protein 1 (TEP1), a complement-like protein previously implicated in bacterial phagocytosis (Levashina et al., 2001). TEP1 is expressed by hemocytes specifically after infection and localizes to the surface of ookinetes. The TEP1 ligand is unknown, but it seems to be mostly ookinete-specific; oocysts are very weakly recognized, and sporozoites are not recognized at all by this PRR. Recognition occurs between 24 and 48 h post-infection, and ookinetes that have been detected by TEP1 display morphological indicators of death. RNAi-mediated knockdown of *TEP1* causes a drastic increase in live oocysts (up to 7-fold), suggesting that TEP1 is responsible for setting off a powerful killing mechanism. Evidence in two different mosquito genetic backgrounds supports either lysis or lysis and melanization as the disposal mechanism (Blandin et al., 2004).

Like TEP1, a leucine-rich repeat-containing protein called LRIM1 is an antagonist to invading *P. berghei* parasites; oocysts increase almost 4-fold when *LRIM1* is silenced with RNAi. Expression of this PRR is induced upon malarial infection and specifically increases in the midgut during ookinete invasion of the epithelium, although it is expressed more highly in the rest of the mosquito at other times (Osta et al., 2004). LRIM1 is suspected to be a second PRR that is responsible for initiating ookinete killing during invasion, independent of melanization.

Other PRRs have been shown to play a role as agonists to parasite infection: Two infection-inducible C-type lectins, CTL4 and CTLMA2, protect the parasite from the melanization response. It is particularly interesting that although LRIM1 does not need either of the C-type lectins to exert its effects, silencing either *CTL4* or *CTLMA2* results in melanization only in the presence of a functional LRIM1 (Osta et al., 2004). Since the melanization triggered in the absence of either C-type lectin does not occur without the LRIM1-dependent killing mechanism in place, lysis and melanization must occur as two separate events, potentially mediated by different sets of PRRs. Also, the protective capacity of these C-type lectins has been shown in *Anopheles* mosquitoes infected with the rodent malaria parasite *P. berghei* but is not seen when *A. gambiae* mosquitoes from the field are infected with isolates of the human malaria parasite *P. falciparum*. The molecular basis for this discrepancy is unknown, but is most likely a parasite-species-specific effect (Cohuet et al., 2006).

A targeted approach employed by Dong et al. (2006b) used RNAi against the gene encoding AgDscam (*A. gambiae* Down syndrome cell adhesion molecule), a molecule with a complex domain organization suggestive of alternative splicing. This molecule has the architecture of a PRR and, like TEP1, promotes phagocytosis of bacteria. AgDscam exons are up-regulated in response to *P. berghei* or *P. falciparum* challenge, with certain exons being favored in the midgut after invasion by either parasite. It is possible that the parasite or infection events are influencing the abundance of specific AgDscam isoforms, creating a distinct repertoire. In particular, different sets of exons are enriched during infection with either the human or rodent *Plasmodium* species, so that the parasites would elicit unique isoform repertoires (Dong et al., 2006b). The adhesive Ig domains making up the extracellular portion of AgDscam, together with the co-localization of AgDscam to ookinetes *in vivo* and *in vitro* (Dong and Dimopoulos, unpublished data), make this molecule an attractive candidate for *Plasmodium* recognition.

Other immunoglobulin domain-encoding genes with potential roles in innate immune defense have also been identified. One of these is a molecule with multiple immunoglobulin domains as well as two kinase domains, named IRID6 (infection responsive with immunoglobulin domain 6). *IRID6* is up-regulated in response to both *P. berghei* and *P. falciparum* infections, and when it is silenced by RNAi, *P. falciparum* oocyst infection is increased by more than 2-fold (Garver et al., 2007). It is possible that a molecule with such architecture provides a link between pathogen recognition and the downstream immune signaling that eventually leads to interference with infection.

There are several other PRRs that affect malarial infection intensity in *A. gambiae*, but their contributions are still under investigation. Three members of the fibrinogen domain immunolectin (FBN) family, FBN8, FBN9, and FBN39, have been shown to limit *P. falciparum* (Dong et al., 2006a). AgMDL1 is also a PRR that is specific for immune defense against *P. falciparum* (Dong et al., 2006). The mammalian homolog of AgMDL1 is MD-2, which is an adaptor protein linking a bound pathogen with the Toll-like receptor 4 to initiate an immune signaling cascade. Despite its known effects on *P. falciparum* infection, this adaptor role has not been verified for the mosquito protein. Finally, the leucine-rich repeat domain-containing proteins APL1/LRRD19 and APL2/LRRD7 have been identified via gene expression analysis as being up-regulated during *Plasmodium* infection and by quantitative trait loci (QTLs) linkage mapping as belonging to a chromosomal region contributing to *P. falciparum* resistance in certain natural mosquito populations (Dong et al., 2006a; Riehle et al., 2006). As exhibited in Toll and LRIM1, leucine-rich repeats are known pattern recognition/pathogen-binding domains.

8.4   SIGNAL AMPLIFICATION CASCADES

Recognition of pathogens by PRRs can be followed by a rapid amplification of the signal to effectively activate defense reactions. This signal amplification is carried out by serine proteases that are in turn controlled by serine protease inhibitors (serpins).

Most *Anopheles* immunity-related serine proteases belong to the clip-domain serine protease (CLIP) family: either the CLIPA type, with a catalytic domain, or the CLIPB type, with a non-catalytic serine protease domain (Christophides et al., 2002). The *A. gambiae* genome predicts 10 CLIPAs and 17 CLIPBs, and gene expression data have shown that several of these factors are transcriptionally up-regulated after *Plasmodium* infection (Dimopoulos et al., 2002; Dong et al., 2006; Volz et al., 2005). The *A. gambiae* immunity-related serine proteases have mainly been studied with regard to their possible involvement in the activation of melanization reactions that kill *Plasmodium* parasites in refractory strains; these are discussed in detail below. SRPN6, a serpin has been shown to inhibit *P. berghei* development through a non-melanizing defense reaction, which largely remains uncharacterized. Specifically, RNAi-mediated depletion of *SRPN6* causes an increase in the number of developing *P. berghei* oocysts in *A. stephensi* (another member of the *Anopheles* genus), and this protein is highly induced in response to invasion of the midgut epithelium. *SRPN6* is also induced as *P. falciparum* traverses the midgut but at a much lower intensity than in the case of *P. berghei*. Although the effect of SRPN6 on *P. falciparum* development has not been determined, species specificity is apparent with regard to the vector; since ookinete lysis is delayed, but not abrogated, in *SRPN6* kd *A. gambiae* mosquitoes (Abraham et al., 2005).

8.5 IMMUNE SIGNALING PATHWAYS

Recognition of the pathogen by certain PRRs (Section 8.3) leads to the activation of intracellular immune signaling pathways that in turn activate the transcription of AMPs and other effector genes (Dimopoulos, 2003; Medzhitov and Janeway, 2002).

The Toll and Imd pathways are the major immune signaling pathways that have been identified and studied in mosquitoes, though there is also evidence for the existence of Janus kinase/signal transducers and activators of transcription (JAK/STAT) pathway in mosquito immune signaling (Barillas-Mury et al., 1999). Putative immune signaling pathways in the malaria vector have been extrapolated from studies in *Drosophila* or *Aedes* and then verified or altered by studying *Anopheles*. This approach has proved useful for establishing the details of the anti-bacterial and anti-fungal immune response in *Anopheles* because of the high degree of similarity among the dipterans. It also is useful for identifying target molecules to assess the contribution of either Toll or Imd to the anti-*Plasmodium* immune response, which is most relevant in *Anopheles* mosquitoes.

In *Aedes*, the Toll pathway functions as an anti-fungal immune response (Shin et al., 2005), while the Imd pathway in *Anopheles* is involved in defense against both Gram-positive and Gram-negative bacteria (Meister et al., 2005). Although the signaling pathways are known to play key roles in anti-parasitic defense, the precise mechanisms and roles of the various pathways in anti-*Plasmodium* defenses are unclear. Studies of *P. berghei* infections have shown that the basal expression of the major anti-parasitic genes in the pre-invasion phases plays crucial roles in the elimination of parasites, a process that is controlled by the NF-κB transcription factors from both pathways (Section 8.11; Frolet et al., 2006). Other studies have shown that the Imd pathway alone is able to limit the parasite infection (Garver and Dimopoulos, unpublished data; Meister et al., 2005).

Although the specifics of how the immune pathways contribute to *Plasmodium* destruction are unclear, some of the effector genes transcribed upon pathway activation are known to be anti-malarial, and manipulation of the pathways has been shown to cause fluctuations in parasite load; therefore, understanding the pathway composition is critical. Thus, a general picture of the mosquito signaling pathways can be drawn, based on our knowledge of *Drosophila*. There is one important difference between *Drosophila* and mosquitoes; however, mosquitoes lack the Dorsal-related immune factor (Dif). In *Drosophila*, Toll and Imd activate two distinct NF-κB transcription factors, Dif/Dorsal and Relish, respectively, while REL1 (the ortholog of Dorsal, also called Gambif1) and REL2 (the ortholog of Relish) play the analogous roles in mosquitoes (Barillas-Mury et al., 1996; Meister et al., 2005).

Activation of the *Drosophila* Toll pathway by Gram-positive bacteria and fungi is mediated by PGRP-SA or GNBP1 (Leclerc and Reichhart, 2004). Proteolytic cleavage of the circulating cytokine-like molecule Spaetzle (Spz) is then induced by a cascade of serine proteases. This cleavage leads to the binding of Spz to the ectodomain of the Toll transmembrane receptor and initiation of the Toll pathway. The intracytoplasmic domain of Toll then interacts with three cytoplasmic proteins, MyD88, Tube, and Pelle, to form a signaling complex below the cell membrane. All three proteins contain a death domain, with MyD88 and Tube acting as adaptor proteins and Tube as a serine–threonine kinase. Activated Tube can then act on the downstream components, Dif/Dorsal–cactus complexes, leading to the degradation of cactus and the subsequent translocation of Dif/Dorsal from the cytoplasm to the nucleus, which activates the expression of the effector genes such as AMPs (Section 8.6; Hoffmann, 2003).

Nine toll genes are encoded in *Drosophila*, and only one (Toll-1) appears functional in the pathway, although Toll-9 has also been found to be linked to immunity (Hoffmann, 2003; Tauszig et al., 2000). There are 10 *toll* genes in *A. gambiae*, but none is the definitive ortholog of *Drosophila* Toll-1. Six *spz* genes have be found in *Drosophila*; all of them have 1:1 orthologs in *A. gambiae* (Christophides et al., 2002). This high level of homology might indicate conserved signaling events occurring via Spz. Recent studies have suggested that Toll-5A and Spz1C play a major role in the fatbody-specific activation of the Toll pathway in *A. aegypti* (Shin et al., 2006). Alternative transcripts of *Rel-1* (the Toll pathway transcription factor) encode two isoforms, Rel-1A and Rel-1B, in *A. aegypti*. *In vitro* studies have suggested that Rel-1B acts cooperatively with Rel-1A to enhance effector gene transcription (Shin et al., 2005).

The *Drosophila* Imd pathway is mainly involved in defense against Gram-negative bacteria. Upstream PRRs for Imd include PGRP-LC and PGRP-LE, which can function cooperatively (Leclerc and Reichhart, 2004); however, the transmembrane receptor for this pathway is unclear. Imd is an adaptor protein with a death domain that associates with FADD (Fas-associated protein with death domain). Downstream of Imd/FADD is TAK1 (transforming growth factor-β activated kinase 1), which is able to activate the IKK (inhibitors of κB kinase) complex, composed of IKK-β and IKK-γ homologs. The IKK complex then directs the site-specific cleavage and activation of the NF-κB transcription factor Relish and induces the expression of AMPs (Hoffmann, 2003; Tanji and Ip, 2005). In addition to TAK1, a caspase homolog called Dredd is also able to work with FADD to cleavage Relish and initiate AMP transcription. The Imd pathway in *Drosophila* can diverge downstream of TAK1 and activate the expression of cytoskeletal proteins through a JNK cascade (c-Jun *N*-terminal kinase) (Silverman et al., 2003). JNK regulates other immune-inducible genes but not the AMPs genes. Negative cross-talk between the JNK and Relish-dependent mechanisms occurs (Park et al., 2004), and cross-talk between the Toll and Imd pathways has been suggested but not definitively established.

As is true for the Toll pathway, all components of the Imd pathway are conserved between *Drosophila* and *A. gambiae* (Christophides et al., 2002), although evidence for a JNK branch in mosquitoes has not yet been found. The activation of the Imd pathway in mosquitoes leads to the translocation of Rel-2 into the nucleus and induces the expression of effector genes. In *A. gambiae*, two *REL2* gene products are generated by alternative splicing: a full-length form (*REL-2F*) and a short form (*REL-2S*) (Meister et al., 2005). REL-2F is implicated in the defense against Gram-positive bacteria, while REL-2S is involved in the defense against the Gram-negative bacteria (Meister et al., 2005). REL-2F has also been found to regulate the intensity of mosquito infection by *P. berghei*.

8.6   EFFECTOR GENES

The Toll and Imd pathways control the expression of both distinct and overlapping subsets of AMPs. In *Drosophila*, the Toll pathway specifically controls the transcription of the gene encoding anti-fungal peptide Drosomycin and immune peptide 2, whereas the Imd pathway activates the transcription of the genes encoding anti-bacterial AMPs diptericin, cecropin (Cec), and attacin (Att). *Defensin* (*Def*) and *metchnikowin* transcription is controlled by both pathways (Tanji and Ip, 2005). Activation of either the Toll or the Imd pathway alone is sufficient to induce the expression of *Cec1*, *Def1*, and *gambicin* in mosquito cell lines (Luna et al., 2006), and *Def1*, *Cec2*, and *gambicin* have been shown to be up-regulated in adult female mosquitoes in response to *Plasmodium* infection. Over-expression of *Cec A* in transgenic *A. gambiae* causes a 60% reduction in *P. berghei* oocysts (Kim et al., 2004) and synthetic Cec-like peptides can inhibit *P. berghei* in *A. albimanus* (Rodriguez et al., 1995). Defensin transcript levels correlate with gametocyte levels of *P. falciparum* and the increased expression peaks at 24 h post-blood meal (Tahar et al., 2002). Despite an increase in transcription during parasite infection (Dimopoulos et al., 1997; Richman et al., 1997; Tahar et al., 2002), loss of the defensin peptide does not affect mosquito viability or oocyst levels during *P. berghei* infection (Blandin et al., 2002), meaning this AMP alone is not an essential part of the anti-*P. berghei* response but may play a lesser role. Gambicin is a novel mosquito AMP that is expressed by the fat body and midgut and is induced by *Plasmodium* infection (Vizioli et al., 2001). *P. berghei* ookinetes are susceptible to gambicin as knock down of the gene causes an increase in viable oocysts (Dong et al., 2006; Vizioli et al., 2002). In addition, *TEP1* and *LRIM1* are markedly reduced in uninfected mosquitoes in which *Rel1* and *Rel2* have been silenced simultaneously, but not in single *Rel1* or *Rel2* knockdowns (Frolet et al., 2006). These results suggest that the two pathways must interact with each other to eliminate an infection (such as with a malarial parasite). Finally, two members of a family of short peptides known as IRSPs are inducible upon *Plasmodium* infection, and RNAi assays have shown that they are antagonistic to malaria. Their size and sequence suggest that these peptides represent a class of AMP that may be specific for malaria, but this possibility is at present only speculative.

8.7　NITRIC OXIDE

Nitric oxide (NO) is a free radical that is produced as an oxidative by-product of NO synthase (NOS) activity. There is a single NOS gene in *Anopheles* that is weakly induced by blood-feeding and strongly induced during *P. berghei* infection at the time of ookinete invasion and oocyst development. This induction then wanes during oocyst development but is elevated again during sporozoite release and migration. *P. falciparum*-responsive NO production is triggered by surface glycosylphosphatidylinositols (GPI) and propagated via MAP kinase signaling (Lim et al., 2005). The induction during non-infectious blood-feeding is thought to be a reaction to gut bacteria that proliferate during feeding, and could therefore prime the mosquito for an NO-based defense against *Plasmodium* in an infected blood bolus. *P. berghei*-induced NOS expression is localized to parasite-rich regions of the midgut and correlates temporally with increased enzymatic activity (Luckhart et al., 1998). Other studies linking *P. berghei* infection to NOS expression, peroxidase activity, protein nitration, and multiple toxic free radicals indicate that oxygen metabolism plays a key role during parasite-induced cell damage or provides a way to limit infection (Han et al., 2000; Kumar et al., 2004; Peterson et al., 2007). To survive, the ookinete must migrate through the midgut epithelium fast enough to avoid these chemical killing mechanisms. In *A. stephensi* infected with *P. berghei*, this competition between parasite and host is described as the 'time bomb model' where the toxic products resulting from nitration are the 'bomb' and peroxidases are the 'detonators' set off by invading ookinetes. Host midgut cells are damaged in the process, apoptosis ensues, and the damaged cells are extruded from the epithelial layer. The integrity of the midgut epithelium is retained via an actin-mediated reconstitution. Those ookinetes that are fast enough avoid the 'bomb' and develop successfully while others are either killed by the toxins or are extruded with the apoptotic cell (Han et al., 2000, 2002; Kumar and Barillas-Mury, 2005). Peroxiredoxins protect the mosquito host cells from NO metabolites that are non-specifically toxic (Peterson and Luckhart, 2006).

8.8   MELANOTIC ENCAPSULATION

Although the pigment protein melanin has several biological functions in mosquitoes, its immune function is to cross-link proteins around invading parasites, both protozoans such as malaria and metazoans such as filariae, creating an inert capsule. Melanization is the downstream effect of a proteolytic cascade leading to the conversion of inactive prophenoloxidase (PPO) to the active phenoloxidase (PO). The conversion process is activated by PPO-activating enzymes (PPAEs), represented by CLIPs. This mechanism was recognized as relevant to *Anopheles–Plasmodium* interactions when melanization was shown to be a crucial contributor to the refractoriness of the L3-5 strain of *A. gambiae* toward invading ookinetes (Volz et al., 2005). These mosquitoes exhibit hyper-melanizing activity toward invading ookinetes of multiple *Plasmodium* species, and this activity is genetically heritable (Collins et al., 1986; Paskewitz et al., 1988). The genes contributing to this refractoriness exist in three QTLs, *Pen1*, *Pen2*, and *Pen3*, although differences exist in contributions of these loci with respect to both *Plasmodium* species and the mosquito subfamily (Zheng et al., 1997, 2003). The exact pathway leading to this phenotype is not fully understood; nevertheless, these mosquitoes are often used in genetic analyses of proteins involved in the signal amplification pathways leading to melanization.

8.8.1   THE PPO-ACTIVATING ENZYME SYSTEM

PPO conversion is spurred by infection-inducible enzymatic activity and not by increased PPO transcription, which is unaffected by malaria, bacteria, or ingestion of blood (Cui et al., 2000). CLIPAs, the clip domain serine proteases that have catalytic domains, can in some insects directly convert PPO to PO. In some systems, including *Anopheles*, CLIPBs, which are non-catalytic co-factors, are also required for PPO conversion. Once converted, PO subsequently initiates the synthesis of melanin, which is used by the host to envelope a pathogen.

While the refractory *A. gambiae* L3-5 strain provides one genetic background for use in teasing apart the melanization process, the susceptible G3 strain mosquitoes can generate a similar melanizing phenotype by silencing the C-type lectin gene encoding CTL4 (see Section 8.3; Osta et al., 2004). Interestingly, manipulation of the clip domain serine protease *CLIPB14* and *CLIPB15* genes has different effects on these two hyper-melanizing genetic backgrounds. In L3-5 mosquitoes, a lack of either gene causes the number of ookinetes to increase, yet these ookinetes are still melanized efficiently. In contrast, in the G3 background with *CTL4* silenced, the absence of both CLIPs causes an increase in viable ookinetes that can mature to oocysts (Volz et al., 2005). These differences suggest that viability and melanization can be uncoupled in the L3-5 strain, with B14 and B15 regulating viability but not melanization. In the CTL4-depleted G3 background, either B14 or B15 is linked to both viability and melanization, or these processes are somehow coupled.

A full screening of all CLIPs has yielded even more observations that point to a complex serine protease network that is strain-specific. As is true for CLIPB14 and CLIPB15, there are major mosquito strain specificities regarding which CLIPs affect distinct melanization and killing events. CLIPA8 is involved in melanization but not killing in the L3-5 strain, yet it affects melanization as well as live oocyst numbers in the G3 *CTL4*-silenced background. CLIPA2 and CLIPA5 are thought to synergistically inhibit melanization, and the contributions of these CLIPs do not differ between the two mosquito strains tested. Parasite morphology and triple knockdown of *A8*, *A2*, and *A5* suggest that the massive and unusual melanization promoted by *A2* and *A5* silencing is the ookinete killing mechanism in these silenced mosquitoes; thus, a separate lysis event cannot be identified. CLIPA7 has a moderate effect on melanization in L3-5 mosquitoes and no effect in G3 mosquitoes, while four other CLIPA serine proteases have no observable effect on either strain. Four CLIPBs (B3, B4, B8, and B17) enhance parasite melanization, although to different degrees, in the G3 *CTL4* knockdown background, and they are not implicated in lysis. Strain comparisons reveal that while B8 has no effect in L3-5 mosquitoes, B4 and B17 seem to promote melanization synergistically (double knockdown results in a 52-fold reduction in melanized ookinetes). B3 is the only CLIPB that promotes melanization in the refractory strain but not in G3 after *CTL4* knockdown. Seven CLIPBs showed no observable effect on the melanizing process (Volz et al., 2006).

These data complicate early conclusions about the dynamics of lysis and melanization and how these mechanisms come together to eliminate ookinetes. Those CLIPs affecting both lysis and melanization could be part of a stage in which one is dependent on the other, or the CLIPs could be performing dual roles. The spatial and temporal relatedness of the two events would support either scenario. Clearly, genetic background is a crucial factor in determining which molecules are participating in the PPO cascade; however, all genes and proteins tested thus far are functional in both models. The hyper-melanizing phenotype of L3-5 would suggest there are unknown strain-specific regulators or mutations controlling melanization components, but how this situation translates to native populations or to *P. falciparum* infection is as yet unknown.

Because melanization can be toxic to the mosquito, it requires stringent regulation. Some serpins have been identified as possible regulators of this process. Michel et al. (2005) have shown that depletion of SRPN2 in *A. gambiae* increases Sephadex bead melanization, and *P. berghei* oocyst development is impaired. While these results are suggestive of a role for SRPN2 as a negative regulator of the PPO cascade, knockdown of *SRPN2* during *P. falciparum* infection gives no discernible phenotype, even though SRPN2 can bind and inhibit an insect

PPO-activating protease *in vitro* (Michel et al., 2006). This clearly indicates a selectivity in immune response with respect to the parasite species encountered and also raises questions about the significance of natural vector–parasite interactions and how these co-evolved interactions dictate the immune response. SRPN2 is indeed active in *A. gambiae* and contributes to an immune response but not the response directed toward *P. falciparum*. In contrast to SRPN2, SRPN1 and -3 have no effect on *P. berghei* oocyst development (Michel et al., 2005).

In both *A. gambiae* refractory models, knockdown of *SRPN6* (discussed above) results in an increase in melanization. These data not only point to a major difference in immune pathway utilization within the *Anopheles* genus, but they also suggest that SRPN6 orthologs could be providing a lysis mechanism in one species and a clearance mechanism in another. This possibility is not unlikely, given that the CLIP data also show relatedness between the two events.

8.8.2   THE PO ENZYMATIC SYSTEM

There are multiple PPOs that are conserved among arthropods; mosquitoes have a number of different PPOs, which can be expressed throughout development or exhibit stage specificity; six are expressed by a mosquito cell line that is hemocyte-like (Cui et al., 2000; Lee et al., 1998; Muller et al., 1999). The melanin cross-linking that forms the acellular capsule results from a biochemical pathway that begins with the conversion of PPO to PO (discussed above) and ends with the conversion of tyrosine to melanin and the subsequent encapsulation. PPO is primarily produced by the insect's hemocytes (Cui et al., 2000); thus, encapsulation is effective against pathogens contacting the hemolymph. Ookinetes are exposed to hemocyte-derived soluble factors when they are lodged in the midgut basal lamina after penetration. Indeed, the capsule is first noticeable around the apical end of the parasite, which is facing the hemolymph. Although the spatial and kinetic details of *Plasmodium* encapsulation are at present incomplete, it is known that a filamentous, organelle-free zone comprised of actin polymers synthesized by healthy midgut cells incompletely covers the extracellular ookinete (Gupta et al., 2005; Han et al., 2000, 2002; Paskewitz et al., 1988; Shiao et al., 2006; Vernick et al., 1995; Vlachou et al., 2004), and this zone is almost invariably linked to melanin deposition in L3-5 mosquitoes. TEP1, Frizzled2, and cdc42 are required for zone formation and complete melanization (Shiao et al., 2006).

In addition to pathogens, beads comprised of polysaccharides linked to various ionic functional groups can induce melanization when injected into the hemocoele and have been used as recoverable, controlled foreign targets (Gorman et al., 1997; Li and Paskewitz, 2006; Paskewitz and Riehle, 1994; Warr et al., 2006). Such studies have revealed several key components of melanization: First, that melanization of beads has the same genetic basis as melanization of *P. berghei*, implying that melanization is a non-specific arm of immunity (Gorman et al., 1997). Second, only the beads' glucan-based hydroxyl groups are efficient in stimulating melanization. Replacement of these groups with other functional groups reduces the mosquito's encapsulation capacity (Paskewitz et al., 1998), and a mosquito beta 1,3-glucan binding receptor (GRP) can mediate encapsulation directed toward bacteria and microfilariae, although its involvement in malaria infection has yet to be established (Wang et al., 2004). Also, silencing of either *TEP1* or *LRIM1* (Section 8.3), but not *CTL4*, *CTLMA2*, or other PRRs, causes aberrant bead melanization phenotypes (Warr et al., 2006). Both have previously been shown to affect melanization of *Plasmodium* (Blandin et al., 2004; Osta et al., 2004). In contrast, at least one *A. gambiae* strain produces a molecule known as melanization inhibitory factor (MPF), which is deposited on injected beads and prevents encapsulation (Paskewitz and Riehle, 1998). This factor was recently identified as lysozyme Lys c-1, and its inhibitory activity on the oxidative pathway leading to melanin synthesis was confirmed *in vitro* (Li and Paskewitz, 2006).

8.9   SPOROZOITE-DIRECTED DEFENSES

The majority of the work concerning mosquitoes' anti-*Plasmodium* immune defenses has considered only the parasite stages between blood ingestion and the appearance of oocysts on the gut wall. It is true that many defenses are employed during this time period, and the relative increase or decrease in oocyst infection is an easy, quantitative indicator for identifying parasite agonists and antagonists. However, each live oocyst releases thousands of sporozoites, not all of which succeed in reaching the mosquito's salivary gland. This situation could be a matter of simple kinetics: Only some sporozoites are able to find, enter, and survive in the gland. It could also be the result of active mechanisms initiated by the mosquito to kill parasites at this stage.

Once released by the oocyst, sporozoites move freely in the hemolymph throughout the open circulatory system. Only 10–20% of the sporozoites released from ruptured oocysts actually reach the salivary gland (Korochkina et al., 2006; Rosenberg and Rungsiwongse, 1991); the others die or are killed in the hemolymph. Once in the salivary gland, the sporozoites are rapidly degraded within the first 24 h and continue to degrade – only 1% of the total sporozoite population is detectable in the salivary gland after 7 days (Hillyer et al., 2007). Therefore, two stage-specific anti-sporozoite mechanisms exist – one that is active in the hemocoele between oocyst rupture and gland invasion, and one inside the gland. Phagocytosis does occur in the hemolymph, but it is responsible for only 2% of all sporozoite destruction in the circulation, and only a small number of phagocytes participate. The anti-sporozoite mechanism responsible for the remaining deaths is unknown but is considered to be active, on the basis of data obtained using methods that distinguish between parasites that have been destroyed and those that are simply dead (Hillyer et al., 2007).

The decline in sporozoites that have successfully invaded the gland epithelium points to an immune response within the gland. Salivary gland-specific expression of immune genes following sporozoite invasion indicates that the gland is indeed an immune-competent organ, expressing six identified immune markers that include Defensin, GNBP, and NOS (Dimopoulos et al., 1998). Three of these genes are induced in the salivary gland only after 11 days post-infection, corresponding to sporozoite invasion of the gland's epithelium. Moreover, serial analyses of gene expression (SAGE) studies of the salivary gland transcriptome during malaria parasite infection have identified 37 immune genes expressed within the salivary glands, including up-regulation of *Cec2*, *Def1*, *GNBP*, and *SRPN6* during the period of sporozoite invasion. These genes also exhibit different spatial and temporal patterns of expression in different regions of the salivary glands. Genes associated with oxidative stress, apoptosis, or cytoskeletal reorganization are not differentially expressed, observations that are consistent with previous reports that sporozoite invasion of the salivary glands is not accompanied by the host cell pathology occurring during ookinete invasion of the midgut epithelium (Rosinski-Chupin et al., 2007).

8.10   TRANSCRIPTIONAL IMMUNE RESPONSES TO *PLASMODIUM*

Gene transcription is intimately related to function, and it can therefore be used to study the functional responses of an organism to different stimuli such as infection. Comprehensive genomic analyses of adult female mosquitoes and mosquito midguts during *Plasmodium* infection have identified an expansive panel of genes that are regulated during infection events. These analyses provide two key elements for future work in *Anopheles–Plasmodium* biology: (1) It offers a global view of how the insect is responding to the invading parasite, including which functional gene classes are undergoing the most changes and what kinds of genes appear to be specifically transcribed during infection. This view of infection responsiveness is critical for biologists' understanding of how infection disrupts the biology of the mosquito and how the mosquito can be used to disrupt the biology of the parasite. (2) It provides a qualitative list of the molecules whose basal levels are influenced by infection. Such lists have proven useful in multiple studies as a means of identifying candidate genes for further functional characterization using RNAi. Here, the value is in targeting specific genes and elucidating the molecular mechanisms that contribute to the global outcome.

Evaluation of parasite-induced transcriptional changes in the mosquito has been critical for identifying novel contributors to anti-parasitic responses. Microarrays have not only been used to provide global overviews, which give insight into the mechanisms occurring during infection, but they can also be used as screens to identify and then target crucial molecules contributing to these mechanisms. Array data point to innate immunity as one of the major functional gene classes that are regulated during *P. berghei* or *P. falciparum* invasion; this class includes genes that have previously been implicated in anti-parasitic responses and insect immunity (Dimopoulos et al., 2002; Dong et al., 2006a; Vlachou et al., 2005).

The period of ookinete invasion and subsequent oocyst formation has been the focus of several studies assessing the transcription responses of the mosquito to the parasite (Abraham et al., 2004; Dana et al., 2006; Dong et al., 2006a; Srinivasan et al., 2004; Vlachou et al., 2005; Xu et al., 2005). A comprehensive analysis outlined by Dong et al. (2006a) incorporates mosquito midgut-specific expression and carcass expression in response to *P. berghei*, *P. falciparum*, and strains of either species that are incapable of invading the midgut epithelium. With this approach, expression due to infected blood and parasite invasion can be separated. Of the entire transcriptome, 3.4% responds to *P. falciparum* and 8.1% to *P. berghei*, with limited overlap between the two profiles. Importantly, while *P. berghei* causes a more massive global response, the immune response is more potent toward *P. falciparum*. Also interesting is the large number of genes that are differentially expressed in the absence of midgut invasion, supporting the idea that parasites can be detected before the ookinetes traverse the epithelial layer. Forty-seven immune genes are induced in the midgut during *P. falciparum* invasion, and RNAi indicates that 11 products of these genes are potential *Plasmodium* antagonists, including TEP1, AgMDL1, 3 FBN family members, a CLIP domain serine protease, APOD, IRSP1, IRSP5, LRRD7, and the AMP gambicin (Dong et al., 2006a).

This genomic analysis points to several interesting genes to consider in terms of anti-parasitic activity, but the overall picture raises two overarching issues: First, *P. berghei* is not a perfect model for the mosquito immune response to *P. falciparum*, and several studies have provided gene-specific data that underscore this concern. Second, detection events are occurring prior to the ookinete's crossing of the midgut, yet few studies have taken these earlier time points into account. Studies have shown that members of other classes of genes that are related to immunity are differentially expressed during invasion. These functional classes include genes associated with oxidative stress, apoptosis, and/or cytoskeletal reorganization (Abraham et al., 2004; Vlachou et al., 2005; Xu et al., 2005). The responding gene classes indicate that massive cytoskeletal restructuring occurs as the parasite traverses the epithelium and that a complex immune regulatory mechanism is in place, balancing parasite agonists and antagonists.

Warr et al. (2007) characterized the spatial variation in gene expression in four different compartments of the midgut and documented a high degree of spatial specificity of anti-*Plasmodium* factors in the anterior section of the midgut, similar to that shown for all the mosquito's known AMPs, Cec2, Cec3, Def1, and gambicin. Several studies have also begun to characterize gene and protein expression in the hemolymph and hemocytes (Bartholomay et al., 2004; Baton et al., 2007, unpublished) (Li et al., 2006; Paskewitz and Shi, 2005), which are believed to be a major source of immune effector molecules mediating anti-*Plasmodium* responses (Abraham et al., 2005; Blandin et al., 2004; Castillo et al., 2006; Frolet et al., 2006; Volz et al., 2005).

8.11   BASAL IMMUNITY AND
ANTI-*PLASMODIUM* DEFENSE

Recently, Frolet and coworkers (2006) introduced the concept of 'basal immunity' into studies of malaria–mosquito interactions. The mosquito's immune response to infection with the malaria parasite can be separated into two conceptually distinct phases, occurring before and after ookinete invasion of the midgut epithelium. The 'pre-invasion' phase is the constitutive, basal level of immune gene activation in the absence of infection, while the 'post-invasion' phase represents the level of inducible immune gene activation resulting after ookinete infection of the midgut epithelium. The pre/and post-invasion distinction is based on the ability to either increase or decrease, respectively, the levels of *P. berghei* oocyst infection by activating or suppressing the Toll and Imd pathways via RNAi. Suppression or activation of the respective pathways results in either up- or down-regulation of various immune genes, including *Tep1* and *LRIM1*. Constitutive (i.e. pre-invasion) levels of immune gene activation determine the susceptibility of the mosquito to ookinete infection, while the inducible (i.e. post-invasion) immune gene activation performs a replenishment function, replacing the immune factors utilized during ookinete invasion.

The replenishment hypothesis highlights the possibility that the post-invasion inducible component of immune gene expression is not necessarily and may not directly be related to anti-*Plasmodium* effector functions. Transcriptional profiling studies of mosquito immune gene expression have thus far focused on those genes exhibiting differential expression after malaria parasite infection. Although this approach is sensible and has been successful in identifying a number of immune genes that may be involved in the regulation of malarial infection, induction of immune gene expression may not necessarily be a reliable indicator of anti-*Plasmodium* function. It may be that the structural changes related to the parasite's crossing of the midgut epithelium cause fluctuations in transcript levels unrelated to anti-*Plasmodium* effector function. It is also possible that changes in transcript abundance reflect quantitative variation in the levels of parasite infection, rather than a qualitatively different response to the parasites themselves. Furthermore, many relevant and important immune genes may not exhibit parasite-induced changes in transcript abundance. This may be especially true for immune genes encoding PRRs possessing a pathogen-sensing function, for which constitutive levels of expression may be a key determinant of the ability of the mosquito to mount an effective anti-*Plasmodium* immune response prior to the establishment of oocyst infection.

Pathogen-induced transcriptional expression of PRR genes appears unlikely to be of use in sensing parasite infections that are already established. In non-malaria-infected adult female mosquitoes, the level of transcriptional expression of genes encoding PRRs does exhibit significant variation among different members and strains of the *A. gambiae* species complex, which vary in their susceptibility to *P. falciparum* infection (Baton et al., 2006a). Although the functional significance of this variation is yet to be determined, it is consistent with the hypothesis that constitutive levels of PRR gene expression are correlated with susceptibility to malarial infection. The extent to which constitutive, as opposed to inducible, expression levels of genes encoding signaling and effector components of the mosquito immune system have a similar importance in determining susceptibility to malaria infection is less certain, although transcripts of a smaller number of these classes of immune genes also exhibit significant variation among non-malaria-infected adult female mosquitoes (Baton et al., 2006a).

Although the functional differences that may exist between the so-called 'pre-invasion' and 'post-invasion' phases of immune gene expression are important, it is unclear exactly how this distinction relates to the concepts of 'basal immunity,' or to 'constitutive' rather than 'inducible' immune gene expression. A number of studies have shown that a variety of vertebrate host and parasite factors, as well as the mosquito's bacterial flora, may influence the expression of components of the mosquito's immune system prior to ookinete invasion of the midgut epithelium. Changes in mosquito immune gene expression do occur following ingestion of uninfected blood (Marinotti et al., 2005, 2006); e.g. vertebrate host cytokines, antibodies, and white blood cells ingested as part of the mosquito bloodmeal can have an effect on the expression of these genes (Luckhart et al., 2003; Vodovotz et al., 2004). Several transcriptional profiling studies have also shown that non-infectious parasites or parasites in asexual blood stages can be present within the mosquito midgut lumen and induce changes in mosquito immune gene expression in the mosquito (Bonnet et al., 2001; Dong et al., 2006a; Tahar et al., 2002; Vlachou et al., 2005).

How parasites induce immunity either prior to or in the absence of epithelial invasion is currently uncertain. However, proinflammatory malaria parasite moieties such as GPI are present in all malaria parasite stages and have been shown to modify immune gene expression through the mosquito's insulin signaling pathway (Lim et al., 2005).

Non-malarial microorganisms, in particular bacteria within the gut lumen, may also alter the mosquito's immune gene expression. A number of previous studies provide evidence that the presence of bacteria within the midgut lumen negatively affects the level of oocyst infection (Beier et al., 1994; Gonzalez-Ceron et al., 2003; Micks and Ferguson, 1961; Pumpuni et al., 1993, 1996), although the mechanism of such suppression is currently unknown (Azambuja et al., 2005). The effect of bacteria on malaria parasite development may occur either directly or indirectly through bacteria-induced changes in the mosquito host. There is significant overlap between the anti-bacterial and anti-*Plasmodium* immune responses (Christophides et al., 2002; Dimopoulos et al., 2002; Dong et al., 2006a; Lambrechts et al., 2004; Lowenberger et al., 1999; Volz et al., 2005); thus, the presence of bacteria may activate the mosquito's anti-microbial immune responses, which then act against the parasite (Lowenberger et al., 1999).

The scenario just described would certainly complicate the concept of a basal level of immune gene expression. Rather than a constitutive level of immune gene transcription, the response may be actively stimulated by a variety of different sources. Such a response could be induced at any time (e.g. a response to bacteria) or after a bloodmeal but independent of the presence of a parasite (e.g. a response to vertebrate blood components). Such an immune response could act against the parasite but would also confound the interpretation of the differential gene expression that has been thought to be caused by malarial infection. Despite these lingering questions, however, the distinction between constitutive and inducible immune gene expression is both useful and important, as is the distinction between immunity elicited by the parasite itself and immunity elicited by bystander molecules and organisms.

8.12   GENETIC VARIATION IN MOSQUITO SUSCEPTIBILITY TO MALARIA

The ability to easily and rapidly select mosquito strains in the laboratory that differ in their level of oocyst infection attests to the common and well-documented existence of genetic variation in the susceptibility of mosquitoes to malarial infection (Al Mashhadani et al., 1980; Collins et al., 1986; Feldmann and Ponnudurai, 1989; Frizzi et al., 1975; Graves and Curtis, 1982; Huff, 1929, 1931; Hurd et al., 2005; Kilama and Craig, 1969; Micks, 1949; Somboon et al., 1999; Thathy et al., 1994; Trager, 1942; van der Kaay and Boorsma, 1977; Vernick et al., 1995; Ward, 1963; Zheng et al., 1997). Some of these studies have found evidence that mosquito susceptibility to malarial infection is controlled by one or only a few genes, each of large effect; other studies have suggested that the genetic basis for the mosquito's susceptibility is polygenic, involving complex interactions between multiple genes, each of relatively small effect. For example, Zhong et al. (2006) identified six QTLs determining susceptibility of different selected laboratory strains of *A. aegypti* to the avian malaria parasite *P. gallinaceum*. Each of the QTLs explained approximately 10–30% of the observed variation in oocyst infection, and there were epistatic interactions between two pairs of these loci.

Several studies have also addressed the complex interactions between the malaria parasite genotype and the mosquito genotype (Collins et al., 1986; Gonzalez-Ceron et al., 2001, 2007; Lambrechts et al., 2005; Zheng et al., 2003). Laboratory mosquito strains selected for susceptibility to one malaria parasite species are not necessarily resistant to infection with other species/strains, and different interacting loci regulate the intensity of infection of different parasite strains/species (Zheng et al., 1997, 2003). For example, different chromosomal regions determine *A. gambiae*'s susceptibility to different strains of the non-human primate malaria *P. cynomolgi* (Zheng et al., 2003). Also, *A. gambiae* strains selected for resistance to *P. cynomolgi* have varying levels of resistance to *P. falciparum* (Collins et al., 1986). Different *Anopheles* species also exhibit varying degrees of susceptibility to different strains of *P. vivax*, another human *Plasmodium* parasite (Gonzalez-Ceron et al., 2001, 2007).

Several recent field studies have also investigated the inheritance of refractoriness in natural mosquito populations (Menge et al., 2006; Niare et al., 2002; Riehle et al., 2006). Menge et al. (2006) identified two QTLs in natural populations of *A. gambiae* that determine the intensity of *P. falciparum* oocyst infection, although each QTL accounted for only approximately 10% of the observed variation in oocyst loads, leaving 80% of the variation in parasite infection unexplained. In contrast, Riehle et al. (2006) identified up to seven distinct *A. gambiae* QTLs, four of which were clustered in the same region of chromosome 2L and accounted for up to 89% of uninfected mosquitoes challenged with *P. falciparum*. This region contains the leucine-rich repeat-containing gene APL1 (Section 8.3) and is considered a potential determinant of mosquito susceptibility to malaria infection in the field (Riehle et al., 2006). It is particularly intriguing that the genes that have been implicated thus far in mosquito susceptibility to malaria infection through genetic mapping studies have all been putative immune genes, suggesting that the efficiency of the mosquito anti-*Plasmodium* defense may be the major determinant of the mosquito's susceptibility to malarial infection.

Other mosquito immune genes, *LRIM1*, *CTL4*, *CTLMA2*, and *SRPN2*, whose relevance has been determined under laboratory conditions, have no effect on oocyst infection in field mosquitoes infected with natural populations of co-indigenous *P. falciparum* (Cohuet et al., 2006; Michel et al., 2006). The absence of an effect of these immune genes on human malarial parasites is not surprising, given that melanization in field populations of *P. falciparum*-infected *A. gambiae* is rare (Schwartz and Koella, 2002) and that laboratory lines of *A. gambiae* selected for refractoriness to non-human malaria parasites are not refractory to human malaria (Collins et al., 1986; Graves and Curtis, 1982; Somboon et al., 1999).

These studies raise the important issue of the extent to which laboratory models are an appropriate substitute for the natural malaria parasite–mosquito vector combinations found in the field (Aguilar et al., 2005). Undoubtedly, model systems provide a powerful and convenient means for characterizing components of parasite–vector interactions, and laboratory studies have facilitated the characterization of mosquito immune responses to malarial infection and identification of polymorphic variants (e.g. TEP1) determining susceptibility to infection (Abraham et al., 2005; Blandin et al., 2004; Dong et al., 2006a, b; Frolet et al., 2006; Luckhart et al., 1998; Meister et al., 2005; Michel et al., 2005; Osta et al., 2004; Riehle et al., 2006; Shiao et al., 2006; Vlachou et al., 2005, 2006; Volz et al., 2005).

The applicability of data derived from model systems to natural human malaria parasites and their natural *Anopheles* vector populations has yet to be determined (Baton et al., 2006b; Cohuet et al., 2006). However, there is already evidence from laboratory studies for variability in the molecular responses of mosquitoes to different malarial species (Dong et al., 2006a, b; Gupta et al., 2005; Tahar et al., 2002). For example, global transcriptomic analyses of *A. gambiae* have identified differences in mosquito gene expression following midgut infection with ookinetes of different malarial species: *P. berghei* triggers differential regulation of more than twice as many genes as *P. falciparum*, while the latter species induces a greater number of putative immune genes (Dong et al., 2006a). However, functional comparison of 11 *A. gambiae* putative immune genes using RNAi suggested that the mosquitoes' defense against *P. falciparum* and *P. berghei* is largely conserved (Dong et al., 2006a). Seven of the eleven genes investigated – *TEP1, APOD, FBN9, FBN8, IRSP1, SPCLIP1*, and *LRRD7* (also known as *APL2*; Riehle et al., 2006) – were found to influence mosquito resistance to both malaria parasite species (Dong et al., 2006a). In contrast, an MD2-like receptor, AgMDL1, and FBN39 specifically regulate resistance to *P. falciparum*, while gambicin and a novel putative short secreted peptide, IRSP5, are specific for defense against *P. berghei* (Dong et al., 2006a). These differences are probably the result of a variety of factors, including differences in the evolutionary history of the malarial species concerned, *A. gambiae* being a natural vector for the human parasite *P. falciparum* but not for the rodent parasite *P. berghei*. The larger number of putative immune genes induced upon *P. falciparum* infection may indicate that the mosquito's immune surveillance system is more capable of sensing this malaria species, or *P. berghei* suppresses the mosquito's immune response. These studies highlight the importance of parasite variability in determining the outcome of malarial infection in the mosquito vector, and they also suggest that laboratory models using appropriate parasite–vector combinations may be able to provide data that can be extrapolated to the field.

8.13   CONCLUSION

Our knowledge of the mosquito's innate immune system has advanced in the last 10 years, enhanced by the completion of the *A. gambiae* genome and the subsequent development of techniques to assess gene expression and target specific genes for manipulation. The *Anopheles* mosquito has become an important model organism for the understanding of how invertebrate immunity functions but, more importantly, its immune system directly interferes with a disease affecting millions of people worldwide making it relevant to public health. The intention is to use the molecules, cells, and pathways of the immune system as targets for vector-based disease control strategies. In the future, research in mosquito immunity could be coupled with the ongoing research in gene drive and genetic delivery, such that *Plasmodium* antagonists can be activated in field vectors. This type of manipulation is years away but, as anti-parasitic immune effectors are elucidated, this goal becomes more attainable. Studying natural mosquito and parasite populations, understanding immune reactions resulting from a variety of vector–parasite species combinations and clarifying the inconsistencies between different species and strains will allow the determination of how universal any immunity-based intervention could be. Many mosquito immunity researchers are currently focusing their efforts on immune systems of field mosquitoes and the natural mosquito reaction to *P. falciparum*. These studies will reveal the data that is most relevant to combating human malaria. In addition to the direct application in malaria control, *Anopheles* defense against *Plasmodium* is often used as a model for investigations of other vector–pathogen interactions. There are many protozoans, metazoans, viruses, and bacteria that are vectored by insects. *A. gambiae* was the first insect vector to have a fully sequenced and publicly available genome (Holt et al., 2002) and, accordingly, has provided arguably the richest data on vector immunity to a human pathogen. Now, the genomes of other insect vectors have become increasingly available and the *Anopheles* immune system represents a good model for the immune systems of these insects. Although biologists' view of the complete network of mosquito defense against *Plasmodium* is far from complete, the principles outlined in this chapter are critical for understanding vector–pathogen interactions and could contribute to future malaria control efforts.

ACKNOWLEDGMENTS

We thank the Dimopoulos lab for fruitful discussions. We thank Dr. Deborah McClellan at the Editing Referral Service, William H. Welch Medical Library, Johns Hopkins University School of Medicine. This work has partly been supported by the NIH/NIAID 1R01AI061576-01A1, the Ellison Medical Foundation, the WHO/TDR, and the Johns Hopkins Malaria Research Institute.

REFERENCES

Abraham, E. G., Islam, S., Srinivasan, P., Ghosh, A. K., Valenzuela, J. G., Ribeiro, J. M., Kafatos, F. C., Dimopoulos, G., and Jacobs-Lorena, M. (2004). Analysis of the *Plasmodium* and *Anopheles* transcriptional repertoire during ookinete development and midgut invasion. *J. Biol. Chem.* **279**, 5573–5580.

Abraham, E. G., Pinto, S. B., Ghosh, A., Vanlandingham, D. L., Budd, A., Higgs, S., Kafatos, F. C., Jacobs-Lorena, M., and Michel, K. (2005). An immune-responsive serpin, SRPN6, mediates mosquito defense against malaria parasites. *Proc. Natl. Acad. Sci. USA* **102**, 16327–16332.

Aguilar, R., Dong, Y., Warr, E., and Dimopoulos, G. (2005). *Anopheles* infection responses; laboratory models versus field malaria transmission systems. *Acta Trop.* **95**, 285–291.

Al Mashhadani, H. M., Davidson, G., and Curtis, C. F. (1980). A genetic study of the susceptibility of *Anopheles gambiae* to *Plasmodium berghei*. *T. Roy. Soc. Trop. Med. H.* **74**, 585–594.

Azambuja, P., Garcia, E. S., and Ratcliffe, N. A. (2005). Gut microbiota and parasite transmission by insect vectors. *Trends Parasitol.* **21**, 568–572.

Barillas-Mury, C., Charlesworth, A., Gross, I., Richman, A., Hoffmann, J. A., Kafatos, F. C. (1996). Immune factor Gambif1, a new rel-family from the human malaria vector, *Anopheles gambiae*. *EMBO J.* **15**, 4691–4701.

Barillas-Mury, C., Han, Y. S., Seeley, D., and Kafatos, F. C. (1999). *Anopheles gambiae* Ag-STAT, a new insect member of the STAT family, is activated in response to bacterial infection. *EMBO J.* **18**, 959–967.

Bartholomay, L. C., Cho, W. L., Rocheleau, T. A., Boyle, J. P., Beck, E. T., Fuchs, J. F., Liss, P., Rusch, M., Butler, K. M., Wu, R. C., Lin, S. P., Kuo, H. Y., Tsao, I. Y., Huang, C. Y., Liu, T. T., Hsiao, K. J., Tsai, S. F., Yang, U. C., Nappi, A. J., Perna, N. T., Chen, C. C., and Christensen, B. M. (2004). Description of the transcriptomes of immune response-activated hemocytes from the mosquito vectors *Aedes aegypti* and *Armigeres subalbatus*. *Infect. Immun.* **72**, 4114–4126.

Baton, L. A., Dong,Y., and Dimopoulos, G. (2006a). Strain- and species-specific comparison of the immune responses of different members of the *Anopheles gambiae* complex to *Plasmodium falciparum* infection. *Amer. J. Trop. Med. Hyg.*, **75**, 311.

Baton, L. A., Dong, Y., Li, J., and Dimopoulos, G. (2006b). "Variation in mosquito immunity to Plasmodium" paper in *Proceedings of the 11th International Congress of Parasitology* – ICOPA XI (August 6–11, 2006, Glasgow, Scotland, United Kingdom), Medimond, International Proceedings, Bologna.

Baton, L. A., Castillo, J., Robertson, A., Warr, E., Strand, M. R., and Dimopoulos, G. (2007). Transcriptomic analysis of mosquito hemocytes during bacterial and *Plasmodium* infection. *Gen. Biol.* (submitted).

Beier, M. S., Pumpuni, C. B., Beier, J. C., and Davis, J. R. (1994). Effects of *para*-aminobenzoic acid, insulin, and gentamicin on *Plasmodium falciparum* development in anopheline mosquitoes (Diptera: Culicidae). *J. Med. Entomol.* **31**, 561–565.

Blandin, S., Moita, L. F., Kocher, T., Wilm, M., Kafatos, F. C., and Levashina, E. A. (2002). Reverse genetics in the mosquito *Anopheles gambiae*: Targeted disruption of the defensin gene. *EMBO Rep.* **3**, 852–856.

Blandin, S., Shiao, S. H., Moita, L. F., Janse, C. J., Waters, A. P., Kafatos, F. C., and Levashina, E. A. (2004). Complement-like protein TEP1 is a determinant of vectorial capacity in the malaria vector *Anopheles gambiae*. *Cell* **116**, 661–670.

Bonnet, S., Prevot, G., Jacques, J. C., Boudin, C., and Bourgouin, C. (2001). Transcripts of the malaria vector *Anopheles gambiae* that are differentially regulated in the midgut upon exposure to invasive stages of *Plasmodium falciparum*. *Cell Microbiol.* **3**, 449–458.

Castillo, J. C., Robertson, A. E., and Strand, M. R. (2006). Characterization of hemocytes from the mosquitoes *Anopheles gambiae* and *Aedes aegypti*. *Insect Biochem. Mol. Biol.* **36**, 891–903.

Christophides, G. K., Zdobnov, E., Barillas-Mury, C., Birney, E., Blandin, S., Blass, C., Brey, P. T., Collins, F. H., Danielli, A., Dimopoulos, G., Hetru, C., Hoa, N. T., Hoffmann, J. A., Kanzok, S. M., Letunic, I., Levashina, E. A., Loukeris, T. G., Lycett, G., Meister, S., Michel, K., Moita, L. F., Muller, H. M., Osta, M. A., Paskewitz, S. M., Reichhart, J. M., Rzhetsky, A., Troxler, L., Vernick, K. D., Vlachou, D., Volz, J., von Mering, C., Xu, J., Zheng, L., Bork, P., and Kafatos, F. C. (2002). Immunity-related genes and gene families in *Anopheles gambiae*. *Science* **298**, 159–165.

Cohuet, A., Osta, M. A., Morlais, I., Awono-Ambene, P. H., Michel, K., Simard, F., Christophides, G. K., Fontenille, D., and Kafatos, F. C. (2006). *Anopheles* and *Plasmodium*: From laboratory models to natural systems in the field. *EMBO Rep.* **7**, 1285–1289.

Collins, F. H., Sakai, R. K., Vernick, K. D., Paskewitz, S., Seeley, D. C., Miller, L. H., Collins, W. E., Campbell, C. C., and Gwadz, R. W. (1986). Genetic selection of a *Plasmodium*-refractory strain of the malaria vector *Anopheles gambiae*. *Science* **234**, 607–610.

Cui, L., Luckhart, S., and Rosenberg, R. (2000). Molecular characterization of a prophenoloxidase cDNA from the malaria mosquito *Anopheles stephensi*. *Insect Mol. Biol.* **7**, 127–137.

Dana, A. N., Hillenmeyer, M. E., Lobo, N. F., Kern, M. K., Romans, P. A., and Collins, F. H. (2006). Differential gene expression in abdomens of the malaria vector mosquito, *Anopheles gambiae*, after sugar feeding, blood feeding and *Plasmodium berghei* infection. *BMC Genom.* **7**, 119.

Dimopoulos, G. (2003). Insect immunity and its implication in mosquito–malaria interactions. *Cell Microbiol.* **5**, 3–14.

Dimopoulos, G., Richman, A., Muller, H. M., and Kafatos, F. C. (1997). Molecular immune responses of the mosquito *Anopheles gambiae* to bacteria and malaria parasites. *Proc. Natl. Acad. Sci. USA* **94**, 11508–11513.

Dimopoulos, G., Seeley, D., Wolf, A., and Kafatos, F. C. (1998). Malaria infection of the mosquito *Anopheles gambiae* activates immune-responsive genes during critical transition stages of the parasite life cycle. *EMBO J.* **17**, 6115–6123.

Dimopoulos, G., Christophides, G. K., Meister, S., Schultz, J., White, K. P., Barillas-Mury, C., and Kafatos, F. C. (2002). Genome expression analysis of *Anopheles gambiae*: Responses to injury, bacterial challenge, and malaria infection. *Proc. Natl. Acad. Sci. USA* **99**, 8814–8819.

Dong, Y., Aguilar, R., Xi, Z., Warr, E., Mongin, E., and Dimopoulos, G. (2006a). *Anopheles gambiae* immune responses to human and rodent *Plasmodium* parasite species. *PLoS Pathog.* **2**, e52.

Dong, Y., Taylor, H. E., and Dimopoulos, G. (2006b). AgDscam, a hypervariable immunoglobulin domain-containing receptor of the *Anopheles gambiae* innate immune system. *PLoS Biol.* **4**, e229.

Feldmann, A. M., and Ponnudurai, T. (1989). Selection of *Anopheles stephensi* for refractoriness and susceptibility to *Plasmodium falciparum*. *Med. Vet. Entomol.* **3**, 41–52.

Frizzi, G., Rinaldi, A., and Bianchi, U. (1975). Genetic studies on mechanisms influencing the susceptibility of anopheline mosquitoes to plasmodial infection. *Mosquito News* **35**, 505–508.

Frolet, C., Thoma, M., Blandin, S., Hoffmann, J. A., and Levashina, E. A. (2006). Boosting NF-κB-dependent basal immunity of *Anopheles gambiae* aborts development of *Plasmodium berghei*. *Immunity* **25**, 677–685.

Garver, L. S., Xi, Z., and Dimopoulos, G. (2007). Immunoglobulin superfamily members play an important role in the mosquito immune system. *Comp. Dev. Immunol.* (in press).

Garver, L. S. and Dimopoulos, G. (2007). Silencing of immune pathway regulators render *Anopheles* mosquitoes resistant to human malaria (in preparation).

Gonzalez-Ceron, L., Rodriguez, M. H., Santillan, F., Chavez, B., Nettel, J. A., Hernandez-Avila, J. E., and Kain, K. C. (2001). *Plasmodium vivax*: Ookinete destruction and oocyst development arrest are responsible for *Anopheles albimanus* resistance to circumsporozoite phenotype VK247 parasites. *Exp. Parasitol.* **98**, 152–161.

Gonzalez-Ceron, L., Santillan, F., Rodriguez, M. H., Mendez, D., and Hernandez-Avila, J. E. (2003). Bacteria in midguts of field-collected *Anopheles albimanus* block *Plasmodium vivax* sporogonic development. *J. Med. Entomol.* **40**, 371–374.

Gonzalez-Ceron, L., Rodriguez, M. H., Chavez-Munguia, B., Santillan, F., Nettel, J. A., and Hernandez-Avila, J. E. (2007). *Plasmodium vivax*: Impaired escape of Vk210 phenotype ookinetes from the midgut blood bolus of *Anopheles pseudopunctipennis*. *Exp. Parasitol.* **115**, 59–67.

Gorman, M. J., Severson D. W., Cornel, A. J., Collins, F. H., and Paskewitz, S. M. (1997). Mapping a quantitative trait locus involved in melanotic encapsulation of foreign bodies in the malaria vector, *Anopheles gambiae*. *Genetics* **3**, 965–971.

Graves, P. M., and Curtis, C. F. (1982). Susceptibility of *Anopheles gambiae* to *Plasmodium yoelii nigeriensis* and *Plasmodium falciparum*. *Ann. Trop. Med. Parasit.* **76**, 633–639.

Gupta, L., Kumar, S., Han, Y. S., Pimenta, P. F., and Barillas-Mury, C. (2005). Midgut epithelial responses of different mosquito–*Plasmodium* combinations: The actin cone zipper repair mechanism in *Aedes aegypti*. *Proc. Natl. Acad. Sci. USA* **102**, 4010–4015.

Han, Y. S., and Barillas-Mury, C. (2002). Implications of time bomb model of ookinete invasion of midgut cells. *Insect Biochem. Mol. Biol.* **32**, 1311–1316.

Han, Y. S., Thompson, J., Kafatos, F. C., and Barillas-Mury, C. (2000). Molecular interactions between *Anopheles stephensi* midgut cells and *Plasmodium berghei*: The time bomb theory of ookinete invasion of mosquitoes. *EMBO J.* **19**, 6030–6040.

Hillyer, J. F., Barreau, C, and Vernick, K. D. (2007). Efficiency of salivary gland invasion by malaria sporozoites is controlled by rapid sporozoite destruction in the mosquito haemocoel. *Int. J. Parasitol.* **37**, 673–681.

Hoffmann, J. A. (2003). The immune response of *Drosophila*. *Nature* **426**, 33–38.

Holt, R. A., et al. (2002). The genome sequence of the malaria mosquito *Anopheles gambiae*. *Science* **298**, 129–149.

Huff, C. G. (1929). The effects of selection upon susceptibility to bird malaria in *Culex pipiens* Linn. *Ann. Trop. Med. Parasit.* **23**, 427–442.

Huff, C. G. (1931). The inheritance of natural immunity to *Plasmodium cathemerium* in two species of *Culex*. *J. Prev. Med.* **5**, 249–259.

Hurd, H., Taylor, P. J., Adams, D., Underhill, A., and Eggleston, P. (2005). Evaluating the costs of mosquito resistance to malaria parasites. *Evol. Int. J. Org. Evol.* **59**, 2560–2572.

Kilama, W. L., and Craig, G. B. (1969). Monofactorial inheritance of susceptibility to *Plasmodium gallinaceum* in *Aedes aegypti*. *Ann. Trop. Med. Parasit.* **63**, 419–432.

Kim, W., Koo, H., Richman, A. M., Seeley, D., Vizioli, J., Klocko, A. D., and O'Brochta, D. A. (2004). Ectopic expression of a cecropin transgene in the human malaria vector mosquito *Anopheles gambiae* (Diptera: Culicidae): Effects on susceptibility to *Plasmodium*. *J. Med. Entomol.* **41**, 447–455.

Korochkina, K., Barreau, C., Pradel, G., Jeffery, E., Li, J., Natarajan, R., Shabanowitz, J., Hunt, D., Frevert, U., and Vernick, K. D. (2006). A mosquito-specific protein family includes candidate receptors for malaria sporozoite invasion of salivary glands. *Cell Microbiol.* **8**, 163–175.

Kumar, S., and Barillas-Mury, C. (2005). Ookinete-induced midgut peroxidases detonate the time bomb in anopheline mosquitoes. *Insect Biochem. Mol. Biol.* **35**, 721–737.

Kumar, S., Christophides, G. K., Cantera, R., Charles, B., Han, Y. S., Meister, S., Dimopoulos, G., Kafatos, F. C., and Barillas-Mury, C. (2003). The role of reactive oxygen species on *Plasmodium* melanotic encapsulation in *Anopheles gambiae. Proc. Natl. Acad. Sci. USA* **100**, 14139–14144.

Kumar, S., Gupta, L., Han, Y. S., Barillas-Mury, C. (2004). Inducible peroxidases mediate nitration of *Anopheles* midgut cells undergoing apoptosis in response to *Plasmodium* invasion. *J. Biol. Chem.* **270**, 53475–53482.

Lambrechts, L., Vulule, J. M., and Koella, J. C. (2004). Genetic correlation between melanization and antibacterial immune responses in a natural population of the malaria vector *Anopheles gambiae. Evol. Int. J. Org. Evol.* **58**, 2377–2381.

Lambrechts, L., Halbert, J., Durand, P., Gouagna, L. C., and Koella, J. C. (2005). Host genotype by parasite genotype interactions underlying the resistance of anopheline mosquitoes to *Plasmodium falciparum. Malaria J.* **4**, 3.

Leclerc, V., and Reichhart, J. M. (2004). The immune response of *Drosophila melanogaster. Immunol. Rev.* **198**, 59–71.

Lee, W. J., Ahmed, A., della Torre, A., Kobayashi, A., Ashida, M., and Brey, P. T. (1998). Molecular cloning and chromosomal localization of a prophenoloxidase cDNA from the malaria vector *Anopheles gambiae. Insect Mol. Biol.* **7**, 41–50.

Levashina, E. A., Moita, L. F., Blandin, S., Vriend, G., Lagueux, M., and Kafatos, F. C. (2001). Conserved role of a complement-like protein in phagocytosis revealed by dsRNA knockout in cultured cells of the mosquito, *Anopheles gambiae. Cell* **104**, 709–718.

Li, B., and Paskewitz, S. M. (2006). A role for lysozyme in melanization of Sephadex beads in *Anopheles gambiae. J. Insect Physiol.* **52**, 936–942.

Lim, J., Gowda, D. C., Krishnegowda, G., and Luckhart, S. (2005). Induction of nitric oxide synthase in *Anopheles stephensi* by *Plasmodium falciparum*: Mechanism of signaling and the role of parasite glycosylphosphatidylinositols. *Infect. Immun.* **73**, 2778–2789.

Lowenberger, C. A., Kamal, S., Chiles, J., Paskewitz, S., Bulet, P., Hoffmann, J. A., and Christensen, B. M. (1999). Mosquito–*Plasmodium* interactions in response to immune activation of the vector. *Exp. Parasitol.* **91**, 59–69.

Luckhart, S., Vodovotz, Y., Cui, L., and Rosenberg, R. (1998). The mosquito *Anopheles stephensi* limits malaria parasite development with inducible synthesis of nitric oxide. *Proc. Natl. Acad. Sci. USA* **95**, 5700–5705.

Luckhart, S., Crampton, A. L., Zamora, R., Lieber, M. J., Dos Santos, P. C., Peterson, T. M., Emmith, N., Lim, J., Wink, D. A., and Vodovotz, Y. (2003). Mammalian transforming growth factor β1 activated after ingestion by *Anopheles stephensi* modulates mosquito immunity. *Infect. Immun.* **71**, 3000–3009.

Luna, C., Hoa, N. T., Lin, H., Zhang, L., Nguyen, H. L., Kanzok, S. M., and Zheng, L. (2006). Expression of immune responsive genes in cell lines from two different Anopheline species. *Ins. Mol. Biol.* **15**, 721–729.

Marinotti, O., Nguyen, Q. K., Calvo, E., James, A. A., and Ribeiro, J. M. (2005). Microarray analysis of genes showing variable expression following a blood meal in *Anopheles gambiae. Ins. Mol. Biol.* **14**, 365–373.

Marinotti, O., Calvo, E., Nguyen, Q. K., Dissanayake, S., Ribeiro, J. M., and James, A. A. (2006). Genome-wide analysis of gene expression in adult *Anopheles gambiae. Ins. Mol. Biol.* **15**, 1–12.

Medzhitov, R., and Janeway, C. A. (2002). Decoding the patterns of self and nonself by the innate immune system. *Science* **296**, 298–300.

Meister, S., Kanzok, S. M., Zheng, X. L., Luna, C., Li, T. R., Hoa, N. T., Clayton, J. R., White, K. P., Kafatos, F. C., Christophides, G. K., and Zheng, L. (2005). Immune signaling pathways regulating bacterial and malaria parasite infection of the mosquito *Anopheles gambiae. Proc. Natl. Acad. Sci. USA* **102**, 11420–11425.

Menge, D. M., Zhong, D., Guda, T., Gouagna, L., Githure, J., Beier, J., and Yan, G. (2006). Quantitative trait loci controlling refractoriness to *Plasmodium falciparum* in natural *Anopheles gambiae* mosquitoes from a malaria-endemic region in western Kenya. *Genetics* **173**, 235–241.

Michel, K., Budd, A., Pinto, S., Gibson, T. J., and Kafatos, F. C. (2005). *Anopheles gambiae* SRPN2 facilitates midgut invasion by the malaria parasite *Plasmodium berghei*. *EMBO Rep.* **6**, 891–897.

Michel, K., Suwanchaichinda, C., Morlais, I., Lambrechts, L., Cohuet, A., Awono-Ambene, P. H., Simard, F., Fontenille, D., Kanost, M. R., and Kafatos, F. C. (2006). Increased melanizing activity in *Anopheles gambiae* does not affect development of *Plasmodium falciparum*. *Proc. Natl. Acad. Sci. USA* **103**, 16858–16863.

Micks, D. W. (1949). Investigations on the mosquito transmission of *Plasmodium elongatum* Huff, 1930. *J. Natl. Malaria Soc.* **8**, 206–218.

Micks, D. W., and Ferguson, M. J. (1961). Microorganisms associated with mosquitoes: III. Effect of reduction in the microbial flora of *Culex fatiagns* Wiedemann on the susceptibility to *Plasmodium relictum* Grassi and Feletti. *J. Insect Pathol.* **3**, 244–248.

Muller, H. M., Dimopoulos, G., Blass, C., and Kafatos, F. C. (1999). A hemocyte-like cell line established from the malaria vector *Anopheles gambiae* expresses six prophenoloxidase genes. *J. Biol. Chem.* **274**, 11727–11735.

Niare, O., Markianos, K., Volz, J., Oduol, F., Toure, A., Bagayoko, M., Sangare, D., Traore, S. F., Wang, R., Blass, C., Dolo, G., Bouare, M., Kafatos, F. C., Kruglyak, L., Toure, Y. T., and Vernick, K. D. (2002). Genetic loci affecting resistance to human malaria parasites in a West African mosquito vector population. *Science* **298**, 213–216.

Osta, M. A., Christophides, G. K., and Kafatos, F. C. (2004). Effects of mosquito genes on *Plasmodium* development. *Science* **303**, 2030–2032.

Park, J. M., Brady, H., Ruocco, M. G., Sun, H., Williams, D., Lee, S. J., Kato, T. Jr., Richards, N., Chan, K., Mercurio, F., Karin, M., and Wasserman, S. A. (2004). Targeting of TAK1 by the NF-kappa B protein Relish regulates the JNK-mediated immune response in *Drosophila*. *Gene Dev.* **18**, 584–594.

Paskewitz, S. M., and Riehle, M. (1994). Response of *Plasmodium* refractory and susceptible strains of *Anopheles gambiae* to inoculated Sephadex beads. *Dev. Comp. Immunol.* **18**, 369–375.

Paskewitz, S. M., and Riehle, M. (1998). A factor preventing melanization of sephadex CM C-25 beads in *Plasmodium*-susceptible and refractory *Anopheles gambiae*. *Exp. Parasitol.* **90**, 34–41.

Paskewitz, S. M., and Shi, L. (2005). The hemolymph proteome of *Anopheles gambiae*. *Ins. Biochem. Mol. Biol.* **35**, 815–824.

Paskewitz, S. M., Brown, M. R., Lea, O. A., and Collins, F. H. (1988). Ultrastructure of the encapsulation of *Plasmodium cynomolgi* (B strain) on the midgut of a refractory strain of *Anopheles gambiae*. *J. Parasitol.* **74**, 432–437.

Paskewitz, S. M., Schwartz, A. M., and Gorman, M. J. (1998). The role of surface characteristics in eliciting humoral encapsulation of foreign bodies in *Plasmodium*-refractory and -susceptible strains of *Anopheles gambiae*. *J. Insect Physiol.* **44**, 947–954.

Peterson, T. M., and Luckhart, S. (2006). A mosquito 2-Cys peroxiredoxin protects against nitrosative and oxidative stresses associated with malaria parasite infection. *Free Radic. Biol. Med.* **40**, 1067–1082.

Peterson, T. M., Gow, A. J., and Luckhart, S. (2007). Nitric oxide metabolites induced in *Anopheles stephensi* control malaria parasite infection. *Free Radic. Biol. Med.* **42**, 132–142.

Pumpuni, C. B., Beier, M. S., Nataro, J. P., Guers, L. D., and Davis, J. R. (1993). *Plasmodium falciparum*: Inhibition of sporogonic development in *Anopheles stephensi* by Gram-negative bacteria. *Exp. Parasitol.* **77**, 195–199.

Pumpuni, C. B., DeMaio, J., Kent, M., Davis, J. R., and Beier, J. C. (1996). Bacterial population dynamics in three anopheline species: The impact on *Plasmodium* sporogonic development. *Am. J. Trop. Med. Hyg.* **54**, 214–218.

Richman, A. M., Dimopoulos, G., Seeley, D., and Kafatos, F. C. (1997). *Plasmodium* activates the innate immune response of *Anopheles gambiae* mosquitoes. *EMBO J.* **16**, 6114–6119.

Riehle, M. M., Markianos, K., Niare, O., Xu, J., Li, J., Toure, A. M., Podiougou, B., Oduol, F., Diawara, S., Diallo, M., Coulibaly, B., Ouatara, A., Kruglyak, L., Traore, S. F., and Vernick, K. D. (2006). Natural malaria infection in *Anopheles gambiae* is regulated by a single genomic control region. *Science* **312**, 577–579.

Rodriguez, M. C., Zamudio, F., Torres, J. A., Gonzalez-Ceron, L., Possani, L. D., and Rodriguez, M. H. (1995). Effect of a cecropin-like synthetic peptide (Shiva-3) on the sporogonic development of *Plasmodium berghei*. *Exp. Parasitol.* **80**, 596–604.

Rosenberg, R., and Rungsiwongse, J. (1991). The number of sporozoites produced by individual malaria oocysts. *Am. J. Trop. Med. Hyg.* **45**, 574–577.

Rosinski-Chupin, I., Briolay, J., Brouilly, P., Perrot, S., Gomez, S. M., Chertemps, T., Roth, C. W., Keime, C., Gandrillon, O., Couble, P., and Brey, P. T. (2007). SAGE analysis of mosquito salivary gland transcriptomes during *Plasmodium* invasion. *Cell Microbiol.* **9**, 708–724.

Schwartz, A., and Koella, J. C. (2002). Melanization of *Plasmodium falciparum* and C-25 Sephadex beads by field-caught *Anopheles gambiae* (Diptera: Culicidae) from southern Tanzania. *J. Med. Entomol.* **39**, 84–88.

Shiao, S. H., Whitten, M. M., Zachary, D., Hoffmann, J. A., and Levashina, E. A. (2006). Fz2 and cdc42 mediate melanization and actin polymerization but are dispensable for *Plasmodium* killing in the mosquito midgut. *PLoS Pathog.* **2**, e133.

Shin, S. W., Kokoza, V., Bian, G., Cheon, H. M., Kim, Y. J., and Raikhel, A. S. (2005). REL1, a homologue of *Drosophila* dorsal, regulates toll antifungal immune pathway in the female mosquito *Aedes aegypti*. *J. Biol. Chem.* **280**, 16499–16507.

Shin, S. W., Bian, G., and Raikhel, A. S. (2006). A toll receptor and a cytokine, Toll5A and Spz1C, are involved in toll antifungal immune signaling in the mosquito *Aedes aegypti*. *J. Biol. Chem.* **281**, 39388–39395.

Silverman, N., Zhou, R., Erlich, R. L., Hunter, M., Bernstein, E., Schneider, D., and Maniatis, T. (2003). Immune activation of NF-κB and JNK requires *Drosophila* TAK1. *J. Biol. Chem.* **278**, 48928–48934.

Somboon, P., Prapanthadara, L., and Suwonkerd, W. (1999). Selection of *Anopheles dirus* for refractoriness and susceptibility to *Plasmodium yoelii nigeriensis*. *Med. Vet. Entomol.* **13**, 355–361.

Srinivasan, P., Abraham, E. G., Ghosh, A. K., Valenzuela, J., Ribeiro, J. M., Dimopoulos, G., Kafatos, F. C., Adams, J. H., Fujioka, H., and Jacobs-Lorena, M. (2004). Analysis of the *Plasmodium* and *Anopheles* transcriptomes during oocyst differentiation. *J. Biol. Chem.* **279**, 5581–5587.

Tahar, R., Boudin, C., Thiery, I., and Bourgouin, C. (2002). Immune response of *Anopheles gambiae* to the early sporogonic stages of the human malaria parasite *Plasmodium falciparum*. *EMBO J.* **21**, 6673–6680.

Tanji, T., and Ip, Y. T. (2005). Regulators of the Toll and Imd pathways in the *Drosophila* innate immune response. *Trends Immunol.* **26**, 193–198.

Tauszig, S., Jouanguy, E., Hoffmann, J. A., and Imler, J. L. (2000). Toll-related receptors and the control of antimicrobial peptide expression in *Drosophila*. *Proc. Natl. Acad. Sci. USA* **97**, 10520–10525.

Thathy, V., Severson, D. W., and Christensen, B. M. (1994). Reinterpretation of the genetics of susceptibility of *Aedes aegypti* to *Plasmodium gallinaceum*. *J. Parasitol.* **80**, 705–712.

Trager, W. (1942). A strain of the mosquito *Aedes aegypti* selected for susceptibility to the avian malaria parasite *Plasmodium lophurae*. *J. Parasitol.* **28**, 457–465.

van der Kaay, H. J., and Boorsma, L. (1977). A susceptible and refractive strain of *Anopheles atroparvus* van Thiel to infection with *Plasmodium berghei*. *Acta Leidensia* **45**, 13–19.

Vaughan, J. A. (2007). Population dynamics of *Plasmodium* sporogony. *Trends Parasitol.* **23**, 63–70.

Vernick, K. D., Fujioka, H., Seeley, D. C., Tandler, B., Aikawa, M., and Miller, L. H. (1995). *Plasmodium gallinaceum*: A refractory mechanism of ookinete killing in the mosquito, *Anopheles gambiae*. *Exp. Parasitol.* **80**, 583–595.

Vizioli, J., Bulet, P., Hoffmann, J. A., Kafatos, F. C., Muller, H. M., and Dimopoulos, G. (2001). Gambicin: a novel immune responsive antimicrobial peptide from the malaria vector *Anopheles gambiae*. *Proc. Natl. Acad. Sci. USA* **98**, 12630–12635.

Vlachou, D., Schlegelmilch, T., Christophides, G. K., and Kafatos, F. C. (2005). Functional genomic analysis of midgut epithelial responses in *Anopheles* during *Plasmodium* invasion. *Curr. Biol.* **15**, 1185–1195.

Vlachou, D., Shlegelmilch, T., Runn, E., Mendes, A., and Kafatos, F. C. (2006). The developmental migration of *Plasmodium* in mosquitoes. *Curr. Opin. Genet. Dev.* **16**, 384–391.

Vlachou, D., Zimmerman, T., Cantera, R., Janse, C. J., Waters, A. P., and Kafatos, F. C. (2004). Real-time, *in vivo* analysis of malaria ookinete locomotion and mosquito midgut invasion. *Cell Microbiol.* **6**, 671–685.

Vodovotz, Y., Zamora, R., Lieber, M. J., and Luckhart, S. (2004). Cross-talk between nitric oxide and transforming growth factor-beta1 in malaria. *Curr. Mol. Med.* **4**, 787–797.

Volz, J., Osta, M. A., Kafatos, F. C., and Muller, H. M. (2005). The roles of two clip domain serine proteases in innate immune responses of the malaria vector *Anopheles gambiae*. *J. Biol. Chem.* **280**, 40161–40168.

Volz, J., Muller, H. M., Zdanowicz, A., Kafatos, F. C., and Osta, M. A. (2006). A genetic module regulates the melanization response of *Anopheles* to *Plasmodium*. *Cell Microbiol.* **8**, 1392–1405.

Wang, X., Fuchs, J. F., Infanger L. C., Rocheleau, T. A., Hillyer, J. F., Chen, C. C., and Christensen, B. M. (2004). Mosquito innate immunity: Involvement of beta 1,3-glucan recognition protein in melanotic encapsulation immune responses in *Armigeres subalbatus*. *Mol. Biochem. Parasitol.* **139**, 65–73.

Ward, R. A. (1963). Genetic aspects of the susceptibility of mosquitoes to malarial infections. *Exp. Parasitol.* **13**, 328–341.

Warr, E., Lambrechts, L., Koella, J. C., Bourgouin, C., and Dimopoulos, G. (2006). *Anopheles gambiae* immune responses to Sephadex beads: Involvement of anti-*Plasmodium* factors in regulating melanization. *Ins. Biochem. Mol. Biol.* **36**, 769–778.

Warr, E., Aguilar, R., Dong, Y., Mahairaki, V., and Dimopoulos, G. (2007). Spatial and sex-specific dissection of the *Anopheles gambiae* midgut transcriptome. *BMC Genom.* **8**, 37.

Xu, X., Dong, Y., Abraham, E. G., Kocan, A., Srinivasan, P., Ghosh, A. K., Sinden, R. E., Ribeiro, J. M., Jacobs-Lorena, M., Kafatos, F. C., and Dimopoulos, G. (2005). Transcriptome analysis of *Anopheles stephensi–Plasmodium berghei* interactions. *Mol. Biochem. Parasitol.* **142**, 76–87.

Zheng, L., Cornel, A. J., Wang, R., Erfle, H., Voss, H., Ansorge, W., Kafatos, F. C., and Collins, F. H. (1997). Quantitative trait loci for refractoriness of *Anopheles gambiae* to *Plasmodium cynomolgi* B. *Science* **276**, 425–428.

Zheng, L., Wang, S., Romans, P., Zhao, H., Luna, C., and Benedict, M. Q. (2003). Quantitative trait loci in *Anopheles gambiae* controlling the encapsulation response against *Plasmodium cynomolgi* Ceylon. *BMC Genet.* **4**, 16.

Zhong, D., Menge, D. M., Temu, E. A., Chen, H., and Yan, G. (2006). Amplified fragment length polymorphism mapping of quantitative trait loci for malaria parasite susceptibility in the yellow fever mosquito *Aedes aegypti*. *Genetics* **173**, 1337–1345.

9

INSECT IMMUNITY TO VIRUSES

WENDY O. SPARKS, LYRIC C. BARTHOLOMAY AND
BRYONY C. BONNING

*Department of Entomology, Iowa State University,
Ames, IA 50011-3222, USA*

ABSTRACT: Immunity to viruses that replicate in insect hosts is of interest from both basic and applied perspectives in terms of (1) increased understanding of fundamental processes involved in anti-viral immunity and viral counter-measures, (2) use of this knowledge to facilitate the means by which arthropod-borne viruses (arboviruses) of vertebrate hosts can be managed, and (3) exploitation of strategies used by viruses to overcome physiological barriers for pest management purposes. In this chapter, we review the anti-viral mechanisms against viruses that replicate in insects, including genetic resistance, physical and physiological barriers to infection, cellular and subcellular inhibition, humoral immunity and developmental resistance, and finally highlight some practical applications of this knowledge for management of agricultural pests and vector-borne disease.

Abbreviations:

*Ac*EPV	= *Anomala cuprea* EPV
*Ac*MNPV	= *Autographa californica* multiple nucleopolyhedrovirus
BM	= basement membrane
*Bm*NPV	= *Bombyx mori* nucleopolyhedrovirus
BV	= budded virus
CpGV	= *Cydia pomonella* granulovirus
CrPV	= Cricket paralysis virus
DCV	= *Drosophila* C virus
DDCA	= diethyldithio-carbamic acid
DENV	= dengue virus
Dif	= dorsal-related immune factor
DXV	= *Drosophila* X virus
EEEV	= Eastern equine encephalitis virus
EPV	= entomopoxvirus
fgf	= fibroblast growth factor
FHV	= Flock house virus
HaSV	= *Helicoverpa armigera* stunt virus
IAP	= inhibitor of apoptosis proteins
ICTV	= International Committee on Taxonomy of Viruses
IIM	= insect intestinal mucin
Imd	= immune deficiency
IRES	= internal ribosome entry site
LACV	= LaCrosse encephalitis virus
LPS	= lipopolysaccharide
MEB	= midgut escape barrier
MIB	= midgut infection barrier
NPV	= nucleopolyhedrovirus
ODV	= occlusion-derived virus
PIF	= *per os* infectivity factor

PM   = peritrophic membrane
PO   = phenoloxidase
PP-BP = paralytic peptide-binding protein
PRR   = pattern recognition receptor
SEB   = salivary gland escape barrier
SIP   = salivary gland infection barrier
TLR   = Toll-like receptor
WEEV = Western equine encephalomyelitis virus
WNV   = West Nile virus

9.1   INTRODUCTION

Insect–virus interactions represent a proverbial tug-of-war, in which the virus is driven to propagate and disseminate in order to be transmitted to the next host, while the host struggles to retain/regain homeostasis in the midst of virus takeover of the cellular machinery. This review will explore the two extremes of insect–virus interactions: (1) insect viruses that are lethal to the host (e.g. the baculoviruses) and (2) viruses for which transmission is contingent upon the host's survival (e.g. arthropod-borne viruses that are transmitted between vertebrate hosts by a blood-sucking insect). These interactions are the manifestation of complex and long-term co-evolution. Fossil evidence for baculovirus infections, for example, dates back to 100 million years ago (Poinar and Poinar, 2005). Given the ancient relationship between insect viruses and their hosts, it is not surprising to discover both highly conserved innate immune responses across invertebrates and vertebrates, as well as highly specific virus–host interactions.

Typically, insect viruses are ingested with food material, enter the host tissues via the gut epithelium and replicate in this tissue and in some cases in other tissues as well. Baculovirus replication ultimately leads to the death of the host insect followed by lysis of the cadaver and release of virus into the environment. Other viruses have more subtle effects on the host by reducing longevity and fecundity, with both release of virions from infected individuals and vertical transmission playing important roles in virus dissemination. In all cases, the virus has to negotiate a series of barriers for successful infection: (1) the pH and contents of the gut, (2) penetration of the peritrophic membrane (PM) if present, (3) entry into and infection of gut epithelial cells, (4) subcellular immune mechanisms (apoptosis and RNA interference in particular), (5) navigation of the basement membrane (BM) overlying the gut epithelium, (6) cellular immune responses, (7) humoral immunity, and (8) developmental resistance (Fig. 9.1).

Arthropod-borne viruses (arboviruses) are ingested with a blood meal when a hematophagous insect feeds on a viremic host. Of the arboviruses that infect humans, the majority are RNA viruses in the families Togaviridae, Flaviviridae, and Bunyaviridae. Mosquito vectors for medically significant arboviruses are most often in the genus *Aedes* or *Culex*. Within the mosquito host, the virus travels with the blood to the midgut. In the midgut of a susceptible, competent mosquito vector that supports the replication and dissemination of the virus, the virus binds to and enters midgut epithelial cells, replicates, disseminates from the midgut cells and enters the hemocoel. This process is repeated with entry into, replication in, and escape from salivary gland cells, with virus secreted in the saliva when the mosquito feeds on a host. The compatibility of the virus–mosquito association is determined by factors that influence progression through each of these stages (Higgs, 2004). The barriers to arbovirus infection have been termed the midgut infection barrier (MIB), midgut escape barrier (MEB), the salivary gland infection barrier (SIB), and salivary gland escape barrier (SEB; Fig. 9.1).

There are numerous examples where the development of a particular arbovirus is arrested by one of these barriers in specific species or strains of mosquitoes, as reviewed previously (Black et al., 2002; Hardy et al., 1983; Higgs, 2004; Kramer and Ebel, 2003; Weaver et al., 2004). For example, work with susceptible and refractory strains of *Culex* has demonstrated both MIBs and MEBs for Western equine encephalomyelitis virus (WEEV) (Hardy et al., 1983), but mechanisms responsible for MIBs, MEBs, SIBs, and SEBs remain unknown. The best-described anti-viral mechanism directed toward an arbovirus in a mosquito vector is RNA interference, but to what extent this constitutes an infection or escape barrier in the tissues most critical to virus replication and dissemination remains to be seen.

In any of the host–virus interactions considered here, successful virus infection is dependent upon the compatibility of the relationship between the host and virus which is dictated by the genome of each player. In this chapter, we review

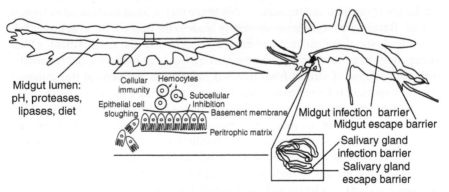

FIGURE 9.1   Known barriers to virus infection of lepidopteran and mosquito hosts. Although four general barriers have been identified in the mosquito, little is known of the anti-viral mechanisms. Cell sloughing and proteolytic action in the midgut are, however, known to be involved in the progression of some arbovirus infections of mosquitoes. See text for further details.

the potential barriers to dissemination faced by insect viruses at each stage of infection.

9.2   GENETIC RESISTANCE TO VIRUSES

Genetic resistance to baculoviruses has been documented against various species of baculovirus, within both laboratory and field insect populations (Fig. 9.2) (Briese, 1986; Eberle and Jehle, 2006; Fuxa, 1993). Evidence in *Cydia pomonella* suggests a non-additive polygenic trait with autosomal inheritance (Eberle and Jehle, 2006). Increasing evidence for host selection of pathogenic genotypes of viruses in field populations of insects has recently been documented (Hitchman et al., 2007).

There is a large body of literature, for a variety of mosquito vector–arbovirus combinations, demonstrating that susceptibility to an arbovirus is genetically dictated. The reader is referred to excellent reviews on this topic (Black et al., 2002; Hardy et al., 1983; Higgs, 2004; Kramer and Ebel, 2003). Briefly, different

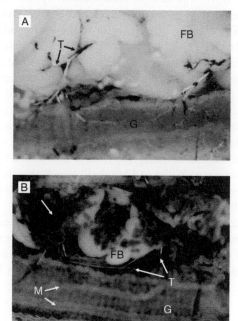

FIGURE 9.2   Impact of genetics on virus propagation within an insect host: Larvae of the semi-permissive host *Ostrinia nubilalis* (A) and the permissive host *Heliothis virescens* (B) dissected 72 h after infection with *Rachiplusia ou* multiple nucleopolyhedrovirus expressing β-galactosidase under the control of the hsp70 promoter. Infection of *H. virescens* is significantly more advanced with extensive lacZ blue staining indicative of infection in the tracheoles (T), fat body (FB) and gut epithelium (G). Infection of *O. nubilalis* is restricted to foci of infection in these tissues (B). The Malpighian tubules (M) are also shown. (See Plate 9.2 for color version of this figure.)

strains of the same mosquito species can vary considerably with respect to susceptibility to infection, and susceptibility can be selectively altered via specific breeding strategies (Hardy et al., 1978; Miller and Mitchell, 1991). Quantitative trait loci mapping has revealed that multiple genes are involved in determining the vector competence of *Ae. aegypti* for the dengue virus 2 (DENV-2), at the level of midgut infection and escape barriers (Black et al., 2002; Bosio et al., 2000), and it is likely that different stages of the infection, i.e. midgut, hemolymph, and salivary glands, will be influenced by a different set of genes.

The potential for studies of phenotypic barriers to arbovirus infection to interface with the genetic bases for those barriers has exploded with the availability of genome sequences and/or functional genomic resources representing three major genera of mosquitoes, i.e. *Aedes*, *Culex*, and *Anopheles* (Lawson et al., 2007). Elegant studies with malaria parasites and *Anopheles gambiae* illustrate the power of combining classical genetic mapping strategies as a means to identify chromosomal regions involved in a phenotypic trait of interest together with genomic sequence and functional genomics studies to pinpoint a gene of interest. In this case, a leucine-rich repeat protein (*Anopheles–Plasmodium*-responsive leucine-rich repeat protein, APL1) was identified and reduced by RNAi, demonstrating an increase in parasite load in mosquitoes deficient in the production of APL1 (Riehle et al., 2006). Adapting this approach to the vector–arbovirus arena and specifically to the *Ae. aegypti*–DENV system, genetic loci with potential involvement in midgut infection and escape barriers have been identified (Bosio et al., 2000), functional genomics studies are being used to explore the response to arbovirus infection (Sanders et al., 2005; Sim et al., 2005) and new tools to continue this work at a larger scale (i.e. with genome-wide arrays) are rapidly becoming available. The means to take a gene of interest and assay transient loss using RNAi (see Reynolds, chapter 12) or viral transducing systems, or heritable loss of function through production of transgenic lines is now also a reality (Shin et al., 2003). Thus we are closer to Miller and Mitchell's prediction that '… inbred mosquito lines will be useful in discovering the molecular basis for flavivirus resistance in *Ae. aegypti*' (Miller and Mitchell, 1991).

The Israeli acute paralysis virus (IAPV) of bees has been correlated with colony collapse syndrome in the United States (Cox-Foster et al., 2007). Honey bees that contained a segment of IAPV stably integrated into their genome were resistant to infection by IAPV, a phenomenon that has also been observed in plants (Maori et al., 2007).

9.3   PHYSICAL AND PHYSIOLOGICAL BARRIERS TO VIRUS INFECTION

Physical and physiological barriers represent the first line of defense of an organism to any invading pathogen, to which viruses are no exception. The invertebrate exoskeleton, impenetrable, waterproof, and cell-less at the outermost layers provide

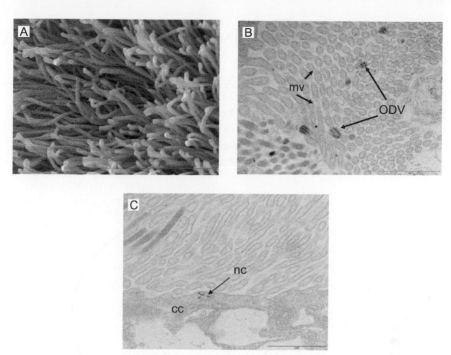

FIGURE 9.3 Baculovirus infection of midgut microvilli of *Heliothis virescens*. (A) Scanning electron micrograph of midgut microvilli, 65 hpi with *Ac*MNPV polyhedra. (B) Transmission electron micrograph of fourth instar midgut microvilli (mv), 1 h post-inoculation with *Ac*MNPV occlusion-derived virus (ODV). ODV can be seen among the microvilli in transverse and longitudinal section. (C) Transmission electron micrograph of fourth instar columnar cell (cc), 1 h post-inoculation with *Ac*MNPV ODV. Nucleocapsids (nc) in the cytoplasm of the cell are indicated. TEMs by Hailin Tang.

an inhospitable environment to an obligate intracellular parasite such as a virus. Of the more than 250 species of invertebrate viruses recognized by the International Committee on Taxonomy of Viruses (ICTV), horizontal transmission is primarily associated with oral ingestion of the virus. Possible infection via the tracheoles has been documented for an iridovirus (Hunter et al., 2003), but does not appear to be the primary mode of transmission for the virus (Williams, 1998). Vertical transmission of insect viruses has also been described, as well as sexual transmission and transmission by parasitoid oviposition. Once ingested, the main target of viral infection is the columnar cells of the midgut epithelia, the only area of the insect that comes into contact with the external environment that does not have a protective layer of cuticle (Tellam et al., 1999). Columnar cells are the predominant cell type found in the rich epithelial lining of the midgut. Characterized by their brush border microvilli and extensive surface area (Fig. 9.3), these cells are responsible for most absorption of nutrients and production of digestive enzymes. Insects utilize multiple innate defenses to protect the columnar cells from viral infection, including the milieu of the gut, the overlying PM, and sloughing of infected cells.

9.3.1   THE GUT: ANTI-VIRAL IMPACT OF pH, PROTEASES, LIPASES, AND DIET

9.3.1.1   pH

Many lepidopteran larvae have a basic gut pH which serves to protect larvae from the potentially deleterious impact of various phytochemicals (Berenbaum, 1980). A midgut pH greater than 12 has been documented for *Acherontia atropos* (Dow et al., 1992). Some insect viruses are activated at high pH, most notably baculoviruses (Fig. 9.4) and entomopoxviruses (EPV). Hence the pH of the insect gut presents the first barrier to infection by these viruses.

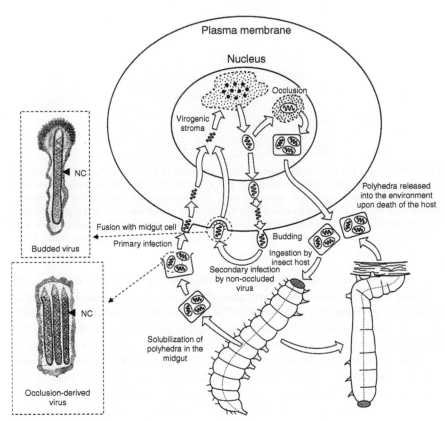

FIGURE 9.4   Life cycle of nucleopolyhedrovirus (NPV). Polyhedra ingested by a lepidopteran larva dissociate in the alkaline environment of the midgut, releasing the infectious occlusion-derived virus (ODV). ODV infect midgut cells and release rod-shaped nucleocapsids (NC) that move into the nucleus. Virus replication ensues within the virogenic stroma in the nucleus. Initially, budded viruses (BV) are produced by the infected cell. BV disseminate the virus within the host insect and initiate infection of other cells and tissues. Subsequently, newly synthesized nucleocapsids are retained within the nucleus and become embedded in a polyhedrin matrix to produce polyhedra. Prior to death the host insect climbs. Disturbance of the fragile cuticle by the elements releases up to 10^{10} polyhedra per cadaver.

9.3.1.2 Proteases

The columnar cells secrete a myriad of proteases into the insect gut lumen. The primary role of these proteases is to break down ingested food, but also serves to degrade the proteins of invading pathogens, including surface proteins of virions that are critical for cellular infection (Tellam et al., 1999). A serine protease purified from larval digestive juice of the silkworm, *Bombyx mori*, has anti-viral activity against *B. mori* nucleopolyhedrovirus (*Bm*NPV: Baculoviridae) under *in vitro* conditions (Nakazawa et al., 2004; Ponnuvel et al., 2003). Addition of a fluorescent whitener allowed infection of *B. mori* by a recombinant *Bm*NPV-expressing phytase. Protease activity in the gut juices showed a dramatic decrease, and the pH was found to be 0.45 lower after 24 h, and 1.34 lower 48 h post-inoculation than the pH of control treatments (pH 9.94 and 9.90, respectively). These results implicate both pH and protease activity as key factors in innate defense, with protease activity also being pH dependent. Proposed mechanisms to account for these observations include destruction of the electric coupling action leading to changes in circulating potassium levels between columnar cells and goblet cells causing a decrease in pH, or changes to the structure of the PM (Wang et al., 2007).

The mosquito, *Aedes triseriatus*, is a vector for LaCrosse encephalitis virus (LACV). LACV infection in the midgut is dependent upon proteolytic activity in the gut to expose the appropriate glycoprotein for binding to the midgut epithelium (Ludwig et al., 1989, 1991). *Ae. aegypti* is a vector of yellow fever and DENV. The replication and dissemination of DENV-2 in the midgut of *Ae. aegypti* significantly decreased when mosquitoes were fed a soybean trypsin inhibitor that decreased expression of late trypsin, at least in part because the virus requires proteolytic processing to be infective to midgut epithelial cells (Molina-Cruz et al., 2005).

9.3.1.3 Lipases

A lipase with strong anti-viral activity against *Bm*NPV was isolated from the digestive juices of the midgut of *B. mori* (Ponnuvel et al., 2003). The lipase was expressed only by the midgut epithelia, and was not expressed during the wandering and molting stages of larval development, possibly indicative of hormonal regulation of gene expression. Expression was not influenced by the presence of *Bm*NPV in the system, but instead appeared to be a constitutive defensive barrier in the gut.

9.3.1.4 Diet

The immunocompetence of herbivorous insects is affected by a variety of phytochemicals and foliar nutrients (Duffey et al., 1995). Larvae of *Heliothis virescens* reared on cotton were more resistant to baculovirus infection than larvae reared on lettuce, a potential mechanism being increased sloughing of midgut cells in larvae reared on cotton (Hoover et al., 2000). The chemical makeup of the diet can have a profound effect on virus survival as observed in a study that isolated distinct *Panolis flammea* NPV (*Pa*NPV: Baculoviridae) genotypes dependent on whether the larvae fed on Scots or lodgepole pine needles (Hodgson et al., 2004). Interestingly, *Helicoverpa* spp. upregulate the expression of gut proteases that are impervious to plant protease inhibitors when feeding on these plants (Volpicella et al., 2003).

Incorporation of selenium (Se) into the diet of *H. virescens* and the resulting increased levels of Se in the plasma correlated with increased resistance to baculovirus infection (*Autographa californica* multiple NPV; *Ac*MNPV) (Shelby and Popham, 2007). Selenium serves as a cofactor for selenoenzymes involved in the stress response, and maintenance of high levels of antioxidants, which may enhance anti-viral defense (Beck et al., 2004).

9.3.2   PERITROPHIC MEMBRANE

9.3.2.1   PM Production

The PM represents a physical barrier between the gut contents and the midgut epithelia found in most insects. The PM is an extracellular mesh composed of chitin, sugars, and protein. Depending on the insect, the PM is produced by all of the cells of the midgut (Type I) or only the anterior cells of the midgut (Type II), and serves as a physical barrier surrounding the gut contents by preventing direct contact with the epithelium. The PM protects the midgut cells from abrasion by food, helps compartmentalize the food for digestion, and serves as an innate defense against viruses and other microbial pathogens (Lehane, 1997). The presence and type of PM produced by an insect is correlated with the microbial content of the diet; insects feeding on more sterile diets such as vertebrate blood, phloem, nectar, or hemolymph usually lack a PM, whereas those that feed on liquid diets such as urine puddles, fecally contaminated waters, muddy pools, and rotting flesh usually produce a PM (Lehane, 1997). Facultative production of PM based on current diet has also been described. Adult hematophagous Diptera producing Type II PM rarely vector arboviruses, in contrast to many other adult Diptera with Type I PM, and mosquito larvae, which produce Type II PM, prove much harder to infect with virus than the adults with a Type I PM (Lehane, 1997). This being said, even the Type I PM in mosquitoes likely does not constitute a barrier because an arbovirus likely binds to and infects midgut epithelial cells quite rapidly following blood ingestion prior to PM formation (Hardy et al., 1983).

9.3.2.2   Pore Size

The porous nature of the PM allows digestive enzymes and nutrients to cross, but prevents many pathogens such as bacteria and large viruses from entering the ectoperitrophic space. The permeability is dependent on the pore size which has been estimated to be in the range of 3–4 nm for *Calliphora erythrocephala* (Diptera), 21–29 nm for Lepidoptera, and 24–36 nm for Orthoptera (Peng et al., 1999). The occlusion-derived virus (ODV; Fig. 9.4) of the baculovirus *Ac*MNPV is estimated to be 190 nm in diameter by 360 nm in length, and the PM of *Trichoplusia ni* larvae were almost impermeable to *Ac*MNPV ODV, whereas

Pseudaletia unipuncta PM allowed passage of more ODV, indicating a larger pore size (Peng et al., 1999). In comparison, the entomopoxvirus particles range in size from 300 by 200 nm to 470 by 300 nm, while the insect dicistroviruses cricket paralysis virus (CrPV) and *Drosophila* C virus (DCV) have small spherical particles between 27 and 30 nm (Goodwin et al., 1991; Moore and Eley, 1991). In contrast to protection of the columnar epithelial cells provided by the PM, the distal ends of the microvilli of *T. ni* can apparently penetrate the pores in the PM, providing direct access to the columnar cells from the endoperitrophic space (Adang and Spence, 1981). In *H. virescens*, baculovirus ODV are about the same size as the microvilli (Fig. 9.3B) but are able to penetrate the PM possibly by enzymatic means (see below).

9.3.2.3 PM Proteins

The peritrophin proteins found in the PM are known for their strong non-covalent interaction with the PM, and resistance to proteases. Two types of modular cysteine-rich domains have been identified, the peritrophin A domain and peritrophin B domain. The cysteines are thought to form disulfide bridges providing protection from proteases (Tellam, 1996). One protein may contain multiple copies of the domain. Thus far, peritrophins have been identified from *Lucilia cuprina*, *An. gambiae*, *Ae. aegypti*, *Chrysoma bezziana*, *Drosophila melanogaster*, and *T. ni* (Devenport, 2005; Elvin et al., 1996; Guo, 2005; Shao, 2005; Shen and Jacobs-Lorena, 1998; Tellam, 1996; Wang and Granados, 1997a, b; Wang, P. et al., 2004b). The peritrophin A domain has been identified only in peritrophic matrix proteins, a few nematode and arthropod chitinases, and baculoviruses (Tellam, 1996). Several of the peritrophins have also been shown to interact with chitin (Elvin et al., 1996; Shen and Jacobs-Lorena, 1998). Insect intestinal mucin (IIM), a peritrophin isolated from *T. ni*, shares similar properties of mammalian mucins expressed by epithelial cells; rich in proline, serine, and threonine, these proteins show high O-linked glycosylation. Another proposed mechanism of the PM is for protection from pathogens through molecular mimicry. The glycoproteins on the PM may mimic those found on the epithelia, and act as decoy sites for binding pathogenic organisms (Lehane, 1997).

9.3.2.4 Viral Proteins that Interact with PM

Enhancin

Many baculovirus genes encode proteins that interact with the PM, further suggesting an important role of the PM in innate insect anti-viral immunity. The baculovirus gene *enhancin* produces a protein that specifically degrades the IIM, allowing faster contact with the midgut epithelial cells, the target of infection (Derksen and Granados, 1988; Lepore et al., 1996; Peng et al., 1999; Slavicek and Popham, 2005). How the enhancin degrades the IIM, which is strongly protected against endogenous proteases such as trypsin and chymotrypsins, is unknown.

Chitinases

Baculoviruses also encode chitinases, which interestingly are closely related to bacterial chitinases (Wang, H. et al., 2004). The chitinases are produced late in infection, and are involved in postmortem liquefaction (Fig. 9.4), rather than acting on the chitin of the PM in early infection. Although chitinase has been detected in polyhedra, a deletion mutant lacking chitinase expression showed no difference in infectivity compared to wild-type AcMNPV (Hawtin et al., 1997). When fed at 1 µg chitinase/g of larval body weight, the baculovirus chitinase perforated the PM and was lethal, and reduction of larval growth was observed when chitinase was applied at sub-lethal doses (Rao et al., 2004). Neither of these outcomes would be beneficial to the baculovirus and may indicate why baculovirus chitinase is not used for penetration of the PM.

Chitin-Binding Proteins

The EPV fusolin protein is found in the spindles and contains a conserved chitin-binding domain (Dall et al., 1993; Hayakawa et al., 1996). A baculovirus homolog, gp37, was also identified (Li et al., 2003). Fusolin causes rapid damage of the PM, and a fusolin from *Pseudosgata separata* EPV enhanced fusion of the *P. unipuncta* MNPV in cultured cells (Hukuhara, 2001; Hukuhara et al., 2001; Mitsuhashi and Miyamoto, 2003). The spindles of *Anomala cuprea* EPV (*Ac*EPV) damage the PM of *A. cuprea*, also enhancing *Ac*EPV infection (Mitsuhashi et al., 2007). The disruption of the PM is thought to be caused by disrupting normal chitin interactions in the PM or by interacting with the chitin in a manner that allows for increased chitinase activity, as observed for bacterial chitin-binding proteins (Vaaje-Kolstad et al., 2005).

Per os *Infectivity Factors*

The baculovirus *per os* infectivity factors, or pif proteins, increase the infectivity of polyhedra. Pif-3 (*Ac*MNPV orf 115) increased oral infectivity, but is not involved in fusion to the midgut cells. The role of Pif-3 is thought to be upstream of this event and it may interact with the PM (Ohkawa et al., 2005). Another baculovirus gene, *Ac*MNPV orf 150, encodes Pif-4, a potentially secreted or type 2 membrane anchored protein containing a single peritrophin A domain (Ayres et al., 1994; Ohkawa et al., 2005). Deletion of the gene encoding this protein, and its homolog orf 126 in *Bm*NPV, resulted in decreased oral virulence (Lapointe et al., 2004; Zhang et al., 2005). The presence of a peritrophin A domain was thought to provide chitin-binding properties, increasing the contact of the baculovirus to the PM. Subsequent studies did not demonstrate chitin binding, but *Ac*MNPV orf150 also contains an integrin-binding domain (Lapointe et al., 2004; Zhang et al., 2005). This protein was found by TEM to associate with ODV envelopes, but Pif-4 was not detected as a component of the ODV envelope (Braunagel et al., 2003; Lapointe et al., 2004). Hence, this protein may have been lost in the purification process or changed in some manner during these analyses. Alternatively, Pif-4 could be packaged into the polyhedra, but not the virions themselves, since the ODV lacking orf150 showed the same infectivity as wild-type ODV, whereas the polyhedra lacking Pif-4 were much less infectious than wild-type polyhedra. This gene also has a homolog in *H. armigera* EPV (Dall et al., 2001).

9.3.3   CELL SLOUGHING

Immediately after ecdysis, some lepidopteran larvae including *T. ni* lack PM and are more susceptible to viral infection. After 3 h, however, the PM is fully formed. There is increasing resistance to viral infection observed throughout an instar (Engelhard and Volkman, 1995). When challenged with baculovirus, all insects were able to clear infected midgut epithelial cells during the molt to fifth instar; thus if secondary infection of other tissues had not occurred, the insect would effectively clear the virus from its system. The increasing resistance throughout the instar was attributed to the age-related rate of establishing and/or sloughing infected epithelial cells (Engelhard and Volkman, 1995). Similar results were described for *H. virescens* (Washburn et al., 1995). Intrahemocoelic injection of virus into both species resulted in similar mortality despite the age of the insect, providing further evidence for the involvement of the midgut epithelia in this developmental resistance (Washburn et al., 1995). Insects fed on cotton were more resistant to baculovirus infection than those fed on artificial diet (Hoover et al., 2000). The proposed mechanism of protection was increased levels of cell sloughing in the cotton feeding insects. Ingestion of cotton has been shown to increase levels of reactive oxygen species that may damage the epithelial cells promoting faster cell turnover, and potentially also inactivate the virus.

Helicoverpa armigera stunt virus (*HaSV*: tetraviridae), which infects only midgut epithelial cells, significantly increased cell sloughing in infected *H. armigera* and virus could be detected in most sloughed cells (Brooks et al., 2002). In the later stages of infection, entire blocks of cells were observed detaching from the BM, and in some places stretches of BM were observed with no epithelial cells, or very degenerate cells which lacked microvilli. Brooks et al. (2002) hypothesized that the *HaSV* infection which can induce apoptosis of cells still within the midgut epithelia may contribute to the increased rates of cell sloughing.

Cell sloughing has also been observed in *Culiseta melanura* and *Culex tarsalis* mosquitoes, in response to infection with Eastern equine encephalitis virus (EEEV), and WEEV respectively (Weaver et al., 1992, 1988). The mechanisms controlling cell sloughing are unknown.

9.3.4   BASEMENT MEMBRANE

Basement membrane, also referred to as basal laminae, are extracellular sheets of proteins that surround tissues, providing structural support, a filtration function, and a surface for cell attachment, migration, and differentiation (Rohrbach and Timpl, 1993). As described for viruses of medical importance (Romoser et al., 2005),

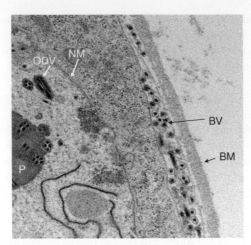

FIGURE 9.5 Transmission electron micrograph of a midgut cell of the tobacco budworm, *Heliothis virescens*, 116 h after infection with *Ac*MNPV, showing accumulation of BV beneath the basement membrane. BM, basement membrane; BV, budded virus; NM, nuclear membrane; ODV, occlusion-derived virus; P, polyhedron. TEM by Hailin Tang.

BMs appear to act as a barrier to dissemination of baculoviruses within infected insects. Budded virus (BVs; Fig. 9.4), the baculovirus phenotype that serves to disseminate infection within an infected host, are too large to freely diffuse through the pores in the BM that surround tissues of the host insect (Reddy and Locke, 1990). Co-injection of BVs and clostridial collagenase, a protease known to degrade BM, resulted in enhanced infection of host tissues (Smith-Johannsen et al., 1986). Ultrastructural studies of infection by the baculovirus *Cydia pomonella* granulovirus (*Cp*GV: Baculoviridae) and *Ac*MNPV revealed a substantial accumulation of BVs in the extracellular spaces between BMs and the plasma membranes of midgut and fat body cells (Hess and Falcon, 1987) (Fig. 9.5). Collectively, these observations suggest that insect BM inhibits the movement of BVs.

The means by which insect viruses negotiate the basement membrane barrier are still unclear. Systemic spread of baculoviruses within the host insect may occur by direct penetration of the basement membrane into the hemocoel, either by an enzymatic process or where the BM is thin (Federici, 1997; Flipsen et al., 1995; Granados and Lawler, 1981). Alternatively, baculoviruses may use the host tracheal system as a conduit to bypass basement membranes and establish systemic infection of host tissues (Engelhard et al., 1994). There is debate over whether one route predominates over the other (Federici, 1997; Volkman, 1997). The identification of virus-encoded proteases with BM-degrading potential in an EPV (Afonso et al., 1999), and granuloviruses (Ko et al., 2000), supports the hypothesis that at least some insect viruses may use enzymatic means to traverse the BM (Liu et al., 2006a). There is also the suggestion that the baculovirus-encoded fibroblast growth factor (*fgf*), which is common to all baculoviruses that infect multiple tissues within an insect as opposed to being restricted to the gut, may function to attract hemocytes to sites of virus infection by chemotaxis. It is possible that a protease from granular cells, used for BM remodeling (Kurata et al., 1991, 1992; Nardi et al., 2001), may facilitate movement of virus across the BM.

9.4   CELLULAR IMMUNITY

One of the most important aspects of immunity is the ability to recognize non-self and respond in an appropriate manner. Extracellular pathogens can be recognized from pattern recognition motifs unique to a group of pathogens, and are then inactivated. Intracellular pathogens pose a more challenging threat to the host, as they are replicating within the host's own tissues. Ideally, only the virally infected tissues are destroyed, with minimal damage to neighboring healthy cells. Examples of cell-based immunity against viruses have been identified, but the precise mechanisms and cell signaling pathways involved are not yet understood.

9.4.1   RECOGNITION OF VIRUS-INFECTED TISSUES

There is high homology between the BM of invertebrates and vertebrates (Fessler and Fessler, 1989). The basement membrane is composed primarily of laminin, collagen IV, and proteoglycans which are known to mediate vertebrate immune cell adhesion and targeting. In insects, the integrity of the BM is important for hemocyte response to both cuticular wounds to facilitate healing, and to endogenous tissues that are developing abnormally. Insects distinguish self from non-self based on the presence or absence of the correct BM structure. In strains of *D. melanogaster* Meigen that produce melanotically encapsulated tumors, the process of melanotic tumor formation is initiated by disruption of the BM overlying the fat body, followed by aggregation of hemocytes around the abnormal surface and melanotic encapsulation of the affected area (Rizki and Rizki, 1974, 1980). Tissue grafts in *D. melanogaster* with mechanically or enzymatically generated BM damage also underwent melanotic encapsulation, but tissue grafts with an undamaged BM preparation were not encapsulated (Rizki and Rizki, 1980). Coating of dishes with a basement membrane preparation (Matrigel) prevented hemocytes of *Pseudoplusia includens* Walker (Lepidoptera: Noctuidae) from spreading on the surface of the dish (Pech et al., 1995), suggesting that the BM can make a foreign surface appear to be self. In addition, antibodies that recognize surface proteins of the insect pathogenic fungus *Nomuraea rileyi* cross-reacted with antigens on the fat body surface of the host insect, the beet armyworm, *Spodoptera exigua* (Pendland and Boucias, 1998; Pendland and Boucias, 2000). Hyphae of *N. rileyi* are able to evade the immune response of *S. exigua* by using molecular mimicry such that they present a surface that resembles the BM of the host insect.

In the few known cases where virus infection results in melanotic encapsulation of the basement membrane overlying the infected cells (Washburn et al.,

1996, 2000), replication of the virus within the underlying cell may result in distension of the basement membrane, which is then recognized as foreign. Disturbance of the basement membrane structure would elicit the immune response in this case, rather than being mediated by direct recognition of the virus. Alternatively, hemocytes may be attracted to the site of infection by recognition of damaged cells or cells undergoing aberrant apoptosis (Zambon et al., 2005).

9.4.2 ENCAPSULATION

Encapsulation of baculovirus-infected cells has been documented in only two species of Lepidoptera, *H. zea* and *Manduca sexta*, which are semi-permissive to AcMNPV (Washburn et al., 1996, 2000). Infected tracheal cells were surrounded by hemocytes and contained capsules of virus-infected cells, the capsules being morphologically identical to those encapsulating bacteria or parasites (Washburn et al., 1996). Suppression of the immune response using either intrahemocoelic injection of diethyldithio-carbamic acid (DDCA) or parasitization by the wasp *Campoletis sonorensis* resulted in an increased number of infected cells. In the semi-permissive host *M. sexta*, hemocyte aggregation, along with melanization, was observed on the tracheal cells as early as 24 h post-inoculation, and mature capsules noted at 72 and 96 h post-infection (hpi) (Washburn et al., 2000). Parasitization of *M. sexta* with *C. congregata*, followed by inoculation with baculovirus resulted in the ability of the virus to spread throughout the host. The ability of AcMNPV to establish primary infection of the midgut cells and secondary infection of the tracheal cells did not differ between the permissive host, *H. virescens* and the semi-permissive host, *H. zea*; however, hemocytes in *H. zea* not only participated in encapsulation of viral foci, but were also resistant to infection by the virus despite endocytosis of BV (Trudeau et al., 2001). The hemocytes play a central role in dissemination of those baculoviruses that infect multiple tissues within the host insect. Baculovirus-infected hemocytes are not able to encapsulate foreign targets (Trudeau et al., 2001). Hence, successful infection of the hemocytes allows not only for virus replication, but also reduces the ability of the host insect to defend itself. *H. zea* is significantly less susceptible to infection by AcMNPV than the closely related *H. virescens*. The physiological basis for this difference appears to be the inability of AcMNPV to replicate in the hemocytes of *H. zea* (Trudeau et al., 2001). Although virus was taken up by the hemocytes of *H. zea*, and the nucleocapsids were detected in the nucleus, replication did not occur. The hemocytes therefore removed virus from the hemolymph while maintaining their ability to encapsulate sites of infection.

Interestingly, the *H. zea* strain that was only semi-permissive to AcMNPV infection was susceptible to the closely related virus *Rachiplusia ou* MNPV (RoMNPV). RoMNPV and AcMNPV are so similar that they are considered to be variants of the same virus (Blissard et al., 2000; Harrison and Bonning, 2003). The difference between AcMNPV and RoMNPV on infection of *H. zea* was the differential ability of RoMNPV to replicate in the hemocytes of *H. zea* (L. Volkman, unpublished data). The genetic basis for this difference between the two viruses remains to be determined.

9.5 SUBCELLULAR INHIBITION

9.5.1 APOPTOSIS

Apoptosis, or programmed cell death, is an effective immune response against viral infections, which has been observed in both invertebrates and vertebrates (Cashio et al., 2005; Siegel, 2006). The most detailed information on apoptosis has been described for *Drosophila*; however, this understanding has its roots in the identification of several baculovirus genes that were found to inhibit apoptosis (Clem, 2005). Apoptosis is activated by caspases, which are normally inactivated in the cell by virus expression of inhibitor of apoptosis proteins (IAPs) that bind to them (Salvesen and Duckett, 2002). The most significant evidence for apoptosis as an effective innate immune response is the decreased virulence of insect viruses that lack genes encoding inhibitors of apoptosis.

The arboviruses often are cytopathic in the vertebrate host, causing apoptotic cell death, and potentially significant disease, but must establish life-long infections in the insect vector in order to be transmitted. Sindbis virus in infected mammalian and mosquito cell cultures illustrate the difference in pathology between the two hosts, wherein apoptosis is more frequently seen in response to infection in mammalian cells (Karpf and Brown, 1998). Recently it has come to light that the arboviruses can induce some apoptosis *in vivo* in mosquito hosts. Illustrative of this, *Culex pipiens quinquefasciatus* infected with West Nile virus (WNV) show ultrastructural evidence of apoptosis in the salivary glands (Girard et al., 2005, 2007).

9.5.1.1 P35

p35 is a universal caspase inhibitor first identified in *Ac*MNPV. Interestingly, with the exception of baculovirus homologs, no homolog of p35 has been identified in any organism sequenced to date (Clem, 2005). When infected with wild-type *Ac*MNPV, Sf-21 cells proceed through the normal infection process producing both BV and ODV in the infected cells; however, infection with *Ac*MNPV lacking a functional *p35* gene induced apoptosis, and resulted in limited BV and no ODV production (Clem et al., 1991). Cellular caspases cleave p35 causing a conformational change and subsequent covalent bond formation between the caspase and p35, rendering the caspase inactive (Clem, 2005). Despite its definition as a universal caspase inhibitor, p35 appears to have a greater affinity to effector caspases, and exhibits low levels of inhibition against apical caspases, such as caspase-9 in humans (Vier et al., 2000), *Drosophila* (Dronc) (Meier et al., 2000), and the lepidopteran homolog of Dronc in *Spodoptera frugiperda*, *Sf*-caspase-X (Clem, 2005; LaCount et al., 2000). Interestingly, a baculovirus p35 homolog, *p49*, found in *Spodoptera litura* NPV is able to inhibit the caspsases Dronc (Jabbour et al.,

2002) and Sf-caspase-X (Zoog et al., 2002) as well (Clem, 2005). The p35 protein may also inhibit apoptosis by quenching free radicals; p35 inhibited cytochrome c release from the mitochondria of oxidant exposed cells blocking caspase activation (Sahdev et al., 2003).

9.5.1.2   IAPs

Further study of baculoviruses identified a new anti-apoptotic protein, inhibitor of apoptosis (IAP), in *Cp*GV (Crook et al., 1993). Infection with virus lacking p35 resulted in apoptosis of infected Sf-21 cells in culture, but cells could be rescued by the addition of DNA from *Cp*GV containing the *iap* gene. Both proteins could prevent apoptosis caused by the addition of actinomycin D to the tissue culture. Subsequent studies identified IAP homologues in *Orygia pseudotsugata* MNPV (Birnbaum et al., 1994). IAPs, but not p35, effectively block activation of pro-Sf-caspase-1 (Seshagiri and Miller, 1997).

Studies on the *in vitro* host range of *Ac*MNPV lacking p35, iap-1 or iap-2 showed that p35 knock-out viruses resulted in apoptosis in permissive cell lines, while viruses lacking functional iap-1 and iap-2 in tissue culture had little effect (Griffiths et al., 1999). Interestingly, *Op*MNPV-IAP1 has 58% amino acid identity with *Cp*GV-IAP, but only 28% with *Ac*MNPV (Griffiths et al., 1999), suggesting different functions, and possibly different origins or evolutionary age.

9.5.1.3   Host Homologs

Cellular homologs of the baculoviral *iap* genes were first identified in *D. melanogaster*, and more recently in the lepidopteran hosts *S. frugiperda* and *T. ni*, suggesting that the baculoviral *iaps* were of host origin, acquired through homologous recombination (Clem, 2005). This hypothesis is further supported by the fact that most baculoviruses have an *iap* homolog, but only one branch has *p35*. Phylogenetic studies suggest the presence of three different *iap* genes in the ancestor of the viral genus NPV, as well as a separate capture event in the *iap* gene of EPV (Hughes, 2002). In addition, the *Cp*GV *iap* clustered with the lepidopteran *iaps*, rather than with the NPV *iaps*. Subsequent study of endogenous IAPs suggests that IAPs have affinity to specific caspases (Deveraux et al., 1997; Roy et al., 1997).

One of the IAPs identified from *O. pseudotsugata*, *Op*-IAP3, has been shown to act as a functional E3 ubiqutin ligase instead of directly inhibiting caspases as with other baculovirus IAPs (Green et al., 2004). Phylogenetic analysis of the *iap-3* genes showed that they cluster with the lepidopteran *iap* genes (Blitvich et al., 2002). This protein was able to ubiquitinate pro-apoptotic cellular proteins such as *Drosophila* protein HID, thus playing an important role in anti-apoptotic activity; however, lepidopteran homologs of HID have not been identified. Op-IAP3 expression may also change the subcellular localization of HID (Vucic et al., 1997, 1998).

The sequence and tissue specificity of an IAP from the mosquito *Ae. triseriatus* was determined based on homology to *Drosophila* IAP 1 (Blitvich et al., 2002).

IAP from *Ae. albopictus* has been shown to have anti-apoptotic activity in verte-brate cells infected with Bluetongue virus (Li et al., 2007). IAP1 sequences from *Cx. pipiens*, *Cx. tarsalis*, *Ae. aegypti*, *Ae. albopictus*, and *Ae. triseriatus* also have been cloned and characterized (Beck et al., 2007). The sequence of IAP1 from *Cx. pipiens* (kindly provided by Dr. Brad Blitvich) was used as a template to design a 60-mer oligonucleotide to be used in a microarray analyses. In a study of salivary glands from *Cx. p. quinquefasciatus*, 14 and 21 days post-infection with WNV, IAP1 was down-regulated (Girard et al., unpublished), which may be evidence for the molecular mechanisms that underlie the extensive apoptosis seen in salivary glands at these time points post-infection with WNV (Girard et al., 2005, 2007).

9.5.1.4 Mode of Action

The mechanism by which a cell responds to virus infection by undergoing apoptosis is not understood. Several factors may be of importance, including the synthesis of viral DNA in the cell, late viral gene expression, and the obstruction of host transcription and translation (Clem, 2001). Addition of the transcription inhibitors actinomycin D, 5,6-dichlorobenzimidazole riboside, or alpha-amanatin to *Sf*-21 cells results in rapid induction of apoptosis (Clem and Miller, 1994). Induction may be a result of the lack of endogenous IAP to inhibit apoptosis. Cells transiently infected with a set of genes necessary for *Ac*MNPV DNA synthesis yielded little DNA unless co-transfected with *p35* or an active *iap* gene such as

---

Drosophila, A Model Organism for the Study of Immunity to Viruses

The genetic resources associated with the use of *Drosophila* as a model organism provide several avenues for investigation of the pathways involved in virus immunity (Cherry and Silverman, 2006). Indeed, *Drosophila* has become a model for the study of immunity to viruses. Mutant lines that are defective in production of key proteins involved in a specific pathway (Toll, Imd, RNAi) can be tested for increased susceptibility to virus infection. Cultured *Drosophila* cells can be tested for the presence of small interfering RNAs (siRNAs) following infection, and can be used to delineate viral sequences encoding suppressors of silencing. Microarrays can be screened to examine the host response to virus infection at the level of transcription. As a result of the great potential of this system, our knowledge of insect immunity to viruses has increased significantly in recent years. As a result of research in *Drosophila*, there is evidence for the involvement of three pathways in anti-viral defense: degradation of viral RNA by RNAi, the Toll pathway, and cytokine-mediated induction of genes via the JAK–STAT (signal transducer and activator of transcription) signaling pathway.

Op-iap or *Cp-iap* (Lu and Miller, 1995). This suggests that expression of proteins necessary for viral DNA synthesis induces apoptosis (Clem, 2001). One of the essential baculoviral proteins, the activator protein immediate early-1 (ie-1), itself induces apoptosis (Prikhod'ko and Miller, 1996). Another immediate early protein, pe38, is also involved in induction of apoptosis (Prikhod'ko and Miller, 1999). Both early and late factors are necessary for maximal induction of apoptosis (LaCount and Friesen, 1997).

9.5.2   RNA INTERFERENCE

RNA interference (RNAi) is the evolutionarily conserved mechanism for gene silencing which is guided by double-stranded RNA (Meister and Tuschl, 2004; see Reynolds, chapter 12). RNAi allows for the targeting of specific mRNAs for destruction (Kavi et al., 2005). dsRNA is produced as an intermediate in the replication of RNA viruses or as a byproduct of symmetrical transcription from DNA viruses. The dsRNA is recognized by the endoribonuclease Dicer-2. The Dicer-2/R2D2 complex initiates RNAi by processing long dsRNA into small interfering RNA (siRNA), and facilitates incorporation of siRNA into the RNA-induced silencing complex (RISC) which contains Argonaute 2 (Ago-2) (Liu et al., 2006b).

RNAi pathways have been shown to protect *D. melanogaster* from infection by several RNA viruses: Flock house virus (FHV: Nodaviridae), Cricket paralysis virus (CrPV: Dicistroviridae) (van Rij et al., 2006; Wang et al., 2006), *Drosophila* C virus (DCV: Dicistroviridae) (van Rij et al., 2006), and *Drosophila* X virus (DXV: Birnaviridae) (Zambon et al., 2006). Enhanced disease susceptibility was found for FHV, DCV, CrPV, and Sindbis virus in mutant flies defective in *dcr-2* or *r2d2* (Galiana-Arnoux et al., 2006; Wang et al., 2006). In addition, infection of cultured *Drosophila* S2 cells with CrPV induced anti-viral silencing as determined by detection of CrPV-specific siRNAs. A suppressor of RNAi was detected within the first 140 codons of CrPV ORF1. Anti-viral silencing was also demonstrated on infection of S2 cells with FHV: silencing was RISC-dependent and inhibited by FHV-encoded B2 protein (Li et al., 2002). Hence, *Drosophila* uses RNAi as an anti-viral defense, but both CrPV and FHV encode suppressors of RNAi.

The nuclease Ago-2, which is the central catalytic component of the RNA-induced silencing complex (RISC), was shown to be essential for anti-viral defense against both DCV and CrPV. Flies defective in Ago-2 expression, showed a significant increase in viral RNA accumulation, a 1000-fold increase in virus titer, and increased mortality rate (van Rij et al., 2006).

The recognition and destruction of double-stranded RNA has recently been recognized as an intracellular anti-viral response against insect viruses of medical importance, and is the subject of a number of arbovirus–vector competence studies. By virtue of their positive-sense RNA genomes, the Togaviridae and Flaviviridae (that make up many of the arboviruses transmitted by mosquitoes) produce negative-stranded RNA during replication. These replicative intermediates can bind positive strand RNA to form double-stranded intermediates that are targets for

RNAi degradation (Sanchez-Vargas et al., 2004). Building on studies of the RISC pathway and its role in response to arboviruses in cell culture (Hoa et al., 2003), *A. gambiae* were infected with a GFP-expressing O'nyong-nyong virus (Alphavirus: Togaviridae) and simultaneously subjected to RNAi-based reduction of RISC elements; increased viral titer and dissemination resulted from silencing Ago-2 (Keene et al., 2004).

Applying this novel strategy to vector–virus systems of considerable public health concern, the potential use of RNA interference in disease control recently was described for *Ae. aegypti* and DENV (Franz et al., 2006). A transgenic line of *Ae. aegypti* was engineered to express an inverted repeat of DENV-2 genomic RNA, driven by a midgut-specific, blood meal triggered carboxypeptidase promoter. Cohorts of a transgenic line (Carb77) were significantly resistant to orally provided DENV-2 as measured by reductions in virus titer, but were not resistant when the DENV-2 was injected, bypassing the midgut. That this reduction was the manifestation of RNAi was evidenced by the specificity of resistance (Carb77 mosquitoes were susceptible to infection with DENV-4), the presence of small interfering RNAs (siRNA) homologous to the inverted repeat, and when *ago2* was silenced in the Carb77 line, DENV-2 virus titers increased. Research is ongoing to engineer anti-DENV effector constructs that produce a high level of resistance to all four DENV serotypes.

Infection of mutant *Drosophila* lines with DXV, which has a double-stranded RNA genome, was used as a screen for mutant lines with greater susceptibility to DXV. Infected flies develop acute anoxia sensitivity and die within 2 weeks of the onset of symptoms. From this screen it was found that the Toll and RNAi pathways, but not the immune deficiency (Imd) pathway, function in *Drosophila* anti-viral defense (Zambon et al., 2005, 2006). Flies with mutations in genes that are essential for the Toll or RNAi pathway show dramatic increases in susceptibility to DXV with increased viral titer and earlier susceptibility to infection. The anti-viral effects mediated by the Toll pathway appear to be systemic (Zambon et al., 2005), while the RNAi effect in *Drosophila* (in contrast to RNAi in plants, worms, and some other insects) occurs only in the cell in which dsRNA is detected and does not spread to neighboring cells (Van Roessel et al., 2002). Hence, RNAi acts on a cell-to-cell basis, slowing the progression of infection, which may provide more time for the Toll-mediated response against virus-infected tissues to be mobilized.

9.6   HUMORAL IMMUNITY

9.6.1   CELL SIGNALING PATHWAYS

The *Drosophila* C virus (DCV: Dicistroviridae), which can infect both flies and cells in culture, has been employed to demonstrate that virus entry occurs via clathrin-mediated endocytosis (Cherry and Perrimon, 2004), and to identify factors essential for growth and replication of DCV (Cherry et al., 2005). Through a genome-wide RNAi screen using 21,000 dsRNAs (representing 91% of the pre-dicted genes in *Drosophila*) the ribosome was identified as limiting for viruses such as DCV with internal ribosome entry sites (IRES).

9.6.1.1 Toll Receptor Pathway

Three different cell signaling pathways have been implicated in immune defense against viruses, the Toll, Imd, and Jak–STAT pathways. The most well known is Toll, a transmembrane receptor, and its mammalian homologs Toll-like receptors (TLR). Interestingly, both insects and mammals contain around 10 copies of Toll receptors, but while all 10 TLRs of mammals have immune-related functions, thus far, only one, Toll, has been associated with immune function in insects (Imler and Zheng, 2004). The other homologs in insects appear to be involved in develop-ment. Another important difference between the insect and mammalian pathways is that the mammalian TLRs interact directly with the pathogen (thus they are pat-tern recognition receptors or PRRs), whereas the Toll receptor in insects requires the ligand Spaetzle, a cysteine knot growth factor, for activation (similar to the cytokine receptors of mammals) (Lemaitre et al., 1996). Pattern recognition of pathogens occurs upstream of Toll, and a signaling cascade activates Spaetzle allowing it to bind to Toll. Six *spaetzle*-related genes have been identified in *D. melanogaster*, and may encode ligands for other TLRs in insects (Parker et al., 2001).

Evidence for inhibition of Toll-mediated reponses in an arboviral infection comes from a microarray study on *Ae. aegypti* exposed to Sindbis virus (Alphavirus: Togaviridae). A homolog of *Drosophila* dorsal-related immune factor (Dif), a transcription factor downstream of the Toll receptor, was represented on the arrays used in this study. *Aedes* Dif was induced 1 day post-infection (p.i.), but induction was no longer evident 4 days p.i., and degradative ubiquitin ligases, critical for the cascade of events that occurs upstream of Dif, also were down-regulated (Sanders et al., 2005).

9.6.1.2 Imd Pathway

The production of anti-microbial peptides in *Drosophila* is regulated by the Toll and Imd signaling pathways (see chapters 1, 5 and 11). These pathways also regulate activation of Dif and Relish respectively, two members of the NF-kappaB family of transcription factors (Brennan and Anderson, 2004). DXV, a double-stranded RNA virus, activates both the Toll and Imd pathways during infection; how-ever, only Toll provided a protective anti-viral response to DXV (Zambon et al., 2005). Mutant flies lacking a functional Dif (an NF-kappaB-like transcription fac-tor of the Toll pathway) had greater virus titers, and died sooner than wild-type flies. Polydnaviruses also encode inhibitors structurally homologous to interferon kappa B proteins known to sequester NF-kappaB in the cytoplasm of mammals (Thoetkiattikul et al., 2005). Imd is a death domain adaptor protein which binds to dFadd and results in activation of the NF-kappaB-like transcription factor Relish. A loss of function Relish mutant had similar outcomes to DXV infection as wild-type flies, indicating that although the Imd pathway was activated by the viral infection in WT flies, the resulting antimicrobial peptides produced had no effect on DXV (Zambon et al., 2005).

9.6.1.3 Jak–STAT Pathway

Infection with *Drosophila* C virus (DCV: Dicistroviridae) induces a set of genes distinct from the Toll and Imd pathways, implicating a third, evolutionarily conserved innate immunity pathway in defense against viruses (Dostert et al., 2005). DCV infection induces transcription of some 150 genes. A subset of these genes is regulated by the kinase Jak and transcription factor STAT. Flies with mutations in the gene *hopscotch*, which encodes the Jak kinase of *Drosophila* (Agaisse and Perrimon, 2004), show increased susceptibility to infection by DCV (Dostert et al., 2005; Wang et al., 2006). However, the most abundant gene product, Vir-1, has no effect on DCV infection (Dostert et al., 2005). Furthermore, many of the Jak–STAT responsive genes were expressed in non-immune, non-infected tissues. These results provide evidence for the presence of a novel unde-scribed anti-viral mechanism in insects (Cherry and Silverman, 2006).

9.6.1.4 Antimicrobial Peptides

Proteins induced in the hemolymph of *Drosophila* adults following injection of DCV have been analyzed by MALDI-TOF mass spectrometry (Sabatier et al., 2003). In contrast to the use of this differential screen for analysis of the impact of bacteria on host proteins for which more than 24 induced peptides were identified, only one peptide, pherokine-2 (Phk-2), was induced on virus infection. Constitutive overexpression of Phk-2 did not protect flies against DCV infection, however, and the function of this protein remains to be determined.

Affymatrix gene chips were used to examine *Drosophila* gene expression 24 h after infection with a variety of microbes, including DCV (Roxstrom-Lindquist et al., 2004). Eleven genes were induced in response to infection by DCV, including several anti-microbial peptides (attacin A, cecropins A1 and A2, and Drosomycin).

9.6.2 PHENOLOXIDASE

Induction of the plasma enzyme phenoloxidase (PO: EC 1.14.18.1) appears to function as a constitutive, innate humoral immune response against viruses and bacteria (see chapter 4). Incubation of *H. zea* single NPV with plasma reduced the infectivity of the virus (Popham et al., 2004; Shelby and Popham, 2006). Indeed PO from the plasma of *H. virescens* exhibits anti-viral activity against a variety of vertebrate viruses (Ourth and Renis, 1993). Baculovirus infection of *H. virescens* results in a significant reduction (approximately 40%) in the pool of pro-PO and PO in the hemolymph by 72 h after infection, which may represent a viral counter-measure against the anti-viral properties of this host enzyme (Huarong Li, unpublished data). Honeybees infected with Varroa mites exhibit reduced expression of immune related genes, including PO, and a simultaneous increase in Deformed wing virus (DWV) replication in the newly emerged adult bees (Yang and Cox-Foster, 2005). Interestingly, mite free, uninfected bees were found to lack an inducible PO response during the 24 hours post emergence, providing further evidence for PO as an anti-viral agent (Yang and Cox-Foster, 2007).

9.6.3 HEMOLIN

Hemolin is a hemolymph protein that appears to be restricted to the Lepidoptera, and is strongly induced by bacterial infection (Faye et al., 1975). Hemolin binds to bacteria, lipopolysaccharide (LPS), and to hemocytes, and may function as an opsonin, or pattern recognition molecule. In an attempt to silence hemolin by injection of larvae of the Chinese oak silk moth, *Antheraea pernyi* with dsRNA, it was found the dsRNA itself, and also baculovirus infection resulted in enhanced hemolin expression (Hirai et al., 2004). Hemolin consists of four Ig domains, similar to many mammalian viral receptors in the immunoglobulin superfamily. Soluble hemolin may bind virus in the hemolymph thereby slowing the progression of virus infection (Hirai et al., 2004).

9.6.4 LECTINS

Examination of larvae of the silkworm, *B. mori*, indicated that hemagglutinating activity increased following infection with a Cypovirus (Reoviridae). A lectin-like protein accumulated at increased levels in the hemolymph of infected larvae (Mori et al., 1989).

9.6.5 EICOSANOIDS

Eicosenoids are oxygenated metabolites of some C20 polyunsaturated fatty acids, and include arachidonic acid. Eicosenoids mediate many cellular actions including cellular reactions to attack by bacteria, fungi, protozoa, and parasitoids, including cell spreading and activation of prophenoloxidase (see chapters 3 and 4). Nine inhibitors of eicosanoid biosynthesis were found to increase susceptibility of larvae of the gypsy moth, *Lymantria dispar* to LdMNPV with 3.5–6.6-fold reductions in LC_{50}s (Stanley and Shapiro, 2007). These results implicate a role for eicosenoids in insect anti-viral defense reactions (see chapters 3, 4, and 9).

9.6.6 PARALYTIC PEPTIDE-BINDING PROTEIN

A protein in *B. mori* called paralytic peptide-binding protein (PP-BP), which was highly homologous to the growth blocking peptide-binding protein, can block the immune responses of paralytic and plasmatocyte spreading activities in the hemocyte immune reaction in *Pseudaletia separata* (Hu et al., 2006). PP-BP was expressed almost exclusively in the hemolymph of *B. mori*, with increased expression following infection with *Bm*NPV suggesting a role in a cellular immune response against *Bm*NPV.

9.7   DEVELOPMENTAL RESISTANCE

Lepidopteran larvae become increasingly resistant to baculovirus infection as they age (Engelhard and Volkman, 1995; Kirkpatrick et al., 1998; Teakle et al., 1986). Two factors relevant to this phenomenon are an age-dependent rate of establishing and/or sloughing-infected midgut cells, and the ability of fourth instar larvae of *T. ni* to clear infection of the midgut epithelium by ecdysis to the fifth instar. Developmental resistance may be hormonally mediated (Engelhard and Volkman, 1995; Hoover et al., 2002). In contrast to other insects, larvae of the gypsy moth, *L. dispar*, appear to have systemic resistance to baculovirus infection. In other species, a systemic component to resistance was only seen in the penultimate instar and not earlier instars (Kirkpatrick et al., 1998; Teakle et al., 1986). This developmental resistance was reduced by the addition of the fluorescent brightener M2R in fifth instar *H. virescens* orally inoculated with baculovirus, although a greater decrease was observed for the earlier instars (Kirkpatrick et al., 1998). The developmental resistance to BV in the fifth instar was also influenced by hormones, with application of the juvenile hormone mimic methoprene resulting in higher mortality of intrahemocoelically inoculated *H. virescens* 36–48 h post-molt. Further study of M2R revealed that it effectively stopped production of the PM (Sheppard and Shapiro 1994; Wang and Granados, 2000). It has also been suggested to inhibit sloughing of virus-infected midgut cells (Hoover et al., 2000; Washburn et al., 1998), and/or to suppress apoptosis thereby allowing virus replication to occur in the midgut. The systemic developmental resistance of gypsy moth larvae may be mediated by an anti-viral immune response, capable of clearing virus from the hemolymph, or may result from hormonally mediated refractoriness to infection (Hoover et al., 2002).

9.8   DISCUSSION

Although significant advances have been made in our understanding of the mechanisms and counter-mechanisms involved in anti-viral immunity, there are still many gaps in our knowledge of insect immunity to viruses. Genomic and proteomic analyses to examine the impact of virus infection on levels of gene transcription and translation provide tremendous potential for investigation of new leads in this area. The potential for this type of analysis has already been demonstrated using viruses that infect *Drosophila*, and similar analyses are underway for examination of the impact of virus infection on mosquitoes.

Given the evidence for immunological specificity and memory in insects repeatedly challenged with bacteria (Sadd and Schmid-Hempel, 2006), whether Down's syndrome cell adhesion molecule (Dscam) is involved in anti-viral immunity will be of particular interest. Dscam is an immunoglobulin-like gene that can encode 38,000 different transcripts in *D. melanogaster* and 31,000 different transcripts in *A. gambiae* via alternative splicing (Dong et al., 2006; Watson et al., 2005).

From a practical standpoint, some of the means by which insect viruses have overcome physiological barriers to infection, such as the use of enzymes that degrade the PM, can potentially be exploited for insect pest control (Liu et al., 2006a). In addition, insect viruses have potential as vectors for virus-induced gene silencing (VIGS), to target genes in the insect for both fundamental study of gene function and for pest management purposes. Increased understanding of virus–host mechanisms will also facilitate development of arbovirus vectors that are incapable of virus transmission. Overall the study of insect immunity to viruses promises to provide many fruitful future avenues for exciting research.

ACKNOWLEDGMENTS

The authors thank Drs. Loy Volkman, Huarong Li, and Brad Blitvich for permission to cite unpublished research. This material is based in part upon work supported by USDA NRI 2003-35302-13558 as well as Hatch Act and State of Iowa funds.

REFERENCES

Adang, M. J., and Spence, K. D. (1981). Surface morphology of peritrophic membrane formation in the cabbage looper, *Trichoplusia ni*. *Cell Tissue Res.* **218**, 141–147.

Afonso, C. L., Tulman, E. R., Lu, Z., Oma, E., Kutish, G. F., and Rock, D. L. (1999). The genome of *Melanoplus sanguinipes* Entomopoxvirus. *J. Virol.* **73**, 533–552.

Agaisse, H., and Perrimon, N. (2004). The roles of JAK/STAT signaling in *Drosophila* immune responses. *Immunol. Rev.* **198**, 72–82.

Ayres, M. D., Howard, S. C., Kuzio, J., Lopez-Ferber, M., and Possee, R. D. (1994). The complete DNA sequence of *Autographa californica* nuclear polyhedrosis virus. *Virology* **202**, 586–605.

Beck, E. T., Blair, C. D., Black IV, W. C., Beaty, B. J. and Blitvich, B. J. (2007). Alternative splicing generates multiple transcripts of the inhibitor of apoptosis protein 1 in *Aedes* and *Culex* spp. mosquitoes. *Insect Biochem. Molec. Biol.* **37**, 1222–1233.

Beck, M., Handy, J., and Levander, O. A. (2004). Host nutritional status: The neglected virulence factor. *Trends Microbiol.* **12**, 417–423.

Berenbaum, M. (1980). Adaptive significance of midgut pH in larval lepidoptera. *Am. Natl.* **115**, 138–146.

Birnbaum, M. J., Clem, R. J., and Miller, L. K. (1994). An apoptosis inhibiting gene from a nuclear polyhedrosis virus encoding a peptide with cys/his sequence motifs. *J. Virol.* **68**, 2521–2528.

Black, W. C., Bennett, K. E., Gorrochotegui-Escalante, N., Barillas-Mury, C. V., Fernandez-Salas, I., de Lourdes Munoz, M., Farfan-Ale, J. A., Olson, K. E., and Beaty, B. J. (2002). Flavivirus susceptibility in *Aedes aegypti*. *Arch. Med. Res.* **33**, 379–388.

Blissard, G., Black, B., Crook, N., Keddie, B. A., Possee, R., Rohrmann, G., Theilmann, D., and Volkman, L. (2000). Family baculoviridae. In *Virus Taxonomy Seventh Report of the International Committee on the Taxonomy of Viruses* (M. H. V. V. Regenmortel, C. M. Fauquet, D. H. L. Bishop, E. B. Carstens, M. K. Estes, S. M. Lemon, J. Maniloff, M. A. Mayo, D. J. McGeoch, C. R. Pringle, and R. B. Wickner, Eds.), pp. 195–202. Springer-Verlag, Vienna/New York.

Blitvich, B. J., Blair, C. D., Kempf, B. J., Hughes, M. T., Black, W. C., Mackie, R. S., Meredith, C. T., Beaty, B. J., and Rayms-Keller, A. (2002). Developmental- and tissue-specific expression of an inhibitor of apoptosis protein 1 homologue from *Aedes triseriatus* mosquitoes. *Insect Mol. Biol.* **11**, 431–442.

Bosio, C. F., Fulton, R. E., Salasek, M. L., Beaty, B. J., and Black IV., W. C. (2000). Quantitative trait loci that control vector competence for dengue-2 virus in the mosquito *Aedes aegypti*. *Genetics* **156**, 687–698.

Braunagel, S. C., Russell, W. K., Rosas-Acosta, G., Russell, D. H., and Summers, M. D. (2003). Determination of the protein composition of the occlusion-derived virus of *Autographa californica* nucleopolyhedrovirus. *Proc. Natl. Acad. Sci. USA* **100**, 9797–9802.

Brennan, C. A., and Anderson, K. V. (2004). *Drosophila*: The genetics of innate immune recognition and response. *Annu. Rev. Immunol.* **22**, 457–483.

Briese, D. T. (1986). Insect resistance to baculoviruses. In *The Biology of Baculoviruses* (B. A. Federici, and R. Granados, Eds.), Vol. 2, pp. 237–263. CRC Press, Boca Raton, FL.

Brooks, E. M., Gordon, K. H., Dorrian, S. J., Hines, E. R., and Hanzlik, T. N. (2002). Infection of its lepidopteran host by the *Helicoverpa armigera* stunt virus (Tetraviridae). *J. Invertebr. Pathol.* **80**, 97–111.

Cashio, P., Lee, T. V., and Bergmann, A. (2005). Genetic control of programmed cell death in *Drosophila melanogaster*. *Semin. Cell Dev. Biol.* **16**, 225–235.

Cherry, S., and Perrimon, N. (2004). Entry is a rate-limiting step for viral infection in a *Drosophila melanogaster* model of pathogenesis. *Nat. Immunol.* **5**, 81–87.

Cherry, S., and Silverman, N. (2006). Host–pathogen interactions in *Drosophila*: New tricks from an old friend. *Nat. Immunol.* **7**, 911–917.

Cherry, S., Doukas, T., Armknecht, S., Whelan, S., Wang, H., Sarnow, P., and Perrimon, N. (2005). Genome-wide RNAi screen reveals a specific sensitivity of IRES-containing RNA viruses to host translation inhibition. *Genes Dev.* **19**, 445–452.

Clem, R. J. (2001). Baculoviruses and apoptosis: The good, the bad, and the ugly. *Cell Death Differ.* **8**, 137–143.

Clem, R. J. (2005). The role of apoptosis in defense against baculovirus infection in insects. *Curr. Top. Microbiol. Immunol.* **289**, 113–129.

Clem, R. J., and Miller, L. K. (1994). Control of programmed cell death by the baculovirus genes *p35* and *iap*. *Mol. Cell. Biol.* **14**, 5212–5222.

Clem, R. J., Fechheimer, M., and Miller, L. K. (1991). Prevention of apoptosis by a baculovirus gene during infection of insect cells. *Science* **254**, 1388–1389.

Cox-Foster, D. L., Conlan, S., Holmes, E. C., Palacios, G., Evans, J. D., Moran, N. A., Quan, P. L., Briese, T., Hornig, M., Geiser, D. M., Martinson, V., Vanengelsdorp, D., Kalkstein, A. L., Drysdale, A., Hui, J., Zhai, J., Cui, L., Hutchison, S. K., Simons, J. F., Egholm, M., Pettis, J. S., Lipkin, W. I. (2007). A metagenomic survey of microbes in honey bee colony collapse disorder. *Science* **318**, 283–287.

Crook, N. E., Clem, R. J., and Miller, L. K. (1993). An apoptosis-inhibiting baculovirus gene with a zinc finger-like motif. *J. Virol.* **67**, 2168–2174.

Dall, D., Sriskantha, A., Vera, A., Lai-Fook, J., and Symonds, T. (1993). A gene encoding a highly expressed spindle body protein of *Heliothis armigera* entomopoxvirus. *J. Gen. Virol.* **74**, 1811–1818.

Dall, D., Luque, T., and O'Reilly, D. (2001). Insect–virus relationships: Sifting by informatics. *Bioessays* **23**, 184–193.

Derksen, A. C. G., and Granados, R. R. (1988). Alteration of a lepidopteran peritrophic membrane by baculoviruses and enhancement of viral infectivity. *Virology* **167**, 242–250.

Devenport, M. F. H., Donnelly-Doman, M., Shen, Z., Jacobs-Lorena, M. (2005). Storage and secretion of Ag-Aper14, a novel peritrophic matrix protein, and Ag-Muc1 from the mosquito *Anopheles gambiae*. *Cell Tissue Res.* **320**, 175–185.

Deveraux, Q. L., Takahashi, R., Salvesen, G. S., and Reed, J. C. (1997). X-linked IAP is a direct inhibitor of cell-death proteases. *Nature* **388**, 300–304.

Dong, Y., Taylor, H. E., and Dimopoulos, G. (2006). AgDscam, a hypervariable immunoglobulin domain-containing receptor of the *Anopheles gambiae* innate immune system. *PLoS Biol.* **4**, e229.

Dostert, C., Jouanguy, E., Irving, P., Troxler, L., Galiana-Arnoux, D., Hetru, C., Hoffmann, J. A., and Imler, J. L. (2005). The Jak–STAT signaling pathway is required but not sufficient for the anti-viral response of *Drosophila*. *Nat. Immunol.* **6**, 946–953.

Dow, J. A., Goodwin, S. F., and Kaiser, K. (1992). Analysis of the gene encoding a 16-kDa proteolipid subunit of the vacuolar H(+)-ATPase from *Manduca sexta* midgut and tubules. *Gene* **122**, 355–360.

Duffey, S., Hoover, K., Bonning, B. C., and Hammock, B. D. (1995). The impact of host plant on the efficacy of baculoviruses. In *Reviews in Pesticide Toxicology* (R. M. Roe, and R. J. Kuhr, Eds.), Vol. 3, pp. 137–275. Toxicology Communications, Inc., Raleigh, NC.

Eberle, K. E., and Jehle, J. A. (2006). Field resistance of codling moth against *Cydia pomonella* granulovirus (*Cp*GV) is autosomal and incompletely dominant inherited. *J. Invertebr. Pathol.* **93**, 201–206.

Elvin, C. M., Vuocolo, T., Pearson, R. D., East, I. J., Riding, G. A., Eisemann, C. H., and Tellam, R. L. (1996). Characterization of a major peritrophic membrane protein, peritrophin-44, from the larvae of *Lucilia cuprina*: cDNA and deduced amino acid sequences. *J. Biol. Chem.* **271**, 8925–8935.

Engelhard, E. K., and Volkman, L. E. (1995). Developmental resistance in fourth instar *Trichoplusia ni* orally inoculated with *Autographa californica* M nuclear polyhedrosis virus. *Virology* **209**, 384–389.

Engelhard, E. K., Kam-Morgan, L. N. W., Washburn, J. O., and Volkman, L. E. (1994). The insect tracheal system: A conduit for the systemic spread of *Autographa californica* M nuclear polyhedrosis virus. *Proc. Natl. Acad. Sci. USA* **91**, 3224–3227.

Faye, I., Pye, A., Rasmuson, T., Boman, H. G., and Boman, I. A. (1975). Insect immunity. 11. Simultaneous induction of antibacterial activity and selection synthesis of some hemolymph proteins in diapausing pupae of *Hyalophora cecropia* and *Samia cynthia*. *Infect. Immun.* **12**, 1426–1438.

Federici, B. A. (1997). Baculovirus pathogenesis. In *The Baculoviruses* (L. K. Miller, Ed.), pp. 33–60. Plenum Press, New York.

Fessler, J. H., and Fessler, L. I. (1989). *Drosophila* extracellular matrix. *Annu. Rev. Cell Biol.* **5**, 309–339.

Flipsen, J. T. M., Martens, J. W. M., Oers, M. M. V., Vlak, J. M., and Lent, J. W. M. V. (1995). Passage of *Autographa californica* nuclear polyhedrosis virus through the midgut epithelium of *Spodoptera exigua* larvae. *Virology* **208**, 328–335.

Franz, A. W. E., Sanchez-Vargas, I., Adelman, Z. N., Blair, C. D., Beaty, B. J., James, A. A., and Olson, K. E. (2006). Engineering RNA interference-based resistance to dengue virus type 2 in genetically modified *Aedes aegypti*. *Proc. Natl. Acad. Sci. USA* **103**, 4198–4203.

Fuxa, J. R. (1993). Insect resistance to viruses. In *Parasites and Pathogens of Insects* (N. E. Beckage, S. N. Thompson, and B. A. Federici, Eds.), Vol. 2, pp. 197–209. Academic Press Inc, New York.

Galiana-Arnoux, D., Dostert, C., Schneemann, A., Hoffmann, J. A., and Imler, J. L. (2006). Essential function *in vivo* for Dicer-2 in host defense against RNA viruses in *Drosophila*. *Nat. Immunol.* **7**, 590–597.

Girard, Y. A., Popov, V., Wen, J., Han, V., and Higgs, S. (2005). Ultrastructural study of West Nile virus pathogenesis in *Culex pipiens quinquefasciatus* (Diptera: Culicidae). *J. Med. Entomol.* **42**, 429–444.

Girard, Y. A., Schneider, B. S., McGee, C. E., Wen, J., Han, V. C., Popov, V., Mason, P. W., and Higgs, S. (2007). Salivary gland morphology and virus transmission during long-term cytopathologic West Nile virus infection in *Culex* mosquitoes. *Am. J. Trop. Med. Hyg.* **76**, 118–128.

Goodwin, R. H., Milner, R. J., and Beaton, C. D. (1991). Entomopoxvirinae. In *Atlas of Invertebrate Viruses* (J. A. Adams, and J. R. Bonami, Eds.), pp. 259–286. CRC Press, Inc., Boca Raton, FL.

Granados, R. R., and Lawler, K. A. (1981). *In vivo* pathway of *Autographa californica* baculovirus invasion and infection. *Virology* **108**, 297–308.

Green, M. C., Monser, K. P., and Clem, R. J. (2004). Ubiquitin protein ligase activity of the anti-apoptotic baculovirus protein Op-IAP3. *Virus Res.* **105**, 89–96.

Griffiths, C. M., Barnett, A. L., Ayres, M. D., Windass, J., King, L. A., and Possee, R. D. (1999). *In vitro* host range of *Autographa californica* nucleopolyhedrovirus recombinants lacking functional *p35*, *iap1* or *iap2*. *J. Gen. Virol.* **80**, 1055–1066.

Guo, W. L. G., Pang, Y., and Wang, P. (2005). A novel chitin-binding protein identified from the peritrophic membrane of the cabbage looper, *Trichoplusia ni*. *Insect Biochem. Mol. Biol.* **35**, 1224–1234.

Hardy, J. L., Apperson, G., Asman, S., and Reeves, W. C. (1978). Selection of a strain of *Culex tarsalis* highly resistant to infection following ingestion of western equine encephalomyelitis virus. *Am. J. Trop. Med. Hyg.* **27**, 313–321.

Hardy, J. L., Houk, E. J., Kramer, L. D., and Reeves, W. C. (1983). Intrinsic factors affecting vector competence of mosquitoes for arboviruses. *Annu. Rev. Entomol.* **28**, 229–262.

Harrison, R. L., and Bonning, B. C. (2003). Comparative analysis of the genomes of *Rachiplusia ou* and *Autographa californica* multiple nucleopolyhedrovirus. *J. Gen. Virol.* **84**, 1827–1842.

Hawtin, R. E., Zarkowska, T., Arnold, K., Thomas, C. J., Gooday, G. W., King, L. A., Kuzio, J. A., and Possee, R. D. (1997). Liquefaction of *Autographa californica* nucleopolyhedrovirus-infected insects is dependent on the integrity of virus-encoded chitinase and cathepsin genes. *Virology* **238**, 243–253.

Hayakawa, T., Xu, J., and Hukuhara, T. (1996). Cloning and sequencing of the gene for an enhancing factor from *Pseudaletia separata* entomopoxvirus. *Gene* **177**, 269–270.

Hess, R. T., and Falcon, L. A. (1987). Temporal events in the invasion of the codling moth, *Cydia pomonella*, by a granulosis virus. *J. Invertebr. Pathol.* **50**, 85–105.

Higgs, S. (2004). How do mosquito vectors live with their viruses? In *Microbe–Vector Interactions in Vector-Borne Diseases* (S. Gillespie, G. Smith, and A. Osborne, Eds.), pp. 103–137. Cambridge (Eng); Cambridge University Press, New York.

Hirai, M., Terenius, O., Li, W., and Faye, I. (2004). Baculovirus and dsRNA induce hemolin, but no antibacterial activity, in *Antheraea pernyi*. *Insect Mol. Biol.* **13**, 399–405.

Hitchman, R. B., Hodgson, D. J., King, L. A., Hails, R. S., Cory, J. S., and Possee, R. D. (2007). Host mediated selection of pathogen genotypes as a mechanism for the maintenance of baculovirus diversity in the field. *J. Invertebr. Pathol.* **94**, 153–162.

Hoa, N. T., Keene, K. M., Olson, K. E., and Zheng, L. (2003). Characterization of RNA interference in an *Anopheles gambiae* cell line. *Insect Biochem. Molec. Biol.* **33**, 949–957.

Hodgson, D. J., Hitchman, R. B., Vanbergen, A. J., Hails, R. S., Possee, R. D., and Cory, J. S. (2004). Host ecology determines the relative fitness of virus genotypes in mixed-genotype nucleopolyhedrovirus infections. *J. Evol. Biol.* **17**, 1018–1025.

Hoover, K., Washburn, J. O., and Volkman, L. E. (2000). Midgut-based resistance of *Heliothis virescens* to baculovirus infection mediated by phytochemicals in cotton. *J. Insect Physiol.* **46**, 999–1007.

Hoover, K., Grove, M. J., and Su, S. (2002). Systemic component to intrastadial developmental resistance in *Lymantria dispar* to its baculovirus. *Biol. Contr.* **25**, 92–98.

Hu, Z. G., Chen, K. P., Yao, Q., Gao, G. T., Xu, J. P., and Chen, H. Q. (2006). Cloning and characterization of *Bombyx mori* PP-BP a gene induced by viral infection. *Yi Chuan Xue Bao* **33**, 975–983.

Hughes, A. L. (2002). Evolution of inhibitors of apoptosis in baculoviruses and their insect hosts. *Infect. Genet. Evolut.* **2**, 3–10.

Hukuhara, T., and Wijonarko, A. (2001). Enhanced fusion of a nucleopolyhedrovirus with cultured cells by a virus enhancing factor from an entomopoxvirus. *J. Invertebr. Pathol.* **77**, 62–67.

Hukuhara, T., Hayakawa, T., and Wijonarko, A. (2001). A bacterially produced virus enhancing factor from an entomopoxvirus enhances nucleopolyhedrovirus infection in armyworm larvae. *J. Invertebr. Pathol.* **78**, 25–30.

Hunter, W. B., Lapointe, S. L., Sinisterra, X. H., Achor, D. S., and Funk, C. J. (2003). Iridovirus in the root weevil *Diaprepes abbreviatus*. *J. Insect Sci.* **3**, 9.

Imler, J. L., and Zheng, L. (2004). Biology of Toll receptors: Lessons from insects and mammals. *J. Leukoc. Biol.* **75**, 18–26.

Jabbour, A. M., Ekert, P. G., Coulson, E. J., Knight, M. J., Ashley, D. M., and Hawkins, C. J. (2002). The p35 relative, p49, inhibits mammalian and *Drosophila* caspases including DRONC and protects against apoptosis. *Cell Death Differ.* **9**, 1311–1320.

Karpf, A. R., and Brown, D. T. (1998). Comparison of Sindbis virus-induced pathology in mosquito and vertebrate cell cultures. *Virology* **240**, 193–201.

Kavi, H. H., Fernandez, H. R., Xie, W., and Birchler, J. A. (2005). RNA silencing in *Drosophila*. *FEBS Lett.* **579**, 5940–5949.

Keene, K. M., Foy, B. D., Sanchez-Vargas, I., Beaty, B. J., Blair, C. D., and Olson, K. E. (2004). RNA interference acts as a natural anti-viral response to O'nyong-nyong virus (Alphavirus; Togaviridae) infection of *Anopheles gambiae*. *Proc. Natl. Acad. Sci. USA* **101**, 17240–17245.

Kirkpatrick, B. A., Washburn, J. O., and Volkman, L. E. (1998). AcMNPV pathogenesis and developmental resistance in fifth instar *Heliothis virescens*. *J. Invertebr. Pathol.* **72**, 63

Ko, R., Okano, K., and Maeda, S. (2000). Structural and functional analysis of the *Xestia c-nigrum* granulovirus matrix metalloproteinase. *J. Virol.* **74**, 11240–11246.

Kramer, L. D., and Ebel, G. D. (2003). Dynamics of flavivirus infection in mosquitoes. *Adv. Virus Res.* **60**, 187–232.

Kurata, S., Kobayashi, H., and Natori, S. (1991). Participation of a 200-kDa hemocyte membrane protein in the dissociation of the fat body at the metamorphosis of *Sarcophaga*. *Dev. Biol.* **146**, 179–185.

Kurata, S., Saito, H., and Natori, S. (1992). The 29-kDa hemocyte proteinase dissociates fat body at metamorphosis of *Sarcophaga*. *Dev. Biol.* **153**, 115–121.

LaCount, D. J., and Friesen, P. D. (1997). Role of early and late replication events in induction of apoptosis by baculoviruses. *J. Virol.* **71**, 1530.

LaCount, D. J., Hanson, S. F., Schneider, C. L., and Friesen, P. D. (2000). Caspase inhibitor P35 and inhibitor of apoptosis Op-IAP block *in vivo* proteolytic activation of an effector caspase at different steps. *J. Biol. Chem.* **275**, 15657–15664.

Lapointe, R., Popham, H. J., Straschil, U., Goulding, D., O'Reilly, D. R., and Olszewski, J. A. (2004). Characterization of two *Autographa californica* nucleopolyhedrovirus proteins, Ac145 and Ac150, which affect oral infectivity in a host-dependent manner. *J. Virol.* **78**, 6439–6448.

Lawson, D., Arensburger, P., Atkinson, P., Besansky, N. J., Bruggner, R. V., Butler, R., Campbell, K. S., Christophides, G. K., Christley, S., Dialynas, E., Emmert, D., Hammond, M., Hill, C. A., Kennedy, R. C., Lobo, N. F., MacCallum, M. R., Madey, G., Megy, K., Redmond, S., Russo, S., Severson, D. W., Stinson, E. O., Topalis, P., Zdobnov, E. M., Birney, E., Gelbart, W. M., Kafatos, F. C., Louis, C., and Collins, F. H. (2007). VectorBase: A home for invertebrate vectors of human pathogens. *Nucleic Acids Res.* **35**, D503–D505.

Lehane, M. J. (1997). Peritrophic matrix structure and function. *Annu. Rev. Entomol.* **42**, 525–550.

Lemaitre, B., Nicolas, E., Michaut, L., Reichhart, J. M., and Hoffmann, J. A. (1996). The dorsoventral regulatory gene cassette spatzle/Toll/cactus controls the potent antifungal response in *Drosophila* adults. *Cell* **86**, 973–983.

Lepore, L. S., Roelvink, P. R., and Granados, R. R. (1996). Enhancin, the granulosis virus protein that facilitates nucleopolyhedrovirus (NPV) infections is a metalloprotease. *J. Invertebr. Pathol.* **68**, 131–140.

Li, H., Li, W. X., and Ding, S. W. (2002). Induction and suppression of RNA silencing by an animal virus. *Science* **296**, 1319–1321.

Li, Q., Li, H., Blitvich, B. J., and Zhang, J. (2007). The *Aedes albopictus* inhibitor of apoptosis 1 gene protects vertebrate cells from bluetongue virus-induced apoptosis. *Insect Mol. Biol.* **16**, 93–105.

Li, Z., Li, C., Yang, K., Wang, L., Yin, C., Gong, Y., and Pang, Y. (2003). Characterization of a chitin-binding protein GP37 of *Spodoptera litura* multicapsid nucleopolyhedrovirus. *Virus Res.* **96**, 113–122.

Liu, S., Li, H., Sivakumar, S., and Bonning, B. C. (2006a). Virus-derived genes for insect resistant transgenic plants. In *Insect Viruses: Biotechnological Applications* (B. C. Bonning, Ed.), Vol. 68, pp. 428–457. Academic Press, San Diego, CA.

Liu, X., Jiang, F., Kalidas, S., Smith, D., and Liu, Q. (2006b). Dicer-2 and R2D2 coordinately bind siRNA to promote assembly of the siRISC complexes. *RNA* **12**, 1514–1520.

Lu, A., and Miller, L. K. (1995). The roles of eighteen baculovirus late expression factor genes in transcription and DNA replication. *J. Virol.* **69**, 975–982.

Ludwig, G. V., Christensen, B. M., Yuill, T. M., and Schultz, K. T. (1989). Enzyme processing of La Crosse virus glycoprotein G1: A bunyavirus–vector infection model. *Virology* **171**, 108–113.

Ludwig, G. V., Israel, B. A., Christensen, B. M., Yuill, T. M., and Schultz, K. T. (1991). Role of La Crosse virus glycoproteins in attachment of virus to host cells. *Virology* **181**, 564–571.

Maori, E., Tanne, E., Sela, I. (2007). Reciprocal sequence exchange between non-retro viruses and hosts leading to the appearance of new host phenotypes. *Virology* **362**, 342–349.

Meier, P., Finch, A., and Evan, G. (2000). Apoptosis in development. *Nature* **407**, 796–801.

Meister, G., and Tuschl, T. (2004). Mechanisms of gene silencing by double-stranded RNA. *Nature* **431**, 343–349.

Miller, B., and Mitchell, C. (1991). Genetic selection of a flavivirus-refractory strain of the yellow fever mosquito *Aedes aegypti*. *Am. J. Trop. Med. Hyg.* **45**, 399–407.

Mitsuhashi, W., and Miyamoto, K. (2003). Disintegration of the peritrophic membrane of silkworm larvae due to spindles of an entomopoxvirus. *J. Invertebr. Pathol.* **82**, 34–40.

Mitsuhashi, W., Kawakita, H., Murakami, R., Takemoto, Y., Saiki, T., Miyamoto, K., and Wada, S. (2007). Spindles of an entomopoxvirus facilitate its infection of the host insect by disrupting the peritrophic membrane. *J. Virol.* **81**, 4235–4243.

Molina-Cruz, A., Gupta, L., Richardson, J., Bennett, K., Black, W. I., and Barillas-Mury, C. (2005). Effect of mosquito midgut trypsin activity on dengue-2 virus infection and dissemination in *Aedes aegypti*. *Am. J. Trop. Med. Hyg.* **72**, 631–637.

Moore, N. F., and Eley, S. M. (1991). Picornaviridae: Picornaviruses of the invertebrates. In *Atlas of Invertebrate Viruses* (J. A. Adams, and J. R. Bonami, Eds.), pp. 371–386. CRC Press, Inc., Boca Raton, FL.

Mori, H., Ohyane, M., Ito, M., Iwamoto, S. I., Matsumoto, T., Sumida, M., and Matsubara, F. (1989). Induction of a hemagglutinating activity in the hemolymph of the silkworm, *Bombyx mori*, infected with cytoplasmic polyhedrosis virus. *J. Invertebr. Pathol.* **54**, 112–116.

Nakazawa, H., Tsuneishi, E., Ponnuvel, K. M., Furukawa, S., Asaoka, A., Tanaka, H., Ishibashi, J., and Yamakawa, M. (2004). Anti-viral activity of a serine protease from the digestive juice of *Bombyx mori* larvae against nucleopolyhedrovirus. *Virology* **321**, 154–162.

Nardi, J. B., Gao, C., and Kanost, M. R. (2001). The extracellular matrix protein lacunin is expressed by a subset of hemocytes involved in basal lamina morphogenesis *J. Insect Physiol.* **47**, 997–1006

Ohkawa, T., Washburn, J. O., Sitapara, R., Sid, E., and Volkman, L. E. (2005). Specific binding of *Autographa californica* M nucleopolyhedrovirus occlusion-derived virus to midgut cells of *Heliothis virescens* larvae is mediated by products of pif genes Ac119 and Ac022 but not by Ac115. *J. Virol.* **79**, 15258–15264.

Ourth, D. D., and Renis, H. E. (1993). Anti-viral melanization reaction of *Heliothis virescens* hemolymph against DNA and RNA viruses *in vitro*. *Compart. Biochem. Physiol. B.* **105**, 719–723.

Parker, J. S., Mizuguchi, K., and Gay, N. J. (2001). A family of proteins related to Spatzle, the Toll receptor ligand, are encoded in the *Drosophila* genome. *Proteins* **45**, 71–80.

Pech, L. L., Trudeau, D., and Strand, M. R. (1995). Effects of basement membranes on the behavior of hemocytes from *Pseudoplusia includens* (Lepidoptera: Noctuidae): Development of an *in vitro* encapsulation assay. *J. Insect Physiol.* **41**, 801.

Pendland, J. C., and Boucias, D. G. (1998). Characterization of monoclonal antibodies against cell wall epitopes of the insect pathogenic fungus, *Nomuraea rileyi*: Differential binding to fungal surfaces and cross-reactivity with host hemocytes and basement membrane components. *Eur. J. Cell Biol.* **75**, 118–127.

Pendland, J. C., and Boucias, D. G. (2000). Comparative analysis of the binding of antibodies prepared against the insect *Spodoptera exigua* and against the mycopathogen *Nomuraea rileyi*. *J. Invertebr. Pathol.* **75**, 107–116.

Peng, J., Zhong, J., and Granados, R. R. (1999). A baculovirus enhancin alters the permeability of a mucosal midgut peritrophic matrix from lepidopteran larvae. *J. Insect Physiol.* **45**, 159–166.

Poinar Jr., G., and Poinar, R. (2005). Fossil evidence of insect pathogens. *J. Invertebr. Pathol.* **89**, 243–250.

Ponnuvel, K. M., Nakazawa, H., Furukawa, S., Asaoka, A., Ishibashi, J., Tanaka, H., and Yamakawa, M. (2003). A lipase isolated from the silkworm *Bombyx mori* shows anti-viral activity against nucleopolyhedrovirus. *J. Virol.* **77**, 10725–10729.

Popham, H. J., Shelby, K. S., Brandt, S. L., and Coudron, T. A. (2004). Potent virucidal activity in larval *Heliothis virescens* plasma against *Helicoverpa zea* single capsid nucleopolyhedrovirus. *J. Gen. Virol.* **85**, 2255–2261.

Prikhod'ko, E. A., and Miller, L. K. (1996). Induction of apoptosis by baculovirus transactivator IE1. *J. Virol.* **70**, 7116–7124.

Prikhod'ko, E. A., and Miller, L. K. (1999). The baculovirus PE38 protein augments apoptosis induced by transactivator IE1. *J. Virol.* **73**, 6691–6699.

Rao, R., Fiandra, L., Giordana, B., Eguileor, M. d., Congiu, T., Burlini, N., Arciello, S., Corrado, G., and Pennacchio, F. (2004). AcMNPV ChiA protein disrupts the peritrophic membrane and alters midgut physiology of *Bombyx mori* larvae. *Insect Biochem. Mol. Biol.* **34**, 1205–1213.

Reddy, J. T., and Locke, M. (1990). The size limited penetration of gold particles through insect basal laminae. *J. Insect Physiol.* **36**, 397–407.

Riehle, M. M., Markianos, K., Niare, O., Xu, J., Li, J., Toure, A. M., Podiougou, B., Oduol, F., Diawara, S., Diallo, M., Coulibaly, B., Ouatara, A., Kruglyak, L., Traore, S. F., and Vernick, K. D. (2006). Natural malaria infection in *Anopheles gambiae* is regulated by a single genomic control region. *Science* **312**, 577–579.

Rizki, R. M., and Rizki, T. M. (1974). Basement membrane abnormalities in melanotic tumor formation of *Drosophila*. *Experientia* **30**, 543–546.

Rizki, R. M., and Rizki, T. M. (1980). Hemocyte responses to implanted tissues in *Drosophila melanogaster* larvae. *Roux's Arch. Dev. Biol.* **189**, 207–213.

Rohrbach, D. H., and Timpl, R. (1993). *Molecular and Cellular Aspects of Basement Membranes*. Academic Press, New York.

Romoser, W. S., Turell, M. J., Lerdthusnee, K., Neira, M., Dohm, D., Ludwig, G., and Wasieloski, L. (2005). Pathogenesis of Rift Valley fever virus in mosquitoes – tracheal conduits and the basal lamina as an extra-cellular barrier. *Arch. Virol. Suppl.* **19**, 89–100.

Roxstrom-Lindquist, K., Terenius, O., and Faye, I. (2004). Parasite-specific immune response in adult *Drosophila melanogaster*: A genomic study. *EMBO Rep.* **5**, 207–212.

Roy, N., Deveraux, Q. L., Takahashi, R., Salvesen, G. S., and Reed, J. C. (1997). The c-IAP-1 and c-IAP-2 proteins are direct inhibitors of specific caspases. *EMBO J.* **16**, 6914–6925.

Sabatier, L., Jouanguy, E., Dostert, C., Zachary, D., Dimarcq, J. L., Bulet, P., and Imler, J. L. (2003). Pherokine-2 and -3. *Eur. J. Biochem.* **270**, 3398–3407.

Sadd, B. M., and Schmid-Hempel, P. (2006). Insect immunity shows specificity in protection upon secondary pathogen exposure. *Curr. Biol.* **16**, 1206–1210.

Sahdev, S., Taneja, T. K., Mohan, M., Sah, N. K., Khar, A. K., Hasnain, S. E., and Athar, M. (2003). Baculoviral p35 inhibits oxidant-induced activation of mitochondrial apoptotic pathway. *Biochem. Biophys. Res. Comm.* **307**, 483–490.

Salvesen, G. S., and Duckett, C. S. (2002). IAP proteins: Blocking the road to death's door. *Nat. Rev. Mol. Cell Biol.* **3**, 401–410.

Sanchez-Vargas, I., Travanty, E., Keene, K., Franz, A., Beaty, B. J., Blair, C., and Olson, K. E. (2004). RNA interference, arthropod-borne viruses, and mosquitoes. *Virus Res.* **102**, 65–74.

Sanders, H. R., Foy, B. D., Evans, A. M., Ross, L. S., Beaty, B. J., Olson, K. E., and Gill, S. S. (2005). Sindbis virus induces transport processes and alters expression of innate immunity pathway genes in the midgut of the disease vector, *Aedes aegypti*. *Insect Biochem. Mol. Biol.* **35**, 1293–1307.

Seshagiri, S., and Miller, L. K. (1997). Baculovirus inhibitors of apoptosis (IAPs) block activation of Sf-caspase-1. *Proc. Natl. Acad. Sci. USA* **94**, 13606.

Shao, L., Devenport, M., Fujioka, H., Ghosh, A., and Jacobs-Lorena, M. (2005). Identification and characterization of a novel peritrophic matrix protein, Ae-Aper50, and the microvillar membrane protein, AEG12, from the mosquito, *Aedes aegypti*. *Insect Biochem. Mol. Biol.* **273**, 17665–17670.

Shelby, K. S., and Popham, H. R. J. (2006). Plasma phenoloxidase of larval *Heliothis virescens* is virucidal. *J. Insect Sci.* **6**, 13.

Shelby, K. S., and Popham, H. J. (2007). Increased plasma selenium levels correlate with elevated resistance of *Heliothis virescens* larvae against baculovirus infection. *J. Invertebr. Pathol.* **95**, 77–83.

Shen, Z., and Jacobs-Lorena, M. (1998). A Type I peritrophic matrix protein from the malaria vector *Anopheles gambiae* binds to chitin. Cloning, expression and characterization. *J. Biol. Chem.* **273**, 17665–17670.

Sheppard, C. A., and Shapiro, M. (1994). Physiological effects of a fluroscent brightener on nuclear polyhedrosis virus-infected *Lymantria dispar* (L.) larvae (Lepidoptera: Lymantriidae). *Biol. Control.* **4**, 404–411.

Shin, S. W., Kokoza, V. A., and Raikhel, A. S. (2003). Transgenesis and reverse genetics of mosquito innate immunity. *J. Exp. Biol.* **206**, 3835–3843.

Siegel, R. M. (2006). Caspases at the crossroads of immune-cell life and death. *Nat. Rev. Immunol.* **6**, 308–317.

Sim, C., Hong, Y. S., Vanlandingham, D. L., Harker, B. W., Christophides, G. K., Kafatos, F. C., Higgs, S., and Collins, F. H. (2005). Modulation of *Anopheles gambiae* gene expression in response to o'nyong-nyong virus infection. *Insect Mol. Biol.* **14**, 475–481.

Slavicek, J. M., and Popham, H. J. (2005). The *Lymantria dispar* nucleopolyhedrovirus enhancins are components of occlusion-derived virus. *J. Virol.* **79**, 10578–10588.

Smith-Johannsen, H., Witkiewicz, H., and Iatrou, K. (1986). Infection of silkmoth follicular cells with *Bombyx mori* nuclear polyhedrosis virus. *J. Invertebr. Pathol.* **48**, 74–84.

Stanley, D., and Shapiro, M. (2007). Eicosanoid biosynthesis inhibitors increase the susceptibility of *Lymantria dispar* to nucleopolyhedrovirus LdMNPV. *J. Invertebr. Pathol.* **95**, 119–124.

Teakle, R. E., Jensen, J. M., and Giles, J. E. (1986). Age-related susceptibility of *Heliothis punctiger* to a commercial formulation of nuclear polyhedrosis disease. *J. Invertebr. Pathol.* **47**, 82–92.

Tellam, R. L. (1996). The peritrophic membrane. In *Biology of the Insect Midgut* (M. J. Lehane, and B. F. Billingsley, Eds.), pp. 86–114. Chapman and Hall, New York.

Tellam, R. L., Wijffels, G., and Willadsen, P. (1999). Peritrophic matrix proteins. *Insect Biochem. Mol. Biol.* **29**, 87–101.

Thoetkiattikul, H., Beck, M. H., and Strand, M. R. (2005). Inhibitor kappaB-like proteins from a polydnavirus inhibit NF-kappaB activation and suppress the insect immune response. *Proc. Natl. Acad. Sci. USA* **102**, 11426–11431.

Trudeau, D., Washburn, J. O., and Volkman, L. E. (2001). Central role of hemocytes in *Autographa californica M* nucleopolyhedrovirus pathogenesis in *Heliothis virescens* and *Helicoverpa zea*. *J. Gen. Virol.* **75**, 996–1003.

Vaaje-Kolstad, G., Horn, S. J., Aalten, D. M. v., Synstad, B., and Eijsink, V. G. (2005). The non-catalytic chitin-binding protein CBP21 from *Serratia marcescens* is essential for chitin degradation. *J. Biol. Chem.* **280**, 28492–28497.

van Rij, R. P., Saleh, M. C., Berry, B., Foo, C., Houk, A., Antoniewski, C., and Andino, R. (2006). The RNA silencing endonuclease Argonaute 2 mediates specific anti-viral immunity in *Drosophila melanogaster*. *Gene. Dev.* **20**, 2985–2995.

Van Roessel, P., Hayward, N. M., Barros, C. S., and Brand, A. H. (2002). Two-color GFP imaging demonstrates cell-autonomy of GAL4-driven RNA interference in *Drosophila*. *Genesis* **34**, 170–173.

Vier, J., Furmann, C., and Hacker, G. (2000). Baculovirus P35 protein does not inhibit caspase-9 in a cell-free system of apoptosis. *Biochem. Biophys. Res. Comm.* **276**, 855–861.

Volkman, L. E. (1997). Nucleopolyhedrovirus interactions with their insect hosts. *Adv. Virus Res.* **48**, 313–348.

Volpicella, M., Ceci, L. R., Cordewener, J., America, T., Gallerani, R., Bode, W., Jongsma, M. A., and Beekwilder, J. (2003). Properties of purified gut trypsin from *Helicoverpa zea*, adapted to proteinase inhibitors. *Eur. J. Biochem.* **270**, 10–19.

Vucic, D., Kaiser, W. J., Harvey, A. J., and Miller, L. K. (1997). Inhibition of reaper-induced apoptosis by interaction with inhibitor of apoptosis proteins (IAPs). *Proc. Natl. Acad. Sci. USA* **94**, 10183–10188.

Vucic, D., Kaiser, W. J., and Miller, L. K. (1998). Inhibitor of apoptosis proteins physically interact with and block apoptosis induced by *Drosophila* proteins HID and GRIM. *Mol. Cell. Biol.* **18**, 3300–3309.

Wang, B., Shang, J., Liu, X., Cui, W., Wu, X., and Zhao, N. (2007). Enhanced effect of fluorescent whitening agent on peroral infection for recombinant baculovirus in the host *Bombyx mori* L. *Curr. Microbiol.* **54**, 5–8.

Wang, H., Wu, D., Deng, F., Peng, H., Chen, X., Lauzon, H., Arif, B. M., Jehle, J. A., and Hu, Z. (2004). Characterization and phylogenetic analysis of the chitinase gene from the *Helicoverpa armigera* single nucleocapsid nucleopolyhedrovirus. *Virus Res.* **100**, 179–189.

Wang, P., and Granados, R. R. (1997a). An intestinal mucin is the target substrate for a baculovirus enhancin. *Proc. Natl. Acad. Sci. USA* **94**, 6977–6982.

Wang, P., and Granados, R. R. (1997b). Molecular cloning and sequencing of a novel invertebrate intestinal mucin cDNA. *J. Biol. Chem.* **272**, 16663–16669.

Wang, P. and Granados, R. R. (2000). Calcofluor disrupts the midgut defense system in insects. *Insect Biochem. Molec. Biol.* **30**, 135–143.

Wang, P., Li, G., and Granados, R. R. (2004). Identification of two new peritrophic membrane proteins from larval *Trichoplusia ni*: Structural characteristics and their functions in the protease rich insect gut. *Insect Biochem. Mol. Biol.* **34**, 215–227.

Wang, X. H., Aliyari, R., Li, W. X., Li, H. W., Kim, K., Carthew, R., Atkinson, P., and Ding, S. W. (2006). RNA interference directs innate immunity against viruses in adult *Drosophila*. *Science* **312**, 452–454.

Washburn, J. O., Kirkpatrick, B. A., and Volkman, L. E. (1995). Comparative pathogenesis of *Autographa californica* M nuclear polyhedrosis virus in larvae of *Trichoplusia ni* and *Heliothis virescens*. *Virology* **209**, 561–568.

Washburn, J. O., Kirkpatrick, B. A., and Volkman, L. E. (1996). Insect protection against viruses. *Nature* **383**, 767.

Washburn, J. O., Kirkpatrick, B. E., Stapelton-Haas, E., and Volkman, L. E. (1998). Evidence that the stilbene-derived optical brightener M2R enhances *Autographa californica* M nuclear polyhedrosis virus of *Trichoplusia ni* and *Heliothis virescens* by preventing sloughing of infected midgut epithelial cells. *Biol. Cont.* **11**, 58–69.

Washburn, J. O., Haas-Stapleton, E. J., Tan, F. F., Beckage, N. E., and Volkman, L. E. (2000). Co-infection of *Manduca sexta* larvae with polydnavirus from *Cotesia congregata* increases susceptibility to fatal infection by *Autographa californica* M. nucleopolyhedrovirus. *J. Insect Physiol.* **46**, 179–190.

Watson, F. L., Puttmann-Holgado, R., Thomas, F., Lamar, D. L., Hughes, M., Kondo, M., Rebel, V. I., and Schmucker, D. (2005). Extensive diversity of Ig-superfamily proteins in the immune system of insects. *Science* **309**, 1874–1878.

Weaver, S. C., Scott, T. W., Lorenz, L. H., Lerdthusnee, K., and Romoser, W. S. (1988). Togavirus-associated pathologic changes in the midgut of a natural mosquito vector. *J. Virol.* **62**, 2083–2090.

Weaver, S. C., Lorenz, L. H., and Scott, T. W. (1992). Pathologic changes in the midgut of *Culex tarsalis* following infection with Western equine encephalomyelitis virus. *Am. J. Trop. Med. Hyg.* **47**, 691–701.

Weaver, S. C., Coffey, L., Nussenzveig, R., Ortiz, D., and Smith, D. (2004). Vector competence. In *Microbe–Vector Interactions in Vector-Borne Disease* (S. Gillespie, G. Smith, and A. Osborne, Eds.), pp. 139–180. Cambridge University Press, Cambridge.

Williams, T. (1998). Invertebrate iridescent viruses. In *The Insect Viruses* (L. K. Miller, and L. A. Ball, Eds.), pp. 31–68. Plenum Press, New York.

Yang, X. and Cox-Foster, D. L. (2005). Impact of an ectoparasite on the immunity and pathology of an invertebreate: evidence for host immunosuppression and viral amplification. *Proc. Natl. Acad. Sci. USA* **102**, 7470–7575.

Yang, X. and Cox-Foster, D. L. (2007). Effects of parasitization by *Varroa destructor* on survivorship and physiological traits of *Apis mellifera* with viral incidence and microbial challenge. *Parasitology* **134**, 405–412.

Zambon, R. A., Nandakumar, M., Vakharia, V. N., and Wu, L. P. (2005). The Toll pathway is important for an anti-viral response in *Drosophila*. *Proc. Natl. Acad. Sci. USA* **102**, 7257–7262.

Zambon, R. A., Vakharia, V. N., and Wu, L. P. (2006). RNAi is an anti-viral immune response against a dsRNA virus in *Drosophila melanogaster*. *Cell Microbiol.* **8**, 880–889.

Zhang, J. H., Ohkawa, T., Washburn, J. O., and Volkman, L. E. (2005). Effects of Ac150 on virulence and pathogenesis of *Autographa californica* multiple nucleopolyhedrovirus in noctuid hosts. *J. Gen. Virol.* **86**, 1619–1627.

Zoog, S. J., Schiller, J. J., Wetter, J. A., Chejanovsky, N., and Friesen, P. D. (2002). Baculovirus apoptotic suppressor P49 is a substrate inhibitor of initiator caspases resistant to P35 *in vivo*. *EMBO J.* **21**, 5130–5140.

10

PARASITOID POLYDNAVIRUSES

AND INSECT IMMUNITY

NANCY E. BECKAGE

Departments of Entomology and Cell Biology and Neuroscience and Center for Disease Vector Research, University of California-Riverside, Riverside, CA 92521, USA

ABSTRACT: The polydnaviruses (PDVs) associated with braconid and ichneumonid parasitoid wasps are unique viruses with very large segmented genomes (>200 kb) that are integrated within the wasp's chromosomal DNA, representing genetic symbionts of the parasitoid. In the first phase of their life cycle, they replicate in the female wasp's ovarian calyx cells. They are released into the lumen of the calyx, and abundant numbers of virions are transferred to the parasitoid's host together with wasp eggs in calyx fluid during oviposition. In the second phase of their life cycle, PDV gene expression occurs in the parasitized lepidopteran host in the absence of viral replication. A primary function of braco- and ichnoviral gene expression in the host is to cause its immunosuppression by targeting host hemocyte function either transiently, by inhibiting hemocyte adhesion behavior, or permanently, by inducing host hemocyte apoptosis, so that encapsulation of the parasitoid egg is prevented. While the PDV does not replicate in the host caterpillar,

PDV gene expression is essential for host immunosuppression and successful parasitism. This chapter reviews the mechanisms whereby PDV gene products interface with the host insect immune system and exert potent effects on its cellular and humoral functions. PDV genomic analysis for braco- and ichnoviruses has demonstrated major differences in these two taxonomic groups and offers exciting potential for development of new insect pest control technologies based upon PDV gene action on host immunity, as well as on host endocrine function and development.

Abbreviations:

AcMNPV	=	*Autographa californica* M nucleopolyhedrovirus
BVs	=	bracoviruses
CcBV	=	*Cotesia congregata* bracovirus
CiBV	=	*Chelonus inanitus* bracovirus
CrV1	=	*Cotesia rubecula* virus protein 1
CsBV	=	*Cotesia sesamiae* bracovirus
CsIV	=	*Campoletis sonorensis* ichnovirus
DOPA	=	dihydroxyphenylalanine
EM	=	electron microscopy
EP1	=	early protein 1
HeIV	=	*Hyposoter exiguae* ichnovirus
HfIV	=	*Hyposoter fugitivus* ichnovirus
IVs	=	ichnoviruses
MdBV	=	*Microplitis demolitor* bracovirus
NF-κB	=	nuclear factor kappa B
PDV	=	polydnavirus
PO	=	phenoloxidase
pp	=	post-parasitization
PSP	=	parasitism-specific protein
PTP	=	protein tyrosine phosphatase
SfIV	=	*Spodoptera frugiperda* ichnovirus
ST	=	(E)-1,2-dihydroxy-2-(isopropyl)-5-(2-phenylethenyl)benzene
TnBV	=	*Toxoneuron nigriceps* bracovirus
TnBV1	=	*Toxoneuron nigriceps* bracovirus virus protein 1
TrIV	=	*Tranosema rostrale* ichnovirus
VLP	=	virus-like particle

10.1   INTRODUCTION TO POLYDNAVIRUS BIOLOGY

Parasitoids are insect parasites of other insect species that ultimately kill their host, either by physiological manipulation and induction of host developmental arrest or by physical consumption of host tissues (Beckage and Gelman, 2004; Pennacchio and Strand, 2006). The pioneering work of George Salt on insect host immune reactions to metazoan parasites and parasitoids in particular (Salt, 1963,

1968, 1970) laid the groundwork for modern studies of host–parasitoid immunological interactions. With genomics and proteomics now in hand, we can focus on
molecular analysis of these relationships based on the dynamic interplay between
virulence mechanisms employed by the parasitoid ('offense') and the resistance
characteristics of the host ('defense').

Wasp endoparasitoids in the hymenopteran families Braconidae and
Ichneumonidae specializing in parasitism of lepidopteran insect hosts have evolved
a unique symbiosis with viruses classified in the viral family Polydnaviridae (Stoltz
et al., 1984; Webb et al., 2005). These viruses are termed PDVs, because they have
some of the largest segmented double-stranded DNA ('polydna') genomes seen
among animal viruses (>200 kb) and are integrated within the genomic DNA of the
wasp. Both male and female wasps carry the PDV genome in their chromosomes,

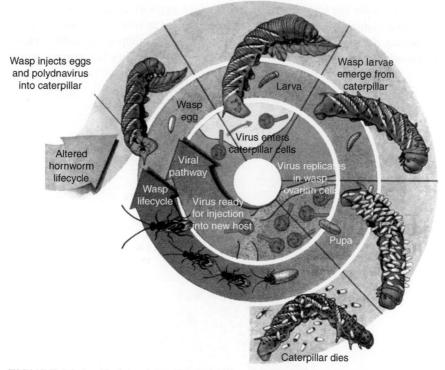

FIGURE 10.1 The interrelationships of the life cycles of insect hosts, parasitoids, and PDVs as
illustrated by tobacco hornworm, *Manduca sexta*, larvae parasitized by the braconid wasp *Cotesia
congregata*. The disrupted life cycle of the parasitized larva is shown to the right. The *C. congregata*
bracovirus (CcBV) replicates in wasp ovarian calyx cells and virions are injected into the caterpillar
host along with the eggs during wasp oviposition. Viral genes are expressed in host *M. sexta* larvae in
the absence of viral replication and induce suppression of the host's immune system. The parasitoids
develop in the host without triggering an encapsulation response, and eventually emerge from the host
to spin pupal cocoons. Replication of the PDV begins during the wasp's late pupal stage, and the female
wasp ecloses as an adult carrying mature virions to parasitize new hosts. The tobacco hornworm larva
with emerged wasps is developmentally arrested in the larval stage. The cycle then begins again. Figure
reprinted from Beckage (1997) with permission from *Scientific American* and Roberto Osti
Illustrations. (See Plate 10.1 for color version of this figure.)

but excision and the viral replication cycle is confined to the female wasp's ovary, specifically in ovarian calyx cells (see Fig. 10.1 for host, parasitoid, and PDV life cycles). Following replication and packaging, the virions are injected in calyx fluid into the host during parasitization along with wasp eggs, and viral genes begin to be expressed within minutes following wasp oviposition. PDV gene expression occurs in host immunoregulatory cells including hemocytes and fat body tissues, as well as in the gut, nervous system, epidermis, and other tissues. While the viruses do not replicate in the host caterpillar, PDV gene expression is required for the successful parasitism of permissive host species by their natural habitual parasitoids.

The presence of virus-like particles (VLPs) in the reproductive tract of the parasitoid *Venturia* (formerly *Nemeritis*) *canescens* was reported by Bedwin (1979a) and their immunoprotective role in preventing encapsulation of the parasitoid egg was proposed (Bedwin, 1979b). Feddersen et al. (1986) showed these VLPs protect the *Venturia* egg by a passive masking of the egg surface against recognition and encapsulation by host blood cells. These particles in *Venturia* are unlike PDVs in that they do not contain nucleic acids and the protein components of the virions are encoded by the wasp genome (Theopold et al., 1994). They function passively

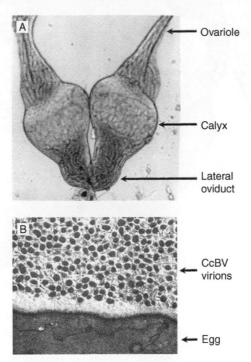

FIGURE 10.2　(A) The appearance of the reproductive tract of *Cotesia congregata* as seen through a light microscope and (B) a transmission electron micrograph of a cross-section through the calyx region of the ovary showing the *C. congregata* PDV virions and the surface of a parasitoid egg. The virus replicates in calyx cells which rupture to release the virions into the lumen of the calyx. In (B) note the abundant numbers of virions in the calyx fluid bathing the parasitoid egg which are injected into the host *M. sexta* larva during parasitization. CcBV: *C. congregata* bracovirus.

to suppress recognition of the parasitoid egg. The *Venturia canescens/Ephestia kuehniella* interaction is unique among host–parasitoid systems in that the cellular defense capacity of the host remains virtually intact following parasitization (Reineke et al., 2006). In contrast, most PDVs, which are the subject of this chapter, function to actively suppress host immunity during parasitism.

Stoltz and Vinson (1979a) and Stoltz et al. (1981) screened a broad range of ichneumonid and braconid taxa for the presence of viruses (which they initially described as baculovirus-like) in parasitoid ovarian calyces (Fig. 10.2A) and calyx fluid (Fig. 10.2B), and found VLPs in a wide range of species belonging to these two taxonomic groups. The fate of the particles, i.e. evidence that the viruses penetrate host caterpillar cells following wasp oviposition, and that viral DNA is released from the viral nucleocapsids and enters host cell nuclei via nuclear pores, was published by Stoltz and Vinson (1979b).

The functional role of these VLPs in braconid and ichneumonid species was demonstrated in experiments by Edson et al. (1981) who showed that parasitoid eggs that were washed free of ovarian calyx fluid containing the virions (as seen in electron micrograms) and then injected into naïve non-parasitized caterpillars were rapidly encapsulated by host hemocytes, whereas washed eggs recombined with calyx fluid (collected from wasp ovarian calyces) that were injected were not attacked. The combination of calyx fluid plus washed parasitoid eggs suppressed the

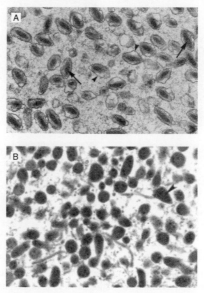

FIGURE 10.3 Transmission electron micrographs showing comparison of the morphology of (A) *Hyposoter exiguae* ichnovirus (HeIV) virions and (B) *Cotesia congregata* bracovirus (CcBV) virions. Note that the HeIV virions in panel (A) have a single lenticular nucleocapsid per envelope (large arrows) and a double envelope (small arrowhead). In contrast, the CcBV virions in panel (B) have multiple nucleocapsids per envelope (large arrowhead) and a single envelope. The genomes of IVs and BVs also differ in their physical organization and genomic sequences. Panel (A) is reprinted from Krell and Stoltz (1980) with permission of Academic Press, and Panel (B) photograph by I. de Buron and N.E. Beckage.

host encapsulation response, mimicking the immunosuppressive effects of natural parasitism, whereas inactivation of the viral DNA destroyed the 'rescue' action of wasp calyx fluid (Edson et al., 1981).

Electron microscopic studies have identified two taxon-species groups of viruses which were later classified as bracoviruses (BVs) (in braconid species) and ichnoviruses (IVs) (in ichneumonid species) in the virus family Polydnaviridae (Stoltz et al., 1984; Webb et al., 2005). BVs and IVs differ in morphology as illustrated in Fig. 10.3. The IVs have one fusiform nucleocapsid per envelope, and bud from the calyx cell in which they replicate into the calyx lumen to generate particles with a double envelope (Fig. 10.3A). In contrast, BVs have multiple nucleocapsids per envelope, and completion of their replication cycle ultimately lyses the ovary cell, producing BV virions having a single envelope (Fig. 10.3B).

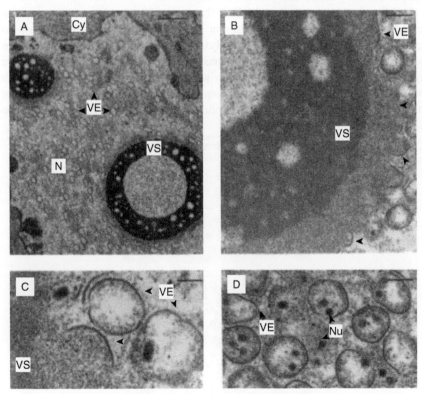

FIGURE 10.4   Transmission electron micrographs showing different stages of *Cotesia congregata* bracovirus (CcBV) replication and assembly in ovarian calyx cells in the pharate adult wasp. (A) calyx cell in which CcBV is undergoing replication in the ring-shaped virogenic stroma, scale bar = 1 μm. In panels (B) and (C) the beginning stages of *de novo* viral envelope synthesis (arrowheads) can be seen at the periphery of the stroma (scale bars = 250 nm (B) and 100 nm (C). In panel (D) packaging of the nucleocapsids into the envelope is underway during the final stages of bracovirus assembly, scale bar = 150 nm. Abbreviations: Cy, cytoplasm; N, nucleus; Nu, nucleocapsid, VE, viral envelope; VS, virogenic stroma. Figure reprinted from Buron and Beckage (1992) with permission of Academic Press.

The replication of *Cotesia congregata* bracovirus (CcBV) in the ovary of the adult wasp (Fig. 10.4) begins earlier in the late pupal stage. Concentric rings of virogenic stroma are seen in calyx cell nuclei (Fig. 10.4A) with beginning stages of virion envelope synthesis and nucleocapsid packaging evident at the periphery (Fig. 10.4B–D). The calyx cell nuclei become packed with virions (Fig. 10.5) and then the cells rupture, releasing the virions into the calyx lumen. Calyx fluid containing an abundance of virions is injected into host along with the parasitoid's eggs into the hemocoel of the host during oviposition. By 2 h post-parasitization (pp), CcBV virions are seen traversing the basement membrane overlying host tissues (Fig. 10.6) and enter host fat body, muscle, and other cells. Hemocytes lack a basal lamina and the virions enter those cells directly from host hemolymph plasma.

Full genome sequences are now available for the *Cotesia congregata* bracovirus (CcBV) (Espagne et al., 2004), *Microplitis demolitor* bracovirus (MdBV) (Webb et al., 2006), *Campoletis sonorensis* ichnovirus (CsIV) (Webb et al., 2006), *Hyposoter fugitivus* ichnovirus (HfIV) (Tanaka et al., 2007), *Tranosema rostrale* ichnovirus (TrIV) (Tanaka et al., 2007), and the banchine PDV of *Glypta fumiferanae*, which may represent a new virus group as it exhibits physical and genomic characteristics of both IVs and BVs (Lapointe et al., 2007). Results of these and ongoing genomic sequencing studies have revealed the presence of multiple complex gene families in both IVs and BVs encoding many closely related protein products whose functional roles are now under intensive investigation. Nevertheless, PDV gene expression has many potent down-regulating effects on the host immune system that are required for the parasitoid to escape the host's immune response, and several classes of PDV genes are proposed to be host immunomodulators (Section 10.3). The roles of PDVs

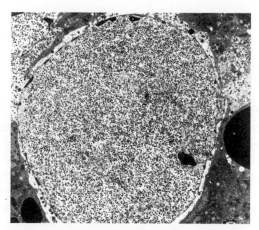

FIGURE 10.5 Transmission electron micrograph of the nucleus of an ovarian calyx cell in an adult female *Cotesia congregata* showing final stages of *C. congregata* bracovirus maturation. Note the high density accumulation of fully formed virions in the nucleus of the calyx cell. The cell later undergoes lysis and releases the mature CcBV virions into the lumen of the calyx preparatory to wasp oviposition. Figure modified from Buron and Beckage (1992).

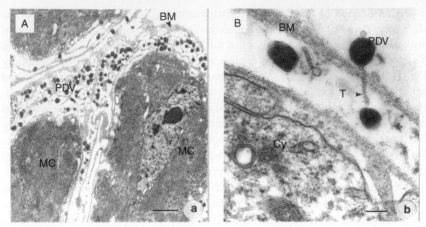

FIGURE 10.6　Movement of *Cotesia congregata* bracovirus (CcBV) virions across the basement membrane (basal lamina) overlying host *Manduca sexta* tissues and transit throughout the hemocoel. In (A), numerous CcBV virions are present beneath the basement membrane overlying muscle and other tissues by 1.5 h post-oviposition. Scale bar = 100 nm. Panel (B) shows the tail of the CcBV virion passing through the basement membrane. Scale bar = 150 nm. Abbreviations: BM, basement membrane; Cy, cytoplasm; MC, muscle cell; PDV = polydnavirions; T, tail of PDV bracovirus virion. Reprinted from Buron and Beckage (1992) with permission of Academic Press.

and PDV gene products in regulation of host immunity are the focus of this chapter. Future prospects for exploiting PDV genes in agriculture, biotechnology, and insect pest management are highlighted in Section 10.5.

10.2　PDVs AS DOWN-REGULATORS OF HOST IMMUNITY

10.2.1　CELLULAR IMMUNITY

10.2.1.1　Alterations in Hemocyte Behavior

PDVs affect both the cellular and humoral arms of the insect host's immune system, and the induced changes in host immune function may be either transient, thus suppressing host immunity temporarily until the parasitoid egg hatches, or permanent and lasting for the duration of parasitism, once parasitization occurs (for reviews see Beckage, 1997, 1998; Kroemer and Webb, 2004; Lavine and Beckage, 1995; Shelby and Webb, 1999; Strand and Pech, 1995a; Webb and Strand, 2005; Webb et al., 2006).

Numerous early studies documented the induction of synthesis of parasitism-specific proteins (PSPs) in insects attacked by parasitoids (reviewed in Beckage, 1993). Many of these PSPs were later found to be PDV gene expression products, some of which are highly abundant in hemolymph of newly parasitized insects (e.g. Harwood et al., 1994). While the functional roles of many of the parasitism-specific PDV-encoded proteins characterized to date in different host–parasitoid combinations remain to be identified, PDV genomic analysis and expression studies, including the application of microarray technologies and other genome-based approaches, are now leading us ever closer to meeting that goal (Section 10.3).

At 30 min pp, PDV gene expression can already be detected in host hemocytes and fat body tissues, and soon thereafter the behavior of host hemocytes is altered to suppress their attachment to biotic targets such as parasitoid eggs (Amaya et al., 2005; Asgari et al., 1997; Le et al., 2003; Strand, 1994; Tian et al., 2007). Encapsulation of abiotic targets such as injected Sephadex beads is often likewise inhibited during parasitism (Ibrahim and Kim, 2006; Lavine and Beckage, 1996), suggestive of a broadly acting suppression of host immunity. Host hemocyte phagocytosis rates are similarly reduced following parasitization due to the action of PDV gene expression (Ibrahim and Kim, 2006; Strand et al., 2006). As alterations in hemocyte behavior in parasitized insects are caused by regulatory disruption of cytoskeletal events such as actin polymerization which are required for hemocyte adhesion (Asgari et al., 1997; Turnbull et al., 2004; chapter 11), or induction of hemocyte apoptosis and clumping (see below), several different PDV genes and gene products act to down-regulate host immunity during successful parasitism. In refractory host species that mobilize an effective defense response against the parasitoid, PDV gene expression can be partially or completely suppressed (Cui et al., 2000; Gitau et al., 2007; Harwood et al., 1998). The latter studies demonstrated that with PDV-encoded transcript and protein production severely diminished, host hemocytes mobilize the encapsulation reaction against the parasitoid resulting in unsuccessful parasitism.

Different hemocyte types may or may not be selectively targeted during parasitism. The appearance of granulocytes and plasmatocytes from a normal non-parasitized *M. sexta* larva is shown in Fig. 10.7. In host–parasitoid systems in which host hemocyte apoptosis is induced, the numbers of hemocytes circulating in host hemolymph exhibit a dramatic reduction following parasitization. Similar decreases in both host granulocyte and plasmatocyte populations are seen in diamondback moth (*Plutella xylostella*) larvae newly parasitized by *Cotesia plutellae* (Ibrahim and Kim, 2006). However, in other host–parasitoid combinations, specific hemocyte morphotypes are selectively targeted for destruction by the PDV. In *Pseudoplusia includens* larvae parasitized by *Microplitis demolitor*, the expression of MdBV gene(s) induces the apoptosis of a single hemocyte class, the granular cells (Pech and Strand, 2000; Strand and Pech, 1995b), while concomitantly inhibiting adhesion behavior in host plasmatocytes.

In some cases BVs or IVs act in concert with wasp factors to suppress host immunoresponsiveness. The CsIV interacts with wasp ovarian proteins to suppress host immunity and induce production of intracellular and secreted proteins that alter hemocyte morphology and behavior (Luckhart and Webb, 1996; Turnbull

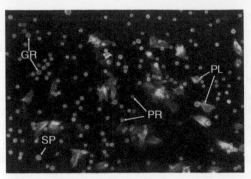

FIGURE 10.7 Light micrograph showing appearance of prohemocytes (PR), plasmatocytes (PL) , granulocytes (GR), and spherule cells (SP) in hemolymph from a non-parasitized fourth instar *Manduca sexta* larva. Note that the plasmatocytes show filopodial extensions resulting from actin polymerization in the cells that spread. The granulocytes are round with accumulations of dense granules. The smallest cells are prohemocytes (see chapter 2). (See Plate 10.7 for color version of this figure.)

et al., 2004). Wasp ovarian proteins and the PDV thus act together to disrupt host immunity, but the nature of the molecular interactions between ovarian proteins and PDV genes or gene products have not yet been identified. Factors present in the wasp's venom gland that are co-injected with PDV into the host during oviposition can also facilitate or even be required for PDV-mediated induction of host immunosuppression in some species (reviewed in Asgari, 2006).

Lack of expression of specific parasitoid PDV gene products in an appropriate time frame following wasp oviposition into refractory host species can result in encapsulation of the parasitoid eggs. Harwood et al. (1998) correlated levels of the CcBV-encoded early protein 1 (EP1) in host hemolymph with susceptibility of different species of sphingid larvae to this parasitoid. The CcBV-encoded EP1 was abundantly expressed in hemolymph of host species that were permissive to *C. congregata* (including the tobacco hornworm, *Manduca sexta*, and the white-lined sphinx, *Hyles lineata*) but lacking or diminished in refractory sphingid host species that successfully encapsulated the parasitoid's eggs, e.g. *Pachysphinx occidentalis* (Harwood et al., 1998). Different stages of encapsulation of *C. congregata* eggs in non-permissive *P. occidentalis* host larvae are shown in Fig. 10.8, beginning with attachment of a small number of cells (Fig. 10.8A and B) to formation of a full melanized capsule completed by 24 h post-parasitization (Fig. 10.8C and D).

Similarly, the CcBV homolog of the *Cotesia rubecula* CrV1 gene is expressed during the very earliest stages of parasitism of tobacco hornworm larvae by *Cotesia congregata*. The CcV1 transcript is detected in host hemocytes and fat body as early as 4 h pp when host hemocytes undergo apoptosis and encapsulation is suppressed (Amaya et al., 2005; Le et al., 2003). Using a novel approach which employs virulent and avirulant biotypes of *C. sesamiae*, a parasitoid of the maize stem borer (*Busseola fusca*), Gitau et al. (2007) demonstrated that levels of expression of the CrV1 homolog in host larvae parasitized by the two biotypes dramatically differed according to their virulence phenotype. High CrV1 homolog expression

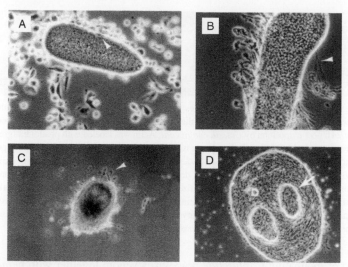

FIGURE 10.8 Light micrographs showing different stages of encapsulation of *Cotesia congregata* eggs in the non-permissive sphingid host species *Pachysphinx occidentalis*. (A) *C. congregata* egg recovered from *P. occidentalis* larva 24 h following oviposition (200 × magnification). Note hemocytes beginning to attach to egg surface (arrowhead). (B) *C. congregata* egg examined 24 h pp of *P. occidentalis*. Note plasmatocytes filopodial extensions beginning to adhere to the egg surface (arrowhead, 400 × magnification). (C) Appearance of a full capsule (100 × magnification) containing a *C. congregata* egg collected in hemolymph from *P. occidentalis* host larva at 48 h pp. Note that melanization of the inner layers of the capsule is in progress. Cells appear to be continuing to be recruited to the outermost layers of the capsule (arrowhead). (D) Appearance of two eggs of *C. congregata* that were fully encapsulated at 72 h pp of *P. occidentalis*. Note density of hemocyte layers comprising the capsule surrounding the eggs (arrowhead, 100 × magnification). Agglutinated masses of encapsulated eggs are sometimes found in melanized foci adhering to the host's internal tissues (fat body, gut, Malpighian tubules, etc.) in non-permissive host species. Abbreviation: pp: post-parasitization. Figure modified from Harwood et al. (1998).

levels in hemocytes and fat body were correlated with the virulence of the *C. sesamiae* strain parasitizing the host. Low or no expression of the CrV1 homolog by *C. sesamiae* bracovirus (CsBV) was detected following parasitization by the avirulent biotype of *C. sesamiae*. Non-synonymous differences in CsV1 gene sequences between virulent and avirulent wasp lineages suggest that differences in *B. fusca* parasitism by *C. sesamiae* may be due to qualitative differences in CsV1–hemocyte interactions in the two lines (Gitau et al., 2007).

In cases where PDV gene expression induces host hemocyte apoptosis, appearance of 'debris' and agglutinated clumps of dead hemocytes and cell remnants are evident microscopically in host hemolymph, which are eventually cleared from circulation. While plasmatocytes from non-parasitized *M. sexta*, larvae have normal actin polymerization and adherence behavior (Fig. 10.9A), the cells of newly parasitized larvae show multiple abnormalities symptomatic of disruption of their function (Fig. 10.9B and C). The appearance of clumps of hemocytes is observed in *M. sexta* hemolymph beginning ~2 h pp *in vivo*. A different *in vitro* approach

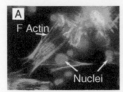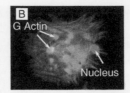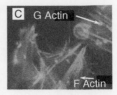

FIGURE 10.9 Appearance of hemocytes from (A) non-parasitized control fourth instar *Manduca sexta* larva compared with (B) and (C) the disrupted appearance of cells collected from parasitized larvae 24 h post-oviposition by *Cotesia congregata*. Actin was labeled with phalloidin (green) and the nuclei were stained with DAPI (red). In non-parasitized larvae (A) note the preponderance of polymerized filamentous actin (F actin) in long filopodial extensions of plasmatocytes which adhere to the slide. In non-parasitized larvae, hemocyte nuclei are round, dense, and centrally located in the cells. In contrast, as seen in (B) and (C), the hemocytes of newly parasitized larvae have disrupted cytoskeletal networks. Note the presence of depolymerized globular actin (G actin) and disrupted ectopic nuclei in hemocytes of newly parasitized larvae. The hemocytes of parasitized larvae are destined to undergo apoptosis (see text). (See Plate 10.9 for color version of this figure.) Photograph modified from Dumpit et al. (submitted).

comprised of incubating *C. plutellae* bracovirus/venom together with hemocytes of non-parasitized diamondback moth larvae in *in vitro* short-term cultures similarly induces cell death of hemocytes (Yu et al., 2007). In addition, Lapointe et al. (2005) demonstrated that expression of the *Toxoneuron nigriceps* bracovirus (TnBV1) gene in a recombinant baculovirus system induces apoptosis in two insect cell lines (High Five cells and Sf21 cells). Expression of TnBV1 induced increased caspase-3-like protease activity and TUNEL staining in cultured cells, two biochemical markers of apoptosis. As both plasmatocytes and granulocytes are key players in encapsulation (Lavine and Strand, 2002; chapter 2), their destruction by apoptosis protects the developing parasitoid egg and larva from encapsulation, ensuring successful parasitoid development.

10.2.1.2 Obligatory Multiparasitism

Early observations documented the interesting phenomenon that some species of parasitoids are able to successfully develop in a particular host species only if the host has been previously parasitized by a different species of wasp for which it is a natural or habitual host (Guzo and Stoltz, 1985). This condition is termed 'obligatory multiparasitism' referring to parasitism of a single host by more than one parasitoid species. This situation contrasts with 'superparasitism', which refers to the deposition of more than one egg of the same species by solitary parasitoids, or more than one clutch of eggs of the same species, in the case of gregarious wasps, into a single host (Dorn and Beckage, 2007; Godfray, 1994).

An example of obligatory multiparasitism is provided by the tussock moth, *Orgyia leucostigma*, which normally encapsulates and destroys eggs of *Hyposoter* species, resulting in unsuccessful parasitism. However, the refractory host *O. leucostigma* can be transformed into a permissive host by previous parasitization by *Cotesia melanoscela*, which is a habitual parasite of this host species in nature. Guzo and Stoltz (1985) successfully reared three different species of *Hyposoter* in host tussock moth larvae that had been previously parasitized with *C. melanoscela*, all of which are normally encapsulated in this host. Additionally they provided experimental evidence that the combination of *C. melanoscela* calyx fluid and venom was the transforming factor responsible for induction of permissiveness to *Hyposoter* species. Thus, the active expression of PDV genes appears species specific to the host–parasitoid combination involved (Gitau et al., 2007; Harwood et al., 1998; Kadash et al., 2003). The existence of obligatory multiparasitism indicates that parasitism of habitual hosts by their natural parasitoid partners 'conditions' the host via PDV-induced immunosuppression, rendering the host suitable for parasitism by novel parasitoid species. Along the same lines, we next examine how PDVs function in increasing host susceptibility to pathogens.

10.2.1.3   Cellular Immunity and PDV-Mediated Enhanced Host Susceptibility to Pathogens

Anecdotal and experimental evidence from the insect pathology literature, as well the author's experiences in laboratory rearing of parasitized insects for a quarter century, reveals that parasitism often enhances susceptibility of the host insect species to fungal, bacterial, and viral infection due to the immunocompromised condition of the parasitized host. *Manduca sexta* larvae parasitized by *C. congregata* are more vulnerable to pathogen infections that have minimal effects on normal larvae, or are kept in a latent state in non-parasitized stocks, rendering them difficult to rear under certain environmental conditions (e.g. high ambient humidity), which foster pathogen proliferation.

As described in chapter 9, antiviral immunity requires both cellular and humoral components. Normally tobacco hornworm larvae are semi-permissive for infection by the *Autographa californica* multiple nucleopolyhedrovirus (AcMNPV) (See Fig. 4 in chapter 9 for an illustration of the AcMNPV life cycle). Oral inoculation of extremely large doses of occluded virus ($>10^5$ occlusions/insect) is required to kill fourth instar *M. sexta* larvae with AcMNPV (Washburn et al., 2000). However, larval susceptibility of *M. sexta* to AcMNPV is greatly enhanced by simultaneous parasitism with *C. congregata*, attributable to the disabling of the cellular immune system induced by CcBV (Washburn et al., 2000). In tobacco hornworm larvae that are parasitized by *C. congregata* as pre-molt fourth instars and orally inoculated with AcMNPV immediately following the molt, or non-parasitized larvae that are injected with four wasp equivalents of CcBV followed by oral inoculation with AcMNPV, the larvae exhibit higher rates of larval mortality and unsuccessful pupation compared to mortality rates in untreated non-parasitized control larvae. Low dosages of BVs or IVs cause a faster speed-of-kill and more rapid and extensive dissemination of the AcMNPV throughout the hemocoel compared to what is seen in non-parasitized control species of larvae that are semi-permissive to AcMNPV infection (Washburn et al., 1996, 2000). The lowered level of host immune capability caused by CcBV-induced apoptosis of host hemocytes, which would normally participate in cellular antiviral responses to AcMNPV infection (Washburn et al., 1996, 2000), contributes to this PDV-mediated increase in susceptibility to viral pathogens.

Similarly, the lepidopteran pest *Spodoptera littoralis* is highly resistant to infection with the occluded form of AcMNPV. Experiments in which the larval immune response of *S. littoralis* was suppressed by injection with the *Chelonus inanitus* bracovirus or co-expression of the CiBV immunosuppressive gene, P-vank-1, demonstrated that the larvae were more susceptible to AcMNPV infection when the PDV gene was co-expressed (Rivkin et al., 2006). Clearly, the discovery that PDVs and their encoded gene products synergize baculovirus infection provides a useful experimental approach to dissecting the responses of the lepidopteran immune system to viruses by using specific PDV immunosuppressive genes to test their functions (Rivkin et al., 2006).

Stoltz and Makkay (2003) observed that overt viral diseases caused by granulosis viruses and others can be induced from apparent latency in host *Trichoplusia ni* larvae following parasitization by the ichneumonid wasp *Hyposoter exiguae*. Viral latency may in some cases be broken through immunosuppressive activity resulting from insect parasitism, lending further support to the idea that PDV-mediated disruption of host immunity renders the host more susceptible to pathogenic infection (Stoltz and Makkay, 2003). Evidence for a parasitism-induced enhancement of host susceptibility to bacterial infection (Matsumoto et al., 1998) and inhibition of production of antimicrobial peptides also exist (Section 10.2.2.2). Thus, the conclusion that the immunosuppressed state of the parasitized insect host evoked by PDV infection contributes to higher rates of pathogen-induced mortality in parasitized larvae is strongly supported by both experimental and observational lines of evidence.

10.2.2 HOST HUMORAL DEFENSES

10.2.2.1 PDVs and the Melanization Pathway

As discussed in chapter 4, melanization by phenoloxidase (PO) enzymes is a critical host immune defense mechanism mobilized during nodulation and encapsulation responses to pathogens and parasites, respectively. Several components of the complex pathway leading to melanin synthesis as well as melanin itself display toxicity to parasites or pathogens. PO activity is associated with both hemocytes (e.g. oenocytoids in lepidopterans or crystal cells in *Drosophila*) and cell-free hemolymph plasma, arguing that it plays a dual role in humoral as well as cell-mediated immunity.

Several early studies documented the occurrence of reduced levels of hemolymph PO activity in parasitized lepidopteran larvae including parasitized *Heliothis virescens* parasitized by *Campoletis sonorensis* (Sroka and Vinson, 1978), *Trichoplusia ni* parasitized by *Hyposoter exiguae* (Stoltz and Cook, 1983) and *Manduca sexta* parasitized by *C. congregata* (Beckage et al., 1990). Rates of hemolymph melanization in non-parasitized *M. sexta* larvae versus those parasitized by *C. congregata* show a dramatic reduction during parasitism (Fig. 10.10A). Beckage et al. (1990) demonstrated that this effect could be mimicked by injection of CcBV into non-parasitized larvae, but inactivated virus had no effect, providing experimental evidence that the observed down-regulation of PO activity was

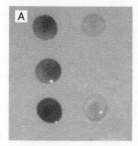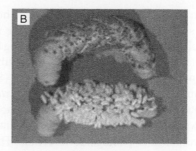

FIGURE 10.10   Panel (A) compares hemolymph melanization rates in non-parasitized control larvae (left) versus parasitized (right) fifth instar *Manduca sexta* larvae. The hemolymph from the parasitized larva was collected on the day *Cotesia congregata* larvae emerged from the host. Hemolymph from the unparasitized larva was collected on day 3 of the fifth instar. Photograph was taken 1.5 h following spotting of the hemolymph samples on Parafilm. Panel (B) shows that when the parasitoid cocoons are stripped from the host (upper larva), localized wound responses and melanization can be seen encircling the sites of parasitoid exit through the integument despite the reduced rates of melanization seen in host hemolymph (A). See text for further details. (See Plate 10.10 for color version of this figure.)

caused by the BV. Interestingly, cuticular melanization around the sites of parasitoid emergence still occurs in the absence of hemolymph melanization (Fig. 10.10B), arguing that hemolymph versus cuticular PO enzymes function independently. An alternative explanation could be that PO moves from the hemolymph to the cuticle to participate in localized host wound responses in the cuticle as the parasitoids exit from the host, depleting the hemolymph pool of PO enzymes.

Beckage et al. (1990) observed a dramatic reduction in hemolymph monophenoloxidase activity in fifth instar *M. sexta* larvae parasitized by *C. congregata* relative to levels of enzyme activity detected in non-parasitized animals. Additionally, CcBV-injected non-parasitized *M. sexta* larvae also had reduced rates indicating the rates of conversion of tyrosine to dihydroxyphenylalanine (DOPA) were inhibited by CcBV. Additionally, more than one level of enzyme inhibition may be operating simultaneously during parasitism. Shelby et al. (2000) found that the immunosuppressive CsIV reduces the protein titer of several key enzymes in the melanization pathway in parasitized fifth instar *Heliothis virescens* larvae, including PO, dopachrome isomerase, and DOPA decarboxylase. Those authors determined that the reductions in enzyme levels were sufficient to cause significant reductions of the corresponding pools of precursor melanization substrates L-DOPA, *N*-acetyldopamine, and *N*-β-alanyl dopamine from millimolar to nanomolar levels in parasitized larvae (Shelby et al., 2000). Thus, in CsIV-infected *H. virescens* larvae, deficiencies in several key enzymes in the melanization pathway result in reduced pools of critical substrates, causing the observed reduction in hemolymph melanization rates in parasitized insects.

Insights into possible mechanisms of PDV-induced PO inhibition can also be gained from recent discoveries in entomopathogenic nematodes which have associated bacterial partners. The insect pathogenic bacterium *Photorhabdus luminescens* is a symbiont of entomopathogenic heterorhabditid nematodes that overcomes the immune responses of the nematode's insect host. This bacterium secretes several antibiotics which prevent putrefaction of the insect cadaver, in which the nematodes develop, by other opportunistic competing bacteria. The *Photorhabdus*-secreted factors include a small molecule with antibiotic activity, ((E)-1,3-dihydroxy-2-(isopropyl)-5-(2-phenylethenyl)benzene) (ST), that acts as an inhibitor of hemolymph PO activity in host *Manduca sexta* larvae (Eleftherianos et al., 2007) (see chapter 12). By shutting down host hemolymph PO activity, ST produced by *Xenorhabdus* inhibits the nodulation/melanization reaction that is normally mobilized in response to pathogenic infection. Thus, this molecule both inhibits an important host defense mechanism while simultaneously inhibiting the growth of microbial competitors due to its antibiotic properties (Eleftherianos et al., 2007). Whether PDVs encode similar inhibitors of PO activity to contribute to the dramatically reduced rates of host hemolymph melanization seen in many species of parasitized insects is a question that obviously merits further study.

10.2.2.2   PDVs and Antibacterial Protein Production

As described above (Section 10.2.1.3), parasitized insects display higher susceptibility to microbial infections including many bacterial pathogens compared to non-parasitized insects of the same age due to suppression of hemocytic responses to infection. Additionally, host humoral immunity to pathogens is likewise suppressed during parasitism. The inducible array of potent antibacterial molecules normally produced by insects in response to bacterial infection include cecropins, attacins, lysozymes, hemolins, defensins, proteases, and protease inhibitors in addition to cytotoxic-free radicals (Gillespie et al., 1997; Hoffmann, 2003; Leclerc and Reichhart, 2004; Waterhouse et al., 2007). Hemocytes, fat body, and gut tissues are the main sources of these molecules, although other tissues can also contribute to their production. How does parasitism and, more specifically, active PDV enhance susceptibility of the host insect to bacterial pathogens?

As both humoral and cellular responses to bacteria participate in antimicrobial immunity, they are dually impacted by parasitism and PDV gene expression in the host. Due to the critical roles of hemocytes in phagocytosis and nodulation, the disruptive effects of parasitism on host hemocyte function obviously reduce the host's ability to counter bacterial pathogenesis and infection. The transient or permanent shutdown in host cell-based immune mechanisms evoked by parasitization has major deleterious impacts on protection from bacterial infection.

Fat body and hemocyte lysozyme transcription normally occur at low levels in non-immunized unparasitized insects, but can be rapidly induced to very high levels following injection of elicitors including peptidoglycans or other bacterial or fungal cell wall components. In *H. virescens* larvae that are newly parasitized by *C. sonorensis* and subjected to an immune challenge (injection of dead *Micrococcus lysodeikticus* cells at 24 and 48 h post-parastization) plasma lysozyme activity is dramatically reduced relative to levels of lysozyme production detected in saline-injected similarly aged non-parasitized control larvae (Shelby et al., 1998). The latter authors demonstrated that parasitization or CsIV injection inhibits plasma lysozyme activity induced by bacterial infection, and that the inhibition is mediated at the post-transcriptional level. Lysozyme transcript levels in fat body tissues of parasitized larvae were indistinguishable from similarly staged and treated non-parasitized controls as shown on Northern blots probed with an 0.8 kb fragment of *H. virescens* lysozyme cDNA (Shelby et al., 1998). Thus, suppression of the humoral antimicrobial response during parasitism is attributable to an inhibition of translation of lysozyme mRNA caused by the PDV, similar to the selective targeting of the PO pathway by CsIV.

10.3   INSIGHTS FROM PDV GENOME BIOLOGY

PDV genome sequencing projects have revealed a wealth of information about potentially significant viral genes that are critical to regulating the host's immunocompetence, endocrine physiology, and development. While complete genome sequences are now available for several PDVs, the function of their viral expression products in regulating host immune function has only just begun to be explored. As noted above (Section 10.1), the sequencing of complete genomes for CcBV (Espagne et al., 2004), MdBV(Webb et al., 2006), CsIV (Webb et al., 2006), HfIV (Tanaka et al., 2007), TrIV (Tanaka et al., 2007) and the banchine PDV genome of *Glypta fumiferanae* (Lapointe et al., 2007) has revealed the existence of complex multi-gene families in PDV genomes. These families encode multiple viral gene products potentially capable of manipulating multiple physiological processes in the host to facilitate successful parasitism by the wasp (*sensu* 'host regulation'). PDV genomes themselves are highly unusual viruses due to the megasize of their segmented viral genomes comprised of multiple circles (20–30) of double-stranded DNA ranging in size from ~5 to ~30 kb. The CcBV genome contains very few virus-like genes as well as a significant proportion of non-coding DNA (Desjardins et al., 2005; Drezen et al., 2006; Espagne et al., 2004). The notably non-virus-like nature of PDV genomes has led to speculation that these viruses represent wasp 'genetic secretions' as opposed to being true viruses (Federici and Bigot, 2003) but this view is clearly a minority opinion among virologists, and the International Committee on Taxonomy of Viruses has recognized both IVs and BVs as belonging to the virus taxon Polydnaviridae (Webb et al., 2005).

The IVs sequenced to date all contain members of six conserved gene families: the repeat element, cysteine motif (cys-motif), viral innexin (vinnexin), viral ankyrin (vankyrin), *N*-family, and polar-residue-rich PDV gene families (Tanaka et al., 2007). The expression of the cys-motif family of genes in CsIV causes immunosuppression in host *Heliothis virescens* larvae, and the cys-motif gene expression products target host hemocytes and are proposed to be immunosuppressors (Cui and Webb, 1998; Cui et al., 1996, 1997). Expression of cys-motif genes can also regulate host developmental programming, providing evidence for another functional role of this IV gene family (Fath-Goodin et al., 2006a).

Kroemer and Webb (2005, 2006) provide an analysis of the IV vankyrin family, another group of potentially important immunosuppressors. Tian et al. (2007) monitored expression of the *Campoletis chloridae* IV vankyrin genes in host *Helicoverpa armigera* larvae and observed two expression peaks in these genes following parasitization. The early peak (at 30 min pp) was followed by a dimuni- tion, then this decline was followed by a second peak in vankyrin expression at 2 days pp, corresponding to the time periods when early protection is needed by the parasitoid egg (at 30 min pp) and later by newly hatched parasitoid larvae (2 days pp) (Tian et al., 2007). They propose that the *Campoletis chloridae* ichnovirus (CcIV) vankyrin gene products prevent host cellular immune responses to the par- asitoid egg and larva by suppressing NF-kappaB signaling cascades that are nor- mally activated to mount an immune reaction and encapsulation. Thoetkiattikul et al. (2005) identified inhibitor κB-like proteins from a PDV that inhibit NF-κB activation and suppress the insect immune response, further implicating this path- way as playing a crucial role in insect defense. Recent studies of PDV-encoded inhibitor κB-like proteins expressed by TnBV provide evidence that these pro- teins cause retention of NF-κB/Rel factors in cell cytoplasm thus disrupting this signaling pathway, contributing to immunosuppression in parasitized host larvae. Falabella et al. (2007) present an analysis of inhibitor NF-κB-like gene family members in BVs belonging to different subfamilies of braconid species.

In BVs such as CcBV, the gene family of cystatins, which encode inhibitors of cysteine protease activity, are proposed to be regulators of lepidopteran host immu- nity (Espagne et al., 2005). The documented roles of cystatins in regulating immu- nity of plants, animals, and other arthropods to parasites and pathogens give credence to this hypothesis (Vray et al., 2002; Zhou et al., 2006). Expression of the CcBV cystatin genes is initiated early after oviposition and is maintained through- out parasitism, suggesting they are required during all stages to protect the devel- oping parasitoid from encapsulation (Espagne et al., 2005). While the latter authors conclude that CcBV cystatins play important roles in host caterpillar physiological deregulation by inhibiting host target proteases in the course of the host–parasite interaction and functioning as immunomodulators, the specific target(s) of CcBV cystatin gene expression remain as yet to be identified.

Another candidate PDV gene family that represent potential regulators of host immunity and hemocyte function during parasitism are the protein tyrosine phos- phatase (PTP) genes (Gundersen-Rindal and Pedroni, 2006; Ibrahim et al., 2007; Provost et al., 2004; Pruijssers and Strand, 2007). *Cotesia plutellae* bracovirus PTP genes are expressed in hemocytes and fat body of host diamondback moth larvae (*Plutella xylostella*) throughout the course of parasitism and affect hemo- cyte PTP activity. This is interpreted to inhibit mobilization of an encapsulation response to the parasitoid egg (Ibrahim and Kim, 2007; Ibrahim et al., 2007). Other hemocyte functions are also targeted by PTP gene products; i.e. in *Microplitis demolitor*, expression of the MdBV PTP genes inhibits phagocytosis behavior in insect immune cells (Pruijssers and Strand, 2007), suggestive of PTPs functioning to suppress hemocyte phagocytosis of pathogens in parasitized larvae.

As both hemocyte adhesion behavior and hemocyte phagocytosis activity are inhibited during parasitism due to the action of PDV genes, the use of *in vitro* cell cultures infected with PDVs offers exciting opportunities to study both of these behaviors directly. Beck and Strand (2003) used RNA interference to silence MdBV genes and identified a candidate viral gene (glc1.8), which when expressed in hemocytes or High Five cells (a cell line derived from *Trichoplusia ni* cells), dramatically inhibited their adhesion behavior. The MdBV infects High Five cells and blocks the ability of these cells to adhere to culture plates, mimicking the response of host hemocytes to infection by MdBV. Co-expression of the glc1.8 gene together with the MdBV PTP genes PTP-H2 and PTP-H3 in the hemocyte-like *Drosophila* S2 cell line results in complete inhibition of their phagocytotic immune response behavior (Pruijssers and Strand, 2007). Hence, both hemocyte adhesion and phagocytosis behaviors are targeted by the PTPs. The PTP gene family may also play a role in regulation of host development during parasitism (Section 10.4).

Using a cDNA microarray approach to study expression of *Spodoptera frugiperda* ichnovirus (SfIV) genes in host *S. frugiperda* hemocytes and fat body, Barat-Houari et al. (2006) found that at 24 h post-injection of SfIV, 4% of the 1750 arrayed host genes showed changes in their transcript levels with a large proportion (74%) showing a reduction in transcript levels. Several of the genes that were found to be modulated by HdIV gene expression are mediators of lepidopteran immune responses, including calreticulin, prophenoloxidase-activating enzyme, and immulectin-2 (Barat-Houari et al., 2006).

Genome-wide expression arrays have similarly been used to selectively identify effects of parasitism on transcript levels in parasitized *Drosophila* larvae, for comparison to transcript levels induced following infection of larvae with microbial pathogens (Wertheim et al., 2005). Interestingly, most genes exhibiting altered expression following parasitoid attack differed from those transcripts whose expression levels were changed in response to microbial infection. Thus, most alterations in gene transcription levels induced by parasitism are not mimicked by pathogenic infection in *Drosophila*.

10.4   PDVs AS REGULATORS OF HOST GROWTH AND DEVELOPMENT

While the immunological roles of PDVs are the main focus of this chapter, PDVs have also been linked to reprogramming of host development and induction of its developmental arrest before the wasps emerge from the host (for reviews see Beckage and Gelman, 2004; Pennacchio and Strand, 2006). In many instances parasitization causes a dramatic reduction in the rate of host food consumption and this effect has been demonstrated to be due to the action of the wasp's PDV by as yet unknown mechanisms, possibly by acting on the nervous system which controls motor movements of the mouthparts and the muscular contractions of the foregut which regulate feeding. Thus, host feeding and developmental pathways are also targeted by PDVs in the parasitized host.

In the *M. sexta–C. congregata* system, non-parasitized larvae injected with as little as 0.001 wasp equivalents of CcBV show arrest in the larval stage, as prepupae, or as larval–pupal intermediates, instead of undergoing normal wandering and pupation (Dushay and Beckage, 1993). These phenotypic effects are indicative of endocrine dysfunction and abnormalities in juvenile hormone and ecdysteroid titers in PDV-injected larvae which prevent normal pupation (Beckage and Gelman, 2004).

PDV-encoded PTPs, whose role in host–parasitoid immunological interactions was explored in Section 10.3, can also influence host endocrine regulation and development. In *Heliothis virescens* larvae parasitized by the braconid *Toxoneuron nigriceps*, parasitism down-regulates prothoracic gland function suppressing ecdysteroid production and ultimately induces host developmental arrest. The TnBV PTPs are rapidly expressed in fat body and hemocytes following parasitization and are also expressed in the host's inactivated prothoracic glands 24 h after parasitoid oviposition (Falabella et al., 2006). As a global inhibition of protein synthesis was not induced by TnBV, which would block ecdysteroidogenesis, Falabella et al. (2006) interpreted their findings as indicating that over-expression of TnBV PTP genes in inactivated prothoracic glands suggests that gland function may be affected by the disruption of phosphorylation of key proteins in hormonal signaling pathways upstream of ecdysone synthesis by the glands. One possibility is that the PDV-encoded PTPs act on the signaling cascade leading to the brain's production of tropic factors (i.e. prothoracicotropic hormone) that normally stimulate the prothoracic glands to secrete ecdysteroids (Falabella et al., 2006).

Thus, PDVs act as endocrine regulators of developmental reprogramming in addition to serving as potent host immunosuppressors during parasitism. Additional clues as to their hormonal role are also provided by studies of egg–larval parasitoids. *Chelonus inanitus* is a solitary egg–larval parasitoid of the lepidopteran *Spodoptera littoralis*. Parasitism induces the host to undergo precocious metamorphosis and arrest occurs when the host reaches the prepupal stage (Grossniklaus-Burgin et al., 1998). In this egg–larval system, the PDV genes are expressed during late stages of parasitism, in contrast to the very early expression of PDV genes typically seen in systems involving larval–larval parasitoids. In combination with playing a metabolic role in inhibiting host feeding and regulating host nutritional physiology (Kaeslin et al., 2005a), the parasitoid's PDV/venom is the inducer of host developmental arrest in the prepupal stage (Grossniklaus-Burgin et al., 1998). Kaeslin et al. (2005b) analyzed the influence of the parasitoid and its PDV on the hemolymph proteome of the host, and detected 11 virally encoded or virally induced proteins at the times corresponding to when arrest was induced. The advantage of using this egg–larval system is that the PDV-encoded proteins are expressed during the last stages of parasitism when host arrest occurs, increasing the likelihood that the arrest-inducing PDV genes and their products can be identified in expression arrays or using other technologies.

In summary, multiple PDV genes are expressed during parasitism to cause immunosuppression and alter developmental programming in the host. Many evolutionary questions remain concerning how these viruses evolved and if they are derived from viral or eukaryotic progenitors (Bezier et al., 2007; Drezen et al., 2006; Dupas et al., 2003; Dupuy et al., 2006; Federici and Bigot, 2003; Whitfield and Asgari, 2003), which will continue to stimulate debate as to their origin for years to come.

10.5   THE FUTURE OF POLYDNAVIROLOGY: IMPLEMENTATION IN AGRICULTURE, PEST MANAGEMENT, AND BIOTECHNOLOGY

A critical consideration is the potential value of application of basic biological and molecular research in immunological host–parasitoid interactions to pest management technology. This subject has particular importance to biological control of insect pests using parasitoids (Hajek, 2004) as we still do not fully appreciate the range of physiological factors that determine the host range of a parasitoid. Host–parasitoid immunological interactions, and the role of the parasitoid's PDV in modulating and suppressing host immunity, are key to understanding the physiological framework of successful parasitism.

As described in this chapter, host–PDV–parasitoid interactions involve a delicate interplay between virulence genes of the PDV which counter the immune resistance genes of the insect host. The information to be gained by continued analysis of strategies of parasitoid/PDV offense versus host defensive response (the host–parasite molecular 'arms race') can be used to improve efficacy of biological control programs that are based on the use of parasitoids and biopesticides for insect pest control.

To optimize rates of parasitism of insect pests in biological control, it is first necessary to distinguish parasitized and non-parasitized larvae in the field. The suppression of hemolymph melanization and coagulation rates during parasitism has been recently developed as a diagnostic tool to determine if larvae of the diamondback moth are parasitized by *Diadegma semiclausum* (Li et al., 2007). This novel technique uses the differential melanization reactions in hemolymph due to immune suppression in parasitized larvae to diagnose parasitism rates. This technique offers potential for application to a variety of pest species attacked by parasitoids to obtain estimates of rates of parasitism without having to dissect the insect to see developing parasitoids or rearing the host larvae until parasitoids emerge, both of which are time consuming and laborious.

Demonstrations that PDV sequences are integrated into lepidopteran chromosomal DNA following infection of different cell lines *in vitro* have been published for *Glyptapanteles indiensis* bracovirus (Gundersen-Rindal and Lynn, 2003) and *Tranosema rostrale* ichnovirus (Doucet et al., 2007). Together with the *in vivo* study by Le et al. (2003) which detected CcBV hybridization to genomic DNA extracted from fat body of parasitized *M. sexta* larvae, but no hybridization of viral DNA with fat body DNA from unparasitized larvae, this evidence using two

experimental approaches supports the conclusion that PDV sequences can be integrated into the chromosomal DNA of host lepidopteran cells, providing a direct
means of regulation of gene expression and transcription. The presence of mobile
elements in the CcBV genome has been confirmed by genomic sequencing
(Drezen et al., 2006; Espagne et al., 2004). The discovery of the mobilization of
PDV sequences to lepidopteran cellular DNA potentially offers promise as a novel
strategy for development of new insect transformation systems based upon PDV
sequence mobilization, and genetic manipulation of vectors of human, animal,
and plant diseases as well as agricultural pests.

Aside from their immunological role, PDV genes and their expression products
have many regulatory roles that were touched on only briefly in this chapter. For
example, ecdysteroid deficiency is a common feature induced during the host's
last instar in many species of parasitized insects (Beckage and Gelman, 2004;
Pennacchio and Strand, 2006). This is in essence makes parasitoids unique among
parasites by invariably causing host developmental arrest and death. In the absence
of ecdysteriods, molting and metamorphosis are impossible, and the host never lives
long enough to reproduce following arrest.

Continued emphasis on identification of PDV genes which when expressed
cause inhibition of insect host immunity, suppress feeding behavior, or act as developmental arrestants, has many potential applications to development of novel
strategies of insect pest control (Beckage and Gelman, 2004, Pennacchio and
Strand, 2006). Examples include expression of targeted PDV genes in plants to
reduce insect feeding damage or cause developmental arrest of the pests (Gill
et al., 2006). Another approach is to generate PDV-hybrid entomopathogens such
as recombinant baculoviruses or bacteria with enhanced pathogen virulence and
faster speed-of-kill based on the immunosuppressive action of PDV gene products
including their hemocyte apoptosis-inducing properties. Applications of polydnavirology to biotechnology include utilization of PDV genes to enhance rates of
protein production in baculovirus expression systems at low cost for pharmaceutical or research-related purposes (Fath-Goodin et al., 2006b).

As we continue to identify more functional roles of PDV-encoded gene products and pinpoint more precisely their roles in manipulating host immunology and
development, our opportunities for exploiting these intriguing symbiotic viruses
in agriculture, forestry, biotechnology, and insect pest management will continue
to expand at a rapid pace. The maturing discipline of polydnavirology clearly provides current and future researchers with exciting opportunities to develop new
tools to address global problems caused by insects that ultimately impact human
health and the environment.

ACKNOWLEDGMENTS

The author expresses her sincere thanks to Ms. Anita Gordillo, Ms. Catherine
Cathers, and Ms. Dyan MacWilliam of the Beckage laboratory at the University

of California-Riverside for critiques of the text, and assistance with the figures and references. Mr. Roberto Osti graciously granted permission to use his artwork (Figure 10.1). I also warmly thank Dr. Donald B. Stoltz (Department of Microbiology and Immunology, Dalhousie University, Halifax, NS, Canada) for introducing me to the subject of polydnavirology early in my academic career. Many of his research findings and insights into the biology of these intriguing viruses have formed the framework of this fascinating and challenging field.

REFERENCES

Amaya, K. E., Asgari, S., Jung, R., Hongskula, M., and Beckage, N. E. (2005). Parasitization of *Manduca sexta* larvae by the parasitoid wasp *Cotesia congregata* induces an impaired host immune response. *J. Insect Physiol.* **51**, 505–512.

Asgari, S. (2006). Venom proteins from polydnavirus-producing endoparasitoids: Their role in host–parasite interactions. *Arch. Insect Biochem. Physiol.* **61**, 146–156.

Asgari, S., Schmidt, O., and Theopold, U. (1997). A polydnavirus-encoded protein of an endoparasitoid wasp is an immune suppressor. *J. Gen. Virol.* **78**, 3061–3070.

Barat-Houari, M., Hilliou, F., Jousset, F. X., Sofer, L., Deleury, E., Rocher, J., Ravallec, M., Galibert, L., Delobel, P., Feyereisen, R., Fournier, P., and Volkoff, A. N. (2006). Gene expression profiling of *Spodoptera frugiperda* hemocytes and fat body using cDNA microarray reveals polydnavirus-associated variations in lepidopteran host genes transcript levels. *BMC Genom.* **7**, 160.

Beck, M., and Strand, M. R. (2003). RNA interference silences *Microplitis demolitor* bracovirus genes and implicates glc1.8 in disruption and adhesion in infected host cells. *J. Virol.* **314**, 521–535.

Beckage, N. E. (1993). Games parasites play: The dynamic roles of proteins and peptides in the relationship between parasite and host. In *Parasites and Pathogens of Insects Vol. 1: Parasites* (N. E. Beckage, S. N. Thompson, and B. A. Federici, Eds.), pp. 25–57. Academic Press, San Diego, CA.

Beckage, N. E. (1997). The parasitic wasp's secret weapon. *Sci. Am.* **277**, 82–87.

Beckage, N. E. (1998). Modulation of immune responses to parasitoids by polydnaviruses. *Parasitology* **116**, S57–S64.

Beckage, N. E., and Gelman, D. B. (2004). Wasp parasitoid disruption of host development: Implications for new biologically based strategies for insect control. *Annu. Rev. Entomol.* **49**, 299–330.

Beckage, N. E., Metcalf, J. S. D., Nesbit, D., Schleifer, K. W., Zetlan, S. R., and Buron, I. de (1990). Host hemolymph monophenoloxidase activity in parasitized *Manduca sexta* larvae and evidence of inhibition by wasp polydnavirus. *Insect Biochem.* **20**, 285–294.

Bedwin, O. (1979a). The particulate basis of the resistance of a parasitoid to the defence reactions of its insect host. *Proc. Roy. Soc. Lond. B. Biol. Sci.* **205**, 267–270.

Bedwin, O. (1979b). An insect glycoprotein: A study of the particles responsible for the resistance of a parasitoid's egg to the defence reactions of its insect host. *Proc. Roy. Soc. Lond. B Biol. Sci.* **205**, 271–286.

Bezier, A., Herbiniere, J., Serbielle, C., Lesobre, J., Wincher, P., Huguet, E., and Drezen, J. M. (2007). Bracovirus gene products are highly divergent from insect proteins. *Arch. Insect Biochem. Physiol.* (in press).

Buron, I. de., and Beckage, N. E. (1992). Characterization of a polydnavirus (PDV) and virus-like filamentous particle (VLFP) in the braconid wasp *Cotesia congregata* (Hymenoptera: Braconidae). *J. Invertebr. Pathol.* **59**, 315–327.

Cui, L., and Webb, B. A. (1998). Relationships between polydnavirus genomes and viral gene expression. *J. Insect Physiol.* **44**, 785–793.

Cui, L., Soldevila, A. L., and Webb, B. A. (1996). Isolation and characterization of a member of the cysteine-rich gene family from *Campoletis sonorensis* polydnavirus. *J. Gen. Virol.* **77**, 797–809.

Cui, L., Soldevila, A. L., and Webb, B. A. (1997). Expression and hemocyte-targeting of a *Campoletis sonorensis* polydnavirus cysteine-rich gene in *Heliothis virescens* larvae. *Arch. Insect Biochem. Physiol.* **36**, 251–271.

Cui, L., Soldevila, A. L., and Webb, B. A. (2000). Relationships between polydnavirus gene expression and host range of the parasitoid wasp *Campoletis sonorensis*. *J. Insect Physiol.* **46**, 1397–1407.

Desjardins, C., Eisen, J. A., and Nene, V. (2005). New evolutionary frontiers from unusual virus genomes. *Genome Biol.* **68**, 212.

Dorn, S., and Beckage, N. E. (2007). Superparasitism in gregarious hymenopteran parasitoids: Ecological, behavioural and physiological perspectives. *Physiol. Entomol.* **32**, 199–211.

Doucet, D., Levasseur, A., Beliveau, C., Lapointe, R., Stoltz, D., and Cusson, M. (2007). *In vitro* integration of an ichnovirus genome segment into the genomic DNA of lepidopteran cells. *J. Gen. Virol.* **88**, 105–113.

Drezen, J. M., Bézier, A., Lesobre, J., Huguet, E., Cattolico, L., Periquet, G., and Dupuy, C. (2006). The few virus-like genes of *Cotesia congregata* bracovirus. *Arch. Insect Biochem. Physiol.* **61**, 110–122.

Dumpit, R. F., Nakanishi, R., Acierto, D., Haroian, N., Powell, E., Bachant, J., and Beckage, N. E. (2007). Cellular studies of *Cotesia congregata* bracovirus-induced host hemocyte apoptosis in *Manduca sexta*. Submitted.

Dupas, S., Turnbull, M. W., and Webb, B. A. (2003). Diversifying selection in a parasitoid's symbiotic virus among genes involved in inhibiting host immunity. *Immunogenetics* **55**, 351–361.

Dupuy, C., Huguet, E., and Drezen, J. M. (2006). Unfolding the evolutionary story of polydnaviruses. *Virus Res.* **117**, 81–89.

Dushay, M. S., and Beckage, N. E. (1993). Dose-dependent separation of *Cotesia congregata*-associated polydnavirus effects on *Manduca sexta* larval development and immunity. *J. Insect Physiol.* **39**, 1029–1040.

Edson, K. M., Vinson, S. B., Stoltz, D. B., and Summers, M. D. (1981). Virus in a parasitoid wasp: Suppression of the cellular immune response in the parasitoid's host. *Science* **211**, 582–583.

Eleftherianos, I., Boundy, S., Joyce, S. A., Aslam, S., Marshall, J. W., Cox, R. J., Simpson, T. J., Clarke, D. J., ffrench-Constant, R. H., and Reynolds, S. E. (2007). An antibiotic produced by an insect-pathogenic bacterium suppresses host defenses through phenoloxidase inhibition. *Proc. Natl. Acad. Sci. Am.* **104**, 2419–2424.

Espagne, E., Dupuy, C., Huguet, E., Cattolico, L., Provost, B., Martins, N., Poirie, M., Periquet, G., and Drezen, J. M. (2004). Genome sequence of a polydnavirus: Insights into symbiotic virus evolution. *Science* **306**, 286–289.

Espagne, E., Douris, V., Lalmanach, G., Provost, B., Cattolico, L., Lesobre, J., Kurata, S., Iatrou, K., Drezen, J. M., and Huguet, E. (2005). A virus essential for insect host–parasite interactions encodes cystatins. *J. Virol.* **79**, 9765–9776.

Falabella, P., Caccialupi, P., Varricchio, P., Malva, C., and Pennacchio, F. (2006). Protein tyrosine phosphatases of *Toxoneuron nigriceps* bracovirus as potential disrupters of host prothoracic gland function. *Arch. Insect Biochem. Physiol.* **61**, 157–169.

Falabella, P., Varricchio, P., Provost, B., Espagne, E., Ferrarese, R., Grimaldi, A., de Equileor, M., Fimiani, G., Ursini, M. V., Malva, C., Drezen, J. M., and Pennacchio, F. (2007). Characterization of the IkappaB-like gene family in polydnaviruses associated with wasps belonging to different braconid subfamilies. *J. Gen. Virol.* **88**, 92–104.

Fath-Goodin, A., Gill, T. A., Martin, S. B., and Webb, B. A. (2006a). Effect of *Campoletis sonorensis* ichnovirus cys-motif proteins on *Heliothis virescens* larval development. *J. Insect Physiol.* **52**, 576–585.

Fath-Goodin, A., Kroemer, J. A., Martin, S. B., Reeves, K., and Webb, B. A. (2006b). Polydnavirus genes that enhance the baculovirus expression vector system. *Adv. Virus Res.* **68**, 75–90.

Feddersen, I., Sander, K., and Schmidt, O. (1986). Virus-like particles with host protein-like determinants protect an insect parasitoid from encapsulation. *Experientia* **42**, 1278–1281.

Federici, B. A., and Bigot, Y. (2003). Origin and evolution of polydnaviruses by symbiogenesis of insect DNA viruses in endoparasitic wasps. *J. Insect Physiol.* **49**, 419–432.

Gill, T. A., Fath-Goodin, A., Maiti, I. I., and Webb, B. A. (2006). Potential uses of cys-motif and other polydnavirus genes in biotechnology. *Adv. Virus Res.* **68**, 393–426.

Gillespie, J. P., Kanost, M. R., and Trenczek, T. (1997). Biological mediators of insect immunity. *Annu. Rev. Entomol.* **42**, 611–643.

Gitau, C. W., Gundersen-Rindal, D., Pedroni, M., Mbugi, P. J., and Dupas, S. (2007). Differential expression of the CrV1 haemocyte inactivation-associated polydnavirus gene in the African maize stem borer *Busseola fusca* (Fuller) parasitized by two biotypes of the endoparasitoid *Cotesia sesamiae* (Cameron). *J. Insect Physiol.* **53**, 676–684.

Godfray, H. C. J. (1994). *Parasitoids: Behavioral and Evolutionary Ecology.* Princeton University Press, Princeton, NJ.

Grossniklaus-Burgin, C., Pfister-Wilhelm, R., Meyer, R., Treiblmayr, K., and Lanzrein, B. (1998). Physiological and endocrine changes associated with polydnavirus/venom in the parasitoid–host system *Chelonus inanitus–Spodoptera littoralis. J. Insect Physiol.* **44**, 305–321.

Gundersen-Rindal, D. E., and Lynn, D. E. (2003). Polydnavirus integration in lepidopteran host cells *in vitro. J. Insect Physiol.* **49**, 453–462.

Gundersen-Rindal, D. E., and Pedroni, M. J. (2006). Characterization and transcriptional analysis of protein tyrosine phosphatase genes and an ankyrin repeat gene of the parasitoid *Glyptapanteles indiensis* polydnavirus in the parasitized host. *J. Gen. Virol.* **87**, 311–322.

Guzo, D., and Stoltz, D. B. (1985). Obligatory multiparasitism in the tussock moth, *Orgyia leucostigma. Parasitology* **90**, 1–10.

Hajek, A. (2004). *Natural Enemies: An Introduction to Biological Control.* Cambridge University Press, Cambridge, UK.

Harwood, S. H., Grosovsky, A. J., Cowles, E. A., Davis, J. W., and Beckage, N. E. (1994). An abundantly expressed hemolymph glycoprotein isolated from newly parasitized *Manduca sexta* larvae is a polydnavirus gene product. *Virology* **205**, 381–392.

Harwood, S. H., McElfresh, J. S., Nguyen, A., Conlan, C. A., and Beckage, N. E. (1998). Production of early expressed parasitism-specific proteins in alternate sphingid hosts of the braconid wasp *Cotesia congregata. J. Invertebr. Pathol.* **71**, 271–279.

Hoffmann, J. A. (2003). The immune response of *Drosophila. Nature* **426**, 33–38.

Ibrahim, A. M., and Kim, Y. (2006). Parasitism by *Cotesia plutellae* alters the hemocyte population and immunological function of the diamondback moth, *Plutella xylostella. J. Insect Physiol.* **52**, 943–950.

Ibrahim, A. M., and Kim, Y. (2007). Transient expression of protein tyrosine phosphatases encoded in *Cotesia plutellae* bracovirus inhibits insect cellular immune responses. *Naturwissenschaften.* (in press).

Ibrahim, A. M., Choi, J. Y., Je, Y. H., and Kim, Y. (2007). Protein tyrosine phosphatases encoded in *Cotesia plutellae* bracovirus: Sequence analysis, expression profile, and a possible biological role in host immunosuppression. *Dev. Comp. Immunol.* **31**, 978–990.

Kadash, K., Harvey, J. A., and Strand, M. R. (2003). Cross-protection experiments with parasitoids in the genus *Microplitis* (Hymenoptera: Braconidae) suggests a high level of specificity in their associated bracoviruses. *J. Insect Physiol.* **49**, 473–482.

Kaeslin, M., Pfister-Wilhelm, R., and Lanzrein, B. (2005a). Influence of the parasitoid *Chelonus inanitus* and its polydnavirus on host nutritional physiology and implications for parasitoid development. *J. Insect Physiol.* **51**, 1330–1339.

Kaeslin, M., Pfister-Wilhelm, R., Molina, D., and Lanzrein, B. (2005b). Changes in the haemolymph proteome of *Spodoptera littoralis* induced by the parasitoid *Chelonus inanitus* or its polydnavirus and physiological implications. *J. Insect Physiol.* **51**, 975–988.

Krell, P. J., and Stoltz, D. B. (1980). Virus-like particles in the ovary of an ichneumonid wasp: Purification and preliminary characterization. *Virology* **101**, 408–418.

Kroemer, J. A., and Webb, B. A. (2004). Polydnavirus genes and genomes: Emerging gene families and new insights into polydnavirus replication. *Ann. Rev. Entomol.* **49**, 431–456.

Kroemer, J. A., and Webb, B. A. (2005). IkappaB-related vankyrin genes in the *Campoletis sonorensis* ichnovirus: Temporal and tissue-specific patterns of expression in parasitized *Heliothis virescens* lepidopteran hosts. *J. Virol.* **79**, 7617–7628.

Kroemer, J. A., and Webb, B. A. (2006). Divergences in protein activity and cellular localization within the *Campoletis sonorensis* ichnovirus vankyrin family. *J. Virol.* **80**, 12219–12228.

Lapointe, R., Wilson, R., Vilaplana, L., O'Reilly, D. R., Falabella, P., Douris, V., Bernier-Cardou, M., Pennacchio, F., Iatrou, K., Malva, C., and Olszewski, J. A. (2005). Expression of a *Toxoneuron nigriceps* polydnavirus-encoded protein causes apoptosis-like programmed cell death in lepidopteran insect cells. *J. Gen. Virol.* **86**, 963–971.

Lapointe, R., Tanaka, K., Barney, W. E., Whitfield, J. B., Banks, J. C., Beliveau C., Stoltz, D., Webb, B. A., and Cusson, M. (2007). Genomic and morphological features of a banchine polydnavirus: Comparison with bracoviruses and ichnoviruses. *J. Virol.* **81**, 6491–6501.

Lavine, M. D., and Beckage, N. E. (1995). Polydnaviruses: Potent mediators of host insect immune dysfunction. *Parasitol. Today* **11**, 368–378.

Lavine, M. D., and Beckage, N. E. (1996). Temporal pattern of parasitism-induced immunosuppression in *Manduca sexta* larvae parasitized by *Cotesia congregata*. *J. Insect Physiol.* **42**, 41–51.

Lavine, M. D., and Strand, M. R. (2002). Insect hemocytes and their role in immunity. *Insect Biochem. Mol. Biol.* **32**, 1295–1309.

Le, N. T., Asgari, S., Amaya, K., Tan, F. F., and Beckage, N. E. (2003). Persistence and expression of *Cotesia congregata* polydnavirus in host larvae of the tobacco hornworm, *Manduca sexta*. *J. Insect Physiol.* **49**, 533–543.

Leclerc, V., and Reichhart, J. M. (2004). The immune response of *Drosophila*. *Immunol. Rev.* **198**, 59–71.

Li, D., Schellhorn, N., and Schmidt, O. (2007). Detection of parasitism in diamondback moth, *Plutella xylostella* (L.), using differential melanization and coagulation reactions. *Bull. Entomol. Res.* **97**, 399–405.

Luckhart, S., and Webb, B. A. (1996). Interaction of a wasp ovarian protein and polydnavirus in host immune suppression. *Dev. Compar. Immunol* **20**, 1–21.

Matsumoto, H., Noguchi, H., and Hayakawa, Y. (1998). Primary cause of mortality in the armyworm larvae simultaneously parasitized by parasitic wasp and infected with bacteria. *Eur. J. Biochem.* **252**, 299–304.

Pech, L. L., and Strand, M. R. (2000). Plasmatocytes from the moth *Pseudoplusia includens* induce apoptosis of granular cells. *J. Insect Physiol.* **46**, 1565–1573.

Pennacchio, F., and Strand, M. R. (2006). Evolution of developmental strategies in parasitic hymenoptera. *Annu. Rev. Entomol.* **51**, 233–258.

Provost, B., Varricchio, P., Arana, E., Espagne, E., Falabella, P., Huguet, E., La Scaleia, R., Cattolico, L., Poirie, M., Malva, C., Olszewski, J. A., Pennacchio, F., and Drezen, J. M. (2004). Bracoviruses contain a large multigene family coding for protein tyrosine phosphatases. *J. Virol.* **78**, 13090–13103.

Pruijssers, A. J., and Strand, M. R. (2007). PTP-H2 and PTP-H3 from *Microplitis demolitor* bracovirus localize to focal adhesions and are antiphagocytic in insect immune cells. *J. Virol.* **81**, 1209–1219.

Reineke, A., Asgari, S., and Schmidt, O. (2006). Evolutionary origin of *Venturia canescens* virus-like particles. *Arch. Insect Biochem. Physiol.* **61**, 123–133.

Rivkin, H., Kroemer, J. A., Bronshtein, A., Belausov, E., Webb, B. A., and Chejanovsky, N. (2006). Response of immunocompetent and immunosuppressed *Spodoptera littoralis* larvae to baculovirus infection. *J. Gen. Virol.* **87**, 2217–2225.

Salt, G. (1963). The defence reactions of insects to metazoan parasites. *Parasitology* **53**, 527–642.

Salt, G. (1968). The resistance of insect parasitoids to the defence reactions of their hosts. *Biol. Rev. Camb. Philos. Soc.* **43**, 200–232.

Salt, G. (1970). *The Cellular Defence Reactions of Insects*. Cambridge University Press, Cambridge, UK.

Shelby, K. S., and Webb, B. A. (1999). Polydnavirus-mediated suppression of insect immunity. *J. Insect Physiol.* **45**, 507–514.

Shelby, K. S., Cui, L., and Webb, B. A. (1998). Polydnavirus-mediated inhibition of lysozyme gene expression and the antibacterial response. *Insect Mol. Biol.* **7**, 265–272.

Shelby, K. S., Adeyeye, O. A., Okot-Kotber, B. M., and Webb, B. A. (2000). Parasitism-linked block of host plasma melanization. *J. Invertebr. Pathol.* **75**, 218–225.

Sroka, P., and Vinson, S. B. (1978). Phenyloxidase activity in the hemolymph of parasitized and unparasitized *Heliothis virescens*. *Insect Biochem.* **8**, 399–402.

Stoltz, D. B., and Cook, D. I. (1983) Inhibition of host phenoloxidase activity by parasitoid hymenoptera. *Experientia* **39**, 1022–1024.

Stoltz, D. B., and Makkay, A. (2003). Overt viral diseases induced from apparent latency following parasitization by the ichneumonid wasp, *Hyposoter exiguae*. *J. Insect Physiol.* **49**, 483–489.

Stoltz, D. B., and Vinson, S. B. (1979a). Viruses and parasitism in insects. *Adv. Virus Res.* **24**, 125–171.

Stoltz, D. B., and Vinson, S. B. (1979b). Penetration into caterpillar cells of virus-like particles injected during oviposition by parasitoid ichneumonid wasps. *Can. J. Microbiol.* **25**, 207–216.

Stoltz, D. B., Krell, P. J., and Vinson, S. B. (1981). Polydisperse viral DNAs in ichneumonid ovaries: A survey. *Can. J. Microbiol.* **27**, 123–130.

Stoltz, D. B., Krell, P. J., Summers, M. D., and Vinson, S. B. (1984). Polydnaviridae – A proposed family of insect viruses with segmented, double-stranded, circular DNA genomes. *Intervirology* **21**, 1–4.

Strand, M. R. (1994). *Microplitis demolitor* polydnavirus infects and expresses in specific morphotypes of *Pseudoplusia includens* haemocytes. *J. Gen. Virol.* **75**, 3007–3020.

Strand, M. R., and Pech, L. L. (1995a). Immunological basis for compatibility in parasitoid–host relationships. *Annu. Rev. Entomol.* **40**, 31–56.

Strand, M. R., and Pech, L. L. (1995b). *Microplitis demolitor* polydnavirus induces apoptosis of a specific haemocyte morphotype in *Pseudoplusia includens*. *J. Gen. Virol.* **76**, 283–291.

Strand, M. R., Beck, M. H., Lavine, M. D., and Clark, K. D. (2006). *Microplitis demolitor* bracovirus inhibits phagocytosis by hemocytes from *Pseudoplusia includens*. *Arch. Insect Biochem. Physiol.* **61**, 134–145.

Tanaka K., Lapointe, R., Barney, W. E., Makkay, A. M., Stoltz, D., Cusson, M., and Webb, B. A. (2007). Shared and species-specific features among ichnovirus genomes. *Virology* **363**, 26–35.

Theopold, U., Krause, E., and Schmidt, O. (1994). Cloning of a VLP-protein coding gene from a parasitoid wasp *Venturia canescens*. *Arch. Insect Biochem. Physiol.* **26**, 137–145.

Thoetkiattikul, H., Beck, M. H., and Strand, M. R. (2005). Inhibitor kappaB-like proteins from a polydnavirus inhibit NF-kappaB activation and suppress the insect immune response. *Proc. Natl. Acad. Sci. Am.* **102**, 11426–11431.

Tian, S. P., Zhang, J. H., and Wang, C. Z. (2007). Cloning and characterization of two *Campoletis chlorideae* ichnovirus *vankyrin* genes expressed in parasitized host *Helicoverpa armigera*. *J. Insect Physiol.* **53**, 699–707.

Turnbull, M. W., Martin, S. B., and Webb, B. A. (2004). Quantitative analysis of hemocyte morphological abnormalities associated with *Campoletis sonorensis* parasitization. *J. Insect Sci.* **4**, 11.

Vray, B., Hartmann, S., and Hoebeke, J. (2002). Immunomodulatory properties of cystatins. *Cell Mol. Life. Sci.* **59**, 1503–1512.

Washburn, J. O., Kirkpatrick, B. A., and Volkman, L. E. (1996). Insect protection against viruses. *Nature* **383**, 767.

Washburn, J. O., Haas-Stapleton, E. J., Tan, F. F., Beckage, N. E., and Volkman, L. E. (2000). Co-infection of *Manduca sexta* larvae with polydnavirus from *Cotesia congregata* increases susceptibility to fatal infection by *Autographa californica* M nucleopolyhedrovirus. *J. Insect Physiol.* **46**, 179–190.

Waterhouse, R. M., Kriventseva, E. V., Meister, S., Xi, Z., Alvarez, K. S., Bartholomay, L. C., Barillas-Mury, C., Bian, G., Blandin, S., Chrestensen, B. M., Dong, Y., Jiang, H., Kanost, M. R., Koutsos, A. C., Levashina, E. A., Li, J., Ligoxygkis, P., Maccallum, R. M., Mayhew, G. F., Mendes, A., Michel, K., Osta, M. A., Paskewitz, S., Shin, S. W., Vlachou, D., Wang, L., Wei, W., Zheng, L., Zou, Z., Severson, D. W., Raikhel, A. S., Kafatos, F. C., Dimopoulos, G., Zdobnov, E. M., and Christophides, G. K. (2007). Evolutionary dynamics of immune-related genes and pathways in disease-vector mosquitoes. *Science* **316**, 1738–1743.

Webb, B. A., and Strand, M. R. (2005). The biology and genomics of polydnaviruses. In *Comprehensive Molecular Insect Science* (L. I. Gilbert, K. Iatrou, and S. S. Gill, Eds.), pp. 323–360. Elsevier, San Diego, CA.

Webb, B. A., Beckage, N. E., Hayakawa, Y., Lanzrein, B., Stoltz, D. B., Strand, M. R., and Summers, M. D. (2005). Polydnaviridae. In *Virus Taxonomy: VIII Report of the International*

Committee on Taxonomy of Viruses (C. M. Fauquet, M. A. Mayo, J. Maniloff, U. Desselberger, and L. A. Ball, Eds.), pp. 253–259. Elsevier, San Diego, CA.

Webb, B. A., Strand, M. R., Dickey, S. E., Beck, M. H., Higarth, R. S., Barney, W. E., Kadash, K., Kroemer, J. A., Lindstrom, K. G., Rattanadechakul, W., Shelby, K. S., Thoetkiattikul, H., Turnbull, M. W., and Witherell, R. A. (2006). Polydnavirus genomes reflect their dual roles as mutualists and pathogens. *Virology* **347**, 160–174.

Wertheim, B., Kraaijeveld, A. R., Schuster, E., Blanc, E., Hopkins, M., Pletcher, S. D., Strand, M. R., Partridge, L., and Godfray, H. C. (2005). Genome-wide expression in response to parasitoid attack in *Drosophila*. *Genome Biol.* **6**, R94.

Whitfield, J. B., and Asgari, S. (2003). Virus or not? Phylogenetics of polydnaviruses and their wasp carriers. *J. Insect Physiol.* **49**, 397–405.

Yu, R. X., Chen, Y. F., Chen, X. X., Huang, F., Lou, Y. G., and Liu, S. S. (2007). Effects of venom/calyx fluid from the endoparasitic wasp *Cotesia plutellae* on the hemocytes of its host *Plutella xylostella in vitro*. *J. Insect Physiol.* **53**, 22–29.

Zhou, J., Ueda, M., Umemiya, R., Battsetseg, B., Boldbaatar, D., Xuan, X., and Fujisaki, K. (2006). A secreted cystatin from the tick *Haemaphysalis longicornis* and its distinct expression patterns in relation to innate immunity. *Insect Biochem. Mol. Biol.* **36**, 527–535.

11

INSECT IMMUNE RECOGNITION AND SUPPRESSION

OTTO SCHMIDT

Insect Molecular Biology, School of Agriculture, Food and Wine, University of Adelaide, Glen Osmond, SA 5964, Australia

ABSTRACT: Parasitoid–host interactions, where one insect grows up inside another insect, are ideal experimental systems to study immunity. The extreme selective pressure of these relationships, where either one of the two insects is going to perish in the process, is likely to force both insect species to continuously evolve in order to stay in this race for survival. Given that these interactions must have produced the most sophisticated evolutionary strategies involving recognition measures by the host and its counter-measures by the parasitoid, investigations into these systems have indeed uncovered important issues related to fundamental problems of how host insects recognize foreign objects as potentially damaging and how parasitoids in turn overcome the hosts' defense reactions. While we are far from knowing all the answers it is obvious that these highly evolved biological interactions challenge our traditional understanding of how innate immune systems are able to identify highly specific targets with a limited number of recognition molecules.

Abbreviations:

ATP	=	adenosinetriphosphate
CrV	=	*Cotesia rubecula* virus
CrV1	=	CrV-encoded protein 1
Gal	=	galactosamine
GalNAc	=	*N*-acetylgalactosamine
GTP	=	guanosinetriphosphate
HPL	=	*Helix pomatia* lectin
LDL	=	low-density lipoprotein
LM	=	leverage mediated
LPS	=	lipopolysaccharide
PAMP	=	pathogen-associated molecular pattern
PAMS	=	parasitoid-associated molecular structure
PDV	=	polydnavirus
PNA	=	peanut agglutinin
PPO	=	prophenoloxidase
RNAi	=	RNA interference

11.1   RECOGNITION OF INSECT PARASITOIDS BY INSECT HOSTS

Parasitoid–host interactions, where one insect grows up inside another insect, are ideal experimental systems to study immunity. The extreme selective pressure, where either one of the two insects is going to perish in the process, is likely to force both insect species to constantly evolve to stay in this relationship. While this so-called 'Red Queen' effect has produced a diverse range of evolutionary adaptations that makes almost every system analyzed so far unique (Pennacchio and Strand, 2006), molecular analysis of about a dozen parasitoid interactions has also focused our attention to some of the fundamental problems of immune recognition. The mechanisms used either by host insects to recognize the parasitoid or by parasitoids to evade and suppress the hosts' defense reactions provide valuable insights into the workings of innate immune recognition and response.

Given some recent reviews on invertebrate immunity (Kanost et al., 2004; Mylonakis and Aballay, 2005) and on parasitoid–host interactions (Pennacchio and Strand, 2006; Schmidt et al., 2001), this is not an attempt to comprehensively review existing literature, but rather a first step in selectively highlighting some questions that may lead to new perspectives on some of the intractable riddles that are at the core of innate immune recognition in general and parasitoid–host interactions in particular. These issues also apply to the suppression of insect immunity by other organisms, such as Wolbachia (Hise et al., 2007; Huigens et al., 2000), nematodes (Brivio et al., 2004; Ribeiro et al., 1999) or associated bacteria, which may in turn affect parasitoid relationships (Zchori-Fein et al., 2001).

In principle, recognition of a parasitoid by a host insect can be summarized within two possible scenarios, which may be difficult to separate operationally and may in fact not be mutually exclusive. However, the two scenarios illustrate some of the fundamental problems we face in trying to explain innate immune recognition and its suppression by parasitoids.

The first possible scenario is based on an apparent ability of insects to identify sclerotized egg shell and larval cuticle structures inside the hemocoel and to respond with a defense reaction. A corollary of this scenario is that any insect can be expected to respond to artificially or accidentally deposited parasitoid eggs and larvae with an encapsulation response. As a consequence parasitoid species that develop inside a particular host insect avoid or prevent this reaction from occurring using maternal secretions, a concept intuitively accepted since George Salt did his groundbreaking experiments on egg transplantations (Salt, 1970) and Brad Vinson summarized all possible parasitoid strategies to overcome the host defense in a classic review (Vinson, 1990).

The second possible scenario is based on a concept where pathogen-associated molecular patterns (PAMPs) are recognized by host recognition molecules (Janeway, 1989). In this context, the host insect utilizes parasitoid-specific molecular structures and while engaging with the parasitoid acquires recognition molecules that signal the presence of the parasitoid, thereby reducing the risk of parasitization. The most compelling cases for this scenario come from genetic studies in *Drosophila* populations (Dupas et al., 2003), where parasitoid virulence and host resistance resemble the gene-for-gene interactions in plants (Jones and Dangl, 2006). In this context, related insect species not engaged in parasitoid–host interactions lack specific recognition proteins and thus are expected to fail to react toward artificially or accidentally deposited parasitoid eggs or larvae.

As indicated before, closer inspection of the two scenarios suggests that the two recognition mechanisms are not mutually exclusive and common to all multicellular organisms. Nevertheless, in the following we will discuss parasitoid–host interactions in the context of these modes of innate immune recognition to highlight problems and point out underlying principles.

11.1.1 PARASITOID-ASSOCIATED MOLECULAR STRUCTURES

The most widely accepted concept of innate immune recognition is based on PAMP recognition, which invokes conserved molecular patterns that distinguish microbes from other (mostly) eukaryotic organisms and thus allow the host to identify potentially damaging microbes (Janeway and Medzhitov, 2002). The implications of this concept for parasitoid–host interactions are that it allows host insects to identify a parasitoid species (or groups of species) and distinguish it from its own by evolving specific proteins that bind to molecular patterns unique to this group.

It has been argued that adopting this concept for parasitoid–host interactions may not be realistic, given that microbial pathogens and their respective multicellular host organisms have evolved together for much longer compared to the time periods available to the more recently emerged parasitoid–host interactions. However, evolutionary adaptations of specific recognition proteins are chance events that depend on other factors as well, such as the available genetic variety and high selection pressures. But while the extreme selective pressure may argue for a rapid evolution of specific recognition molecules in the host, it may also accelerate processes in the parasitoid to overcome host recognition and defense.

Despite the word pattern, which is more accurately defined as molecular structures (Beutler, 2003), this type of recognition is based on the evolution of unique recognition proteins using 'lock and key' interactions to achieve specificity. 'Lock and key'-type interactions, such as protein-adhesion molecules and lectins, bind to particular antigens or sugar determinants with affinities that are usually inversely proportional to the corresponding specificity. In other words, the number of proteins required to achieve high specificity in recognizing parasitoids is directly proportional to the differentiation capacity (Firrao, 1994).

Genetic and protein diversity can be generated by a number of mechanisms, such as gene duplications producing gene families of potential recognition proteins (Christophides, 2002). Alternatively, the creation of multiple protein variants by alternative splicing (Dong et al., 2006; Watson, 2005) potentially expand the receptor repertoire in innate immune recognition proteins. Interestingly, none of these mechanisms have been observed so far in parasitoid–host interactions and while these mechanisms increase the receptor repertoire to select from, insects lack the cellular capacity to implement anticipatory recognition capabilities. Rather insect hosts engaged in parasitoid relationships are likely to recruit recognition proteins from existing genetic diversities of molecules that happen to bind to parasitoid-specific structures. This suggests that hosts species that are able to specifically recognize a parasitoid are likely to achieve this based on few, if not a single gene product. While this is sufficient to identify a parasitoid, it can potentially be overcome by the parasitoid, using antigen-masking or epitope alterations. Such examples of co-evolution have been found in *Drosophila* wild populations exposed to various parasitoid species (Dupas et al., 2003; Kraaijeveld and

Godfray, 1999; Kraaijeveld et al., 1998) and there are reasons to believe that similar gene-for-gene interactions occur in other parasitoid–host interactions (Carton et al., 2005). To distinguish these from microbial pathogens we refer in the following to parasitoid-associated molecular structures (PAMS) that are recognized by specific host recognition proteins to identify particular wasp species.

11.1.2 RECOGNITION OF SCLEROTIZED STRUCTURES

In contrast to the adaptation of specific recognition capacities using PAMS, which require relative long periods of co-evolution, insects seem to posses an underlying ability to cope with new threats relatively quickly. In fact the artificial or accidental deposition of eggs and larvae into the hemocoel of any insect usually causes an encapsulation reaction by the host insect, whether this has had previous encounters with a parasitoid or not. Parasitoid oocytes not covered by calyx secretions are readily encapsulated in the hemolymph of any insect. It is obvious that this is not a pre-adaptation within insect groups to counter any possible parasitoid attack, since other objects, such as nylon threats (Salt, 1970), sephadex beads (Lavine and Strand, 2001) and other artificial structures (Brehelin et al., 1975) are encapsulated as well. This indicates a general capacity to react to 'foreign' structures in the absence of specific recognition proteins. Interestingly, both artificial objects (Li and Paskewitz, 2006) as well as egg and larval surface structures (Asgari et al., 1998) can be masked by proteins to become protected against the host defense. It is in fact conceivable that recognition of potentially damaging objects is the outcome of dynamic protective coating reactions and immune activation reminiscent of vertebrate coagulation inhibitors on self-surfaces (Esmon, 2004).

One mechanism that could explain why eggs are encapsulated inside the hemocoel is based on the assumption that certain molecular structures elicit a response when introduced from outside or released from hemocyte granules but are 'normally' absent inside the hemocoel. For example, terminal sugar modifications, such as N-acetylgalactosamine (GalNAc), are frequently found on plasma glycoproteins, whereas galactosamine (Gal) modifications are only found outside the hemocoel or inside secretory granules of hemocytes. Moreover terminal Gal-containing glyco-modifications are not detected on intact hemocytes or hemocyte-like cells, but are produced in the presence of ecdysone and found on the surface of cells involved in the clearance of debris during metamorphosis (Theopold et al., 2001) or on the surface of hemocytes or microparticles that become part of defense-related aggregation and coagulation reactions (Theopold and Schmidt, 1997).

This begs the questions, whether an immune response by the host is based on the presence of Gal-containing glycodeterminants on the surface of egg and larval structures. Indeed, when parasitoid egg and oocyte surface structures were analyzed with lectins, both GalNAc- and Gal-containing glycodeterminants were detected (Fig. 11.1). In this context, the parasitoid egg and larva are identified

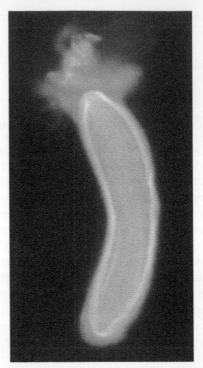

FIGURE 11.1 Mature oocyte of the parthenogenetic wasp *Venturia canescens* stained with a Gal-specific lectin, peanut agglutinin (PNA). The oocyte (or non-fertilized egg) was dissected from the wasp egg reservoir, incubated with FITC-conjugated PNA and inspected under indirect UV light on a confocal microscope. Note the bushy chorion protrusions at the anterior end of the egg shell, which are translucent in the non-stained egg. Encapsulation of the egg usually starts at these heavily stained protrusions (Schmidt and Schuchmann-Feddersen, 1989). (See Plate 11.1 for color version of this figure.)

inside the hemocoel as 'foreign' since the egg chorion and the larval cuticle contain Gal-specific glycomodifications (Korayem et al., 2004; Theopold et al., 2002). Since Gal-specific lectins are present in the hemolymph of lepidopterans (Castro et al., 1987), Gal-containing egg and cuticle surface structures can act as elicitors and cross-link with Gal-containing microparticles (Theopold and Schmidt, 1997) and pro-coagulants (Korayem et al., 2004, 2007) discharged from hemocytes. This scenario describes the host response to the parasitoid as a process based on the difference of hemocoelic and extra-hemocoelic structures rather than the recognition of 'foreigness'. In this model we would expect parasitoid eggs or larvae to become recognized inside the hemocoel of any insect, even of its own species. Thus it is possible that this or similar recognition mechanisms are based on adaptations that are not parasitoid-specific but may have predated parasitoid–host relationships, such as the inherent ability to remove dead cells and clear debris during tissue remodeling and metamorphosis (Theopold et al., 2001).

11.1.3   RECOGNITION OF SYNTHETIC OBJECTS

The real challenge are putative recognition mechanisms based on the observations that insects, or any invertebrate for that matter, are able to respond to objects the organisms have never encountered before? This raises the conundrum of how insects recognize objects, such as nylon threats and sephadex or latex beads, without having a corresponding repertoire of specific binding proteins or the evolutionary time to adapt to it. Are innate recognition capacities available to invertebrates that allow the identification and response to substances, objects and potentially damaging organisms in the absence of anticipatory recognition proteins? This conundrum has its resonance in cell-free defense reactions, where pro-coagulants react independent of cells and cellular signaling pathways, referring to the existence of sensor particles that respond to a wide range of environmental changes, such as osmolarity or oxygen conditions in addition to specific elicitors. The identity and functional roles of these sensors in biological recognition processes have only recently been uncovered.

11.2   CELL-FREE DEFENSE REACTIONS

It has long been known that insects are able to mount humoral immune responses almost independent of hemocytes (Gotz, 1986), such as in cell-free plasma (Theopold et al., 2004) or outside the hemocoel, such as the cuticle (Ashida and Brey, 1995) or gut lumen (Christophides et al., 2004). Moreover, the fact that immune signals are often developed upstream of membrane-bound receptors and used in very different contexts suggests that multiprotein assemblies exist upstream of cell-bound receptors, which are conserved at the mechanistic level and able to carry out its function with more than one combination of protein constituents (Schmidt et al., invited review). For example, the Toll ligand Spaetzle is required for developmental and immune signaling in *Drosophila* (Lemaitre et al., 1996) and is activated to form a dimer by regulatory processes that are reminiscent of proteolytic enzyme cascades activating pro-coagulants (Krem and Cera, 2002) and prophenoloxidase (PPO) (Soderhall and Cerenius, 1998). In fact, the activated Spaetzle ligand resembles arthropod coagulogens at the structural level (Bergner et al., 1997; Delotto and Delotto, 1998).

What is the significance of these extracellular regulatory cascades to recognition processes? It appears that invertebrates with an open circulatory system are using coagulation and melanization reactions not only for wound-healing, but also for the inactivation of pathogens (Theopold et al., 2004) and parasites (Nappi and Christensen, 2005). While these extracellular defense reactions have long escaped our notice due to the difficulty of analyzing covalently linked and sometimes melanized coagulation products at the biochemical level (Bidla et al., 2005), it has become apparent that the regulatory cascades controlling coagulation and melanization (Soderhall and Cerenius, 1998) are part of an ancestral defense reaction that have been adapted to multiple functions in different organisms (Krem and Cera, 2002). But while lipid-containing particles, such as lipophorin (Duvic and Brehelin, 1998; Li et al., 2002) and vitellogenin (Hall et al., 1999), have been known as pro-coagulants in arthropods, we only recently became aware that some plasma components, including immune proteins, become associated with lipid particles under certain conditions (Ma et al., 2006) changing their properties in the presence of elicitors (Schmidt et al., invited review). For example, lipophorin particles in insects interact with exchangeable lipoproteins and other plasma proteins, such as apolipoprotein III (Niere et al., 2001), PPO and its activating proteases (Rahman et al., 2006), imaginal disk growth factors (Ma et al., 2006) and morphogens (Panakova et al., 2005). These modified particles may be involved in multiple processes, including lipid metabolism (Canavoso et al., 2001), immunity (Rahman et al., 2006; Whitten et al., 2004), growth and development (Panakova et al., 2005).

11.2.1   LIPID PARTICLES AS IMMUNE SENSORS

The difficulty in analyzing these particles with traditional analytical approaches is that changes in the environment, such osmolarity, reducing or oxidizing conditions, may affect these particles, causing aggregation into insoluble coagulation products. For example, we noticed that lipid particles of some insects responded to widely used buffers containing Trisamine with aggregation (unpublished data). Conversely, high salt concentrations used in the isolation of lipid particles by low density gradient centrifugation cause most immune proteins to dissociate from lipid particles, such that identification of particles with immune proteins was only detectable in immune-induced insects (Rahman et al., 2006).

In the context of parasitoid eggs deposited inside the hemocoel, the reaction products from oxidative cross-linking of chorion proteins (Li, 1994), a reaction delayed in hydropic eggs (Schmidt et al., 2005) or oxidization of surface structures during egg deposition, may damage host lipid particles in the vicinity of the newly deposited egg. This could cause local coagulation reactions on the egg surface leading to hemocyte attachments to pro-coagulant depositions on foreign surfaces. In fact, one of the four major components of protective virus-like particles coating the surface of *Venturia canescens* eggs is a phospholipid-hydroperoxide glutathione peroxidase-like protein that is able to mask oxidized lipids (Schmidt et al., 2005). This may suspend the coagulation process or turn limited coagulation reactions into a protective layer on the surface of the parasitoid (Kinuthia et al., 1999).

While these properties of lipid particles may have concealed functional involvements other than in lipid metabolism, these observations are compatible with a 'sensor' function of lipid particles. Since some exchangeable proteins, required for lipid uptake and transport, such as apolipoprotein III in insects, register lipid composition (Van der Horst, 2003) and the presence of lipopolysaccharide (LPS) (Whitten et al., 2004), it is conceivable that the first steps in the recognition process involve alterations in the lipid composition of particles in the presence of lipid A or lipid modifications under oxidizing conditions. As a consequence of

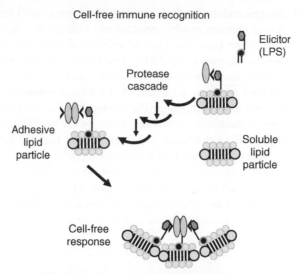

FIGURE 11.2 Cell-free defense reactions involving recognition and aggregation around elicitors or damaging objects. Lipid particles, which can act as circulating sensor particles, are schematically depicted as a disk of lipid bilayer surrounded by ring-shaped apolipoproteins. Associated proteins respond to elicitors (lipopolysaccharide, LPS) or environmental cues by becoming adhesive either directly (Mellroth et al., 2005) or indirectly through regulatory cascades (Krem and Cera, 2002). Adhesive lipid particles aggregate by cross-linking lipid particles around damaging objects or substances.

changes in lipid particle properties, some of the associated proteins can become modified directly (Mellroth et al., 2005; Yu and Kanost, 2002) or indirectly through regulatory cascades (Krem and Cera, 2002) to switch from non-adhesive to adhesive particles (Fig. 11.2).

11.2.2   LIPID PARTICLES AS PRO-COAGULANT

Lipid particles are ideal biological sensors, being continuously secreted and taken up and thus being ubiquitous in and around all cells. These particles contain lipids and glycolipids that are easily oxidized and respond readily to many changes in the environment. Moreover, being similar to cellular membranes they respond to damaging effects and environmental clues by changing their functional properties into adhesive particles that can aggregate around toxins or pathogens (Rahman et al., 2006). Cell-free defense capacities combining afferent and efferent immune functions (Beutler, 2004) have long been implicated with invertebrate hemolymph coagulation reactions. Again, the difficulty with identifying lipid particles involved in cell-free defense using genetic and RNA interference (RNAi) screens is that abrogation of lipid carrier functions are instantly deleterious to the mutant organism hiding any possible immune phenotype.

Although we have yet to uncover the molecular details, the underlying principle of this recognition mechanism is in the responsiveness of lipid particles to environmental changes, and the resulting structural and functional changes of particle properties. These changes involve associations with immune-related plasma proteins, such as regulatory proteases, their inhibitors and zymogens, such as PPO and adhesion proteins (Rahman et al., 2006). These observations in invertebrates are compatible with the existence of cell-free defense reactions, involving coagulation and melanization reactions upstream of cell-bound receptors. The importance of cell-free defense reactions is the notion that an immune signal is developed upstream of membrane-bound receptors by extracellular sensors, such as lipid-containing particles that change properties and become adhesive after exposure to damaging environments and elicitors. Adhesive lipid particles that self-assemble around damaging objects (Rahman et al., 2006) can potentially inactivate potential parasitoids by forming a layer of pro-coagulant that may attract hemocytes. Conversely, the existence of particles with the capacity to self-assemble around objects and substances raises the question of how adhesive particles interact with cells.

11.3   CELLULAR DEFENSE REACTIONS

The internalization of objects and microbes is driven by mechanisms that usually involve a combination of adhesive zipper- (or better) velcro-mechanisms, where cellular uptake reactions are dependent on membrane-bound receptors wrapping the cell membrane around the object, and cellular membrane invaginations dragging the object into the cell (Fig. 11.3). The driving force of the former is based on the adhesive properties of the object and cell surface receptors, and can be described in mechanistic terms, while the latter involves the so-called trigger-mediated endocytosis-like uptake mechanisms (Swanson and Baer, 1995). While the velcro-mechanism is driven by extracellular driving forces, the 'trigger'-mechanism is perceived to be initiated by a signaling process and driven by cytoplasmic forces. However, close inspection of these trigger- or receptor-mediated endocytosis reactions reveals alternative driving forces that resemble velcro-like uptake of sensor particles by adhesive cell surface receptors (Schmidt et al., invited review). For example, addition of soluble adhesion proteins, such as apolipoprotein III, can detach spread cells, while enhancing phagocytosis when immobilized on the surface of bacteria (Whitten et al., 2004). Similar observations, where lectins immobilized on substrate cause cell spreading, whereas addition of soluble lectins causes detachment of spread cells (Glatz et al., 2004), suggest that receptor attachment to external binding sites and de-adhesion by receptor clearance from the cell surface are linked in a dynamic fashion (Schmidt and Schreiber, 2006).

Since attachments to external binding sites and to adhesive lipid particles can involve the same membrane-bound receptors, cell adhesion and de-adhesion can only be described in mechanistic terms, where the internalization of adhesive receptors is viewed in the context of a velcro-like phagocytosis reactions (Fig. 11.3). Given that not much is known about the molecular details of lipid uptake

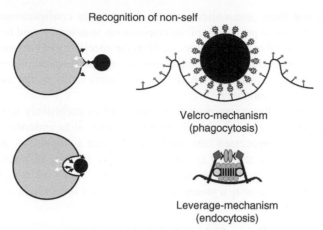

Recognition of non-self

Velcro-mechanism
(phagocytosis)

Leverage-mechanism
(endocytosis)

FIGURE 11.3 Cellular uptake of non-self objects involving attachment and interactions leading to recognition of non-self. Adhesion and uptake of objects and microbes is driven by mechanisms that involve either velcro-mechanisms, where uptake is dependent on adhesive receptors wrapping the cell membrane around the object (black round object covered with lectins as opsonins), or cellular membrane invaginations dragging the object into the cell. Given the size differences of lipid particles and receptors adhesive lipid particles can be regarded as opsonized objects taken up by cells using velcro-like phagocytosis reactions. Since this involves a tilting of membrane-bound receptors around the particle, we called this an LM-uptake reaction (Schmidt and Theopold, 2004). Clustering of lipid particles on the cell surface may drive the uptake of solid and liquid cargo by a cellular clearance reaction based on dynamic adhesion processes on the cell surface (Schmidt and Schreiber, 2006).

and lipid shuttling (Van Hoof et al., 2005) in conjunction with the large size of lipid particles relative to membrane-bound receptors, it is conceivable to consider lipid particles as small objects that are taken up by phagocytosis reactions. In this context, receptors interacting with lipid particles by attaching to associated adhesion proteins assemble around the particle to form velcro-like attachments, where the receptors are tilted around the particle (Fig. 11.3). The outcome is an inverse curvature of the cell membrane, which is the first step toward receptor internalization and endocytosis. Lateral clustering of lipid particle assemblies forms vesicles in proportion to the size of the clusters on the cell surface. Again, like in phagocytosis, this process is affected by a number of extracellular and intracellular factors, such as the presence of adhesive particles and receptor anchorage to cytoplasmic scaffolds and actin fibers.

In principle, any large protein complex that can assemble receptors around objects with hinge-like properties can tilt receptors with the help of oligomeric adhesion proteins. To distinguish velcro-like phagocytosis reactions from cytoplasm-derived mechanisms in other endocytosis reactions, we called this a leverage-mediated (LM) process, which has the capacity to sculpture membranes, internalize objects and substances and dislocate receptors from cytoplasmic anchorage (Schmidt and Theopold, 2004). The importance of this mechanism is that it constitutes an extracellular driving force that is responsive to external cues.

Whether or not these assemblies are able to generate configurational energy depends on the overall properties of the constituents to assemble and rearrange in such a way as to perform an LM process. Thus the specificity of LM-mechanisms is not only determined by the binding properties of the adhesion proteins as part of a 'lock and key' interaction, but also by the functionality of the assembly to produce configurational energy.

Given that the cellular responses were perceived to exclusively rely on intracellular machineries and cytoplasmic energy sources, it is understandable that cellular recognition processes have traditionally been synonymous with signal transduction. However, the very existence of cell-free immune responses suggest that while signal transduction provides immediate instructions for efficient cellular response it is not necessarily a precondition of biological recognition processes.

11.3.1   CELL–CELL INTERACTIONS

In fact our perceptions of cellular response mechanisms to outside cues have been heavily influenced by antibody-receptor (Klein, 1999) and gene-for-gene (Dangl and McDowell, 2006) paradigms, which imply that a multitude of highly specific ligands interact with corresponding membrane receptors to create a signal in the cytoplasm, which allows the cell to respond accordingly. The cell is seen as a box with many button-like receptors on the cell surface, which are pushed by ligand-like proteins to produce a specific response. This signal is understood to be instructive (Fig. 11.4) rather than interactive, constituting a prerequisite for subsequent changes in cell behavior rather than a direct consequence of interactions outside the cell or on the cell surface.

With the exception of zipper-mediated (or velcro-mediated) phagocytosis reactions (see above) cellular activities are perceived to be driven exclusively by cytoplasmic machineries fueled by energy-storing molecules, such as adenosinetriphosphate (ATP) and guanosinetriphosphate (GTP). In fact a conceptual basis

Cellular uptake in the instructive model

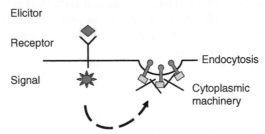

FIGURE 11.4 Instructive model of innate immune recognition and cellular response using the receptor-mediated endocytosis as an example. Elicitors bind to receptors using 'lock and key' interactions. Subsequent structural changes of the elicitor–receptor complex initiate signal pathways in the cytoplasm. In this model the signal is required to activate cytoplasmic machineries for the cell response.

for the instructive model is the assumption that the driving forces that shape a cell are derived from cytoplasmic sources based on the observation that the functional elimination of cytoplasmic components abolish specific cellular processes, such as cell migration, filopodia extension and cellular uptake reactions (Etienne-Manneville and Hall, 2002). But while it is obvious that some of these components are essential for maintaining these activities it is less clear whether these or other cytoplasmic components are sufficient to drive these reactions.

The functional involvement of one and the same or similar receptors and adhesive ligands in opposite cell behaviors, such as cell spreading and detachment (Murphy-Ullrich, 2001), attraction or repulsion of nerve axons (Marquardt et al., 2005), or growth and inhibition of pollen tubes (Lord and Russell, 2002), are difficult to understand in the instructive model, but indicative of a dynamic interplay of extracellular and intracellular driving forces.

11.3.2 LECTIN-MEDIATED HEMOCYTE INACTIVATION

If we accept the existence of extracellular driving forces that mediate adhesion and internalization of objects and substances in a dynamic fashion, it is conceivable that conditions that strongly drive clearance of receptors from the cell surface result in hemocytes that have most adhesive receptors removed from the surface. In line with this assumption, parasitoids may exploit the dynamic balance that exists in the hemolymph, where hemocytes switch between adhesive and de-adhesive cell behavior. If conditions exist in the hemolymph that promote de-adhesion of hemocytes by stimulating cellular uptake reactions, the parasitoid may alter hemolymph conditions to activate extracellular driving forces that may at least temporarily deplete hemocytes of cell surface receptors.

What are the conditions that promote de-adhesion of hemocytes? We and many others have noticed that some oligomeric lectins act as adhesion molecules by promoting spreading on an artificial surface, but when added to spread cells in soluble form act as counter-adhesion molecules by detaching already spread cells (Schmidt et al., invited review). For example, the pioneering work of the Rizki's demonstrated that lectins cause spreading of *Drosophila* cells on a glass surface and cause cell fusion of neighboring cells (Rizki et al., 1975), but the same lectins caused detachment from tissue contacts and associated rearrangements of actin cytoskeleton in fat body cells (Rizki and Rizki, 1983).

Similarly *Drosophila* hemocyte-like cells (and other primary cells as well) will spread more extensively when plated on immobilized lectins (Rogers et al., 2003). In fact, a small population of hemocytes is able to spread or phagocytize in the presence of elicitors (Dean et al., 2004a, b), probably due to the presence of specific adhesive receptors on the cell surface (Nardi et al., 2006). In line with the interactive model increased phagocytosis is expected to cause cells to become round shaped as adhesive receptors are internalized (Dean et al., 2004a). Likewise, in the presence of soluble adhesion proteins, such as lectins, extremely spread cells detach and round up (Glatz et al., 2004).

The importance of lectin-like interactions was confirmed in lectin-mediated responses in ecdysone-treated *Drosophila* hemocyte-like cells, where *mbn-2* cells (displaying GalNAc- and Gal-containing hemomucins (Hellers et al., 1996; Theopold et al., 2001) responded to the presence of two soluble oligomeric lectins, the GalNAc-specific *Helix pomatia* lectin (HPL) and the Gal-specific lectin from *Arachis hypogaea* (peanut agglutinin, PNA), by undergoing lectin-induced uptake reactions (Schmidt et al., invited review). On the assumption that hemomucin is the only glycoprotein receptor on the surface of *Drosophila* cells (Theopold et al., 2001), receptor–lectin interactions leading to adhesion are interdependent with those leading to lectin uptake using the same receptors. The conclusion is that adhesion proteins cause adhesion by interacting with glycoprotein receptors but also mediate cellular uptake reactions causing receptor depletion, which implies that some adhesion proteins act like counter-adhesion molecules (Schmidt et al., invited review), where the two seemingly opposite reactions form a dynamic equilibrium (Schmidt and Schreiber, 2006). While there are more complex explanations the most uncomplicated model is that counter-adhesion proteins interact with lipid particles and by driving LM-uptake reactions remove adhesive receptors from the cell surface (Fig. 11.5).

Using this model we can make some predictions regarding immune suppression by parasitoid factors. Since free lateral receptor movements favour receptor LM-assemblies and internalization, circulating hemocytes require strong receptor anchorage to cytoplasmic scaffolds to maintain receptors on the cell surface, a prerequisite for adhesion which is expected to promote adhesion to extracellular sites or phagocytosis of objects. In line with the model, stabilization of receptors

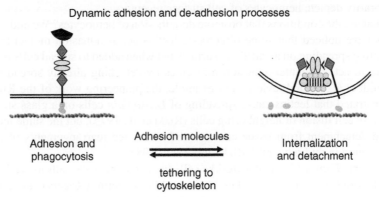

FIGURE 11.5 Dynamic interaction of adhesion (attachment of receptors to external binding sites) and lateral cross-linking with adhesive particles on the cell surface causing receptor internalization (Schmidt and Schreiber, 2006). Since lateral cross-linking of receptors (receptor movements in two dimensional membrane) thermodynamically favored over receptor binding to external binding sites (requiring movements in three dimensions), uptake by LM-mechanisms are favored over adhesion. To retain adhesive properties on the cell surface requires anchorage to cytoplasmic scaffolds, such as actin cytoskeleton. Conversely, destabilization of cytoplasmic scaffolds may enhance macropinocytosis of existing clusters of LM-assemblies, but also prevent the formation of clusters.

on the cell surface can be achieved by receptor anchorage to cytoplasmic scaffold, such as actin cytoskeleton, and by receptor–substrate linkages. Destabilization occurs when receptor anchorage to the actin cytoskeleton is disrupted by the action of molecules, such as cytochalasin D (Glatz et al., 2004) or toxins (Vilcinskas et al., 1997b). In addition, oligomeric adhesion molecules forming LM-assemblies may also be able to dislocate receptor anchorage leading to depolymerization of actin fibers (Glatz et al., 2004).

Likewise, the observation that spread cells lose phagocytic ability in the presence of low concentrations of cytochalasin D (Vilcinskas et al., 1997a) before detaching (Fig. 11.5) is in agreement with this model, given that receptor stabilization at focal adhesion points is provided by receptor linkages to external binding sites in addition to cytoplasmic actin-cytoskeleton anchorage.

In this model hemocytes engaging in adhesive behavior are expected to display adhesive receptors on the cell surface by anchoring their cytoplasmic domains to actin fibers or other cytoplasmic scaffolds. The reinforcement of cytoplasmic anchorage produces characteristic cell shapes, such as odd-shaped hemocytes in lepidopterans known to be involved in spreading and phagocytosis (Dean et al., 2004a) or lamellocytes in *Drosophila* (Meister, 2004), which are known to be involved in encapsulation reactions. Interestingly, the characteristic shapes of lamellocytes are lost in the presence of microtubule inhibitors (Rizki and Rizki, 1990). Moreover, lamellocytes are the target cells for lamellosin, the immune suppressor in larvae parasitised by *Leptopilina heterotoma* or *Leptopilina boulardi*, which target existing lamellocytes rather than preventing lamellocyte differentiation (Rizki and Rizki, 1991, 1992). This again is consistent with the notion that an intact cytoskeleton is required for adhesive hemocyte properties, and if extracellular driving forces exist that internalize anchored surface receptors the result may be a breakdown of fibers leaving the round-shaped cells incapable of making adhesive attachments. Moreover, actin depolymerization precludes vesicle transport to the cell periphery, causing delays in the recovery of adhesive cell properties.

11.4   CELLULAR IMMUNE SUPPRESSION

Several observations suggest that immune suppressors from insect parasitoids resemble counter-adhesion proteins that drive receptor internalization in hemocytes. Although some polydnavirus (PDV) proteins are shared among different parasitoid systems (Whitfield and Asgari, 2003), the general observation is that PDV-encoded gene products involved in immune suppression are remarkably diverse. This includes proteins with conserved cysteine patterns related to conotoxins (Cui and Webb, 1996; Summers and Dib-Hajj, 1995), cysteine-knot motif-containing proteins (Beck and Strand, 2003; Dahlman et al., 2003; Strand et al., 1997) and cystatins (Espagne et al., 2004), coiled-coil-containing proteins (Asgari and Schmidt, 2002; Asgari et al., 1997; Le et al., 2003) and abundantly expressed proteins without any sequence motifs (Harwood et al., 1994, 1998).

Despite the observed diversity of known immune suppressors, there appear to be some commonalities that are interesting to explore in terms of the possible functional mechanisms of these suppressors. Firstly, most of them are secreted proteins that find their target cells after being released into the hemolymph. This is also true for suppressor proteins that are produced in target cells, such as PDV-infected hemocytes. For example, expression of recombinant suppressors lacking signal peptides also lack function (Asgari and Schmidt, 2002). Secondly, suppressors that target cellular immune-related functions appear to affect primarily the cytoskeleton, the most apparent being the destabilization of F-actin (Asgari et al., 1997; Strand and Pech, 1995a; Webb and Luckhart, 1994). Finally, most immune suppressors interact with the hemocyte surface and some are taken up by endocytosis or phagocytosis reactions. Somewhere on the way the suppressor must mediate the destabilization of F-actin in the cytoplasm of hemocytes.

Whether these perceived commonalities among immune suppressor functions are due to a common origin of ancestral genes (Stasiak et al., 2005) or the outcome of convergent evolution (Federici and Bigot, 2003), the maternal factors involved in cellular inactivation of defense functions appear to target the active cytoskeleton. The resulting destabilization mostly through actin depolymerization appears to render hemocytes at least temporarily incapable of phagocytosis (Asgari et al., 1997) and spreading on foreign surfaces (Beck and Strand, 2003; Luckhart and Webb, 1996), with associated changes in the number of hemocytes in circulation (Strand and Pech, 1995a), and increased rates of apoptosis in cases where the suppressor persists in the hemolymph for more than a day (Beckage, 1998; Lavine and Strand, 2002; Pech and Strand, 2000; Strand and Pech, 1995b).

11.4.1 CRV1, A PARADIGM FOR IMMUNE SUPPRESSION

The parasitoid species *Cotesia rubecula* is unique among other parasitoid systems in that virus-infected host tissues produce four major viral proteins in a highly transient fashion (Asgari et al., 1996). Coding sequences of the four major transcripts that can be detected in Northern blots have been cloned and the gene products characterized (Asgari et al., 1997, 2003; Glatz et al., 2003). While all virus-coded proteins appear to have a functional role in immune suppression, a single virus protein CrV1 (*Cotesia rubecula* virus-encoded protein 1) is sufficient to inactivate hemocytes by targeting the functional integrity of the cytoskeleton of host hemocytes (Asgari et al., 1997).

What is the mode of action of CrV1? Since actin-cytoskeleton breakdown is observed together with CrV1-uptake and since cytochalasin D can mimic the immune-suppressed status in hemocytes, it is reasonable to assume that CrV1 mediates actin depolymerization. The question then is how extracellular CrV1 molecules can destabilize F-actin in the cytoplasm of hemocytes. Given that immune suppressors appear to remain stored in endosomal vesicles (Asgari and Schmidt, 2002) suppressor-mediated actin depolymerization must occur as a consequence of a mechanical process that dislocates cytoplasmic receptor domains from actin anchorage during receptor internalization.

11.4.2   POSSIBLE CRV1-UPTAKE REACTIONS

Recombinant CrV1 injected into hemocoel or incubated in whole hemolymph is taken up by hemocytes, but CrV1 does not bind to hemocytes in the absence of hemolymph plasma. These and other observations suggest that hemocytes lack CrV1 receptors but require CrV1 complex formation with a plasma factor, which is then taken up via that factor's receptor. It is known that CrV1 binds to lipophorin forming a complex with the lipid carrier (Asgari and Schmidt, 2002; Glatz et al., 2004). Moreover, CrV1 molecules that lack a coiled-coil structure are precluded from binding to lipophorin and are not taken up by hemocytes (Asgari and Schmidt, 2002; Glatz et al., 2004). This suggests that CrV1 interacts with hemocytes by forming a complex with lipophorin, which is then taken up by the lipophorin or other receptors, such as scavenger receptors, as part of a clearance reaction in the hemolymph of parasitised larvae.

11.4.3   CELLULAR IMMUNE SUPPRESSION: A MODEL

Given that CrV1 interacts with hemocytes after forming a complex with lipophorin, we examined lectin-mediated hemocyte uptake reactions to elucidate the role of actin in these processes. Oligomeric GalNAc- and Gal-specific lectins are known to interact with modified lipophorin to form lipid particle complexes (Rahman et al., 2006). Moreover, lipophorin particles like its vertebrate counterpart low-density lipoprotein (LDL) (Xu et al., 1997) can interact with a number of lipoprotein- and scavenger receptors, including hemomucin (Theopold and Schmidt, 1997) and lipoprotein-like receptors (Acton et al., 1996).

Since lectins resemble immune suppressors that can drive uptake reactions and in the process destabilize actin cytoskeleton (Glatz et al., 2004), the systemic clearance of modified lipid particles inside the hemocoel may provide a model for immune suppression in general. Given that the functional integrity of actin cytoskeleton is more relevant to spreading and adhesion than to pinocytosis (Glatz et al., 2004) and to some degree phagocytosis (Vilcinskas et al., 1997b), the dynamics of lectin-mediated uptake and spreading reactions suggests that extracellular driving forces depend on functional adhesion molecules forming a complex with lipid particles.

As in lectins and other adhesive proteins, where the connective properties depend on oligomerization, dimer formation appears to be a prerequisite for CrV1 uptake (Asgari and Schmidt, 2002). We assume that aggregation of lipophorin particles or receptor interactions with lipid particles are a prerequisite for the uptake. In both scenarios, the uptake of lipid particles may conceptually be compared to velcro-like phagocytosis reactions given the size of lipid particles in relation to cell receptors (Fig. 11.3). Since lipophorin–immune suppressor complexes are taken up by hemocytes, while suppressor proteins alone are not, this suggests that the complex provides binding sites for receptors that are not available by the suppressor protein. This is reminiscent of intact lipid particles being taken up by lipid receptors (Dantuma et al., 1999) but changing adhesive properties under oxidative conditions (Clement-Collin et al., 1999; Dantuma et al., 1999) allows modified particles to be taken up by scavenger receptors. If lipid particles are sensors that become adhesive upon environmental changes, the interaction of the immune suppressor may only require certain modifications of lipid particles to be cleared from the hemolymph by hemocytes.

It will be interesting to uncover the functional properties of other immune suppressors and whether lipid particle-mediated uptake reactions are used by all suppressor systems to inactivate hemocytes. For example, immune suppressors resembling conotoxins (Dib et al., 1993; Nappi et al., 2005; Parkinson et al., 2004) are similar to Kunitz-type protease inhibitors, which could bind to proteases found on lipid particles (Rahman et al., 2006) or to lipoprotein-like receptors (Kasza et al., 1997). The fact that none of the known immune suppressors bind to any of the known hemocyte receptor, could indicate that interactions with lipid particles may be necessary to activate adhesive proteins, such as apolipoprotein III that interact with the hemocyte surface. In this context it is interesting to note that *Microplitis demolitor* not only produce a soluble suppressor but also its cell-bound receptor, which is essential for de-adhesion (Beck and Strand, 2003). Although there are other explanations, the mucin-like receptor could be involved in uptake reactions similar to lectin-mediated uptake and de-adhesion reactions described above.

11.5   CONCLUSIONS

Taken together, these observations provide a conceptual basis for immune suppression, involving the uptake of extracellular suppressor–lipophorin complexes and in the process driving destabilization of actin cytoskeleton. Given the strong expression and secretion of suppressor proteins in PDV-infected fat body cells and hemocytes facilitated by nested PDV gene arrangements (Webb and Cui, 1998), the massive secretion of immune suppressors into the hemocoel has the capacity to create systemic modifications of lipid particles, which in turn are taken up and cleared by hemocytes. Thus the parasitoid exploits a natural cellular clearance mechanism by depleting adhesive receptors from the cell surface thereby precluding hemocyte spreading and phagocytosis (Fig. 11.6).

Although the observed similarities between CrV1 and lectin uptake reactions indicate similar mechanisms of actin-cytoskeleton inactivation, there are differences between immune suppressors and lectins. For example, injection of recombinant CrV1 into the hemocoel of host larvae causes systemic hemocyte inactivation (Asgari et al., 1997), whereas injection of oligomeric lectins and other substances cause plasma coagulation (Haine et al., 2007) and hemocyte aggregation (Glatz et al., 2003). The difference is that while both CrV1 (Asgari and Schmidt, 2002) and

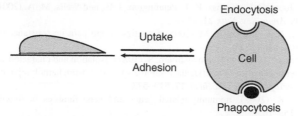

FIGURE 11.6   Schematic depiction of a dynamic relationship between adhesive and uptake reactions. A shift in balance toward internalization of receptors in phagocytosis or endocytosis reactions can deplete the cell surface of receptors leading to loss of adhesive abilities.

lectin (Glatz et al., 2004) mediate uptake reactions, lectins also cause hemocyte agglutination (Kotani et al., 1995) and coagulation (Rahman et al., 2006). What prevents lipophorin–CrV1 complexes to form aggregates is not known and remains to be determined.

REFERENCES

Acton, S., Rigotti, A., Landschulz, K. T., Xu, S., Hobbs, H. H., and Krieger, M. (1996). Identification of scavenger receptor SR-BI as a high density lipoprotein receptor. *Science* **271**, 518–520.

Asgari, S., and Schmidt, O. (2002). A coiled-coil region of an insect immune suppressor protein is involved in binding and uptake by hemocytes. *Insect Biochem. Mol. Biol.* **32**, 497–504.

Asgari, S., Hellers, M., and Schmidt, O. (1996). Host haemocyte inactivation by an insect parasitoid: Transient expression of a polydnavirus gene. *J. Gen. Virol.* **77**, 2653–2662.

Asgari, S., Schmidt, O., and Theopold, U. (1997). A polydnavirus-encoded protein of an endoparasitoid wasp is an immune suppressor. *J. Gen. Virol.* **78**, 3061–3070.

Asgari, S., Theopold, U., Wellby, C., and Schmidt, O. (1998). A protein with protective properties against the cellular defense reactions in insects. *Proc. Natl. Acad. Sci. USA* **95**, 3690–3695.

Asgari, S., Zhang, G., and Schmidt, O. (2003). Polydnavirus particle proteins with similarities to molecular chaperons, heat shock protein 70 and calreticulin. *J. Gen. Virol.* **84**, 1165–1171.

Ashida, M., and Brey, P. T. (1995). Role of the integument in insect defense – Pro-phenol oxidase cascade in the cuticular matrix. *Proc. Natl. Acad. Sci. USA* **92**, 10698–10702.

Beck, M., and Strand, M. R. (2003). RNA interference silences *Microplitis demolitor* bracovirus genes and implicates glc1.8 in disruption of adhesion in infected host cells. *Virology* **314**, 521–535.

Beckage, N. E. (1998). Modulation of immune responses to parasitoids by polydnaviruses. *Parasitol.* **116**, S57–S64.

Bergner, A., Muta, T., Iwanaga, S., Beisel, H. G., Delotto, R., and Bode, W. (1997). Horseshoe crab coagulogen is an invertebrate protein with a nerve growth factor-like domain. *Biol. Chem.* **378**, 283–287.

Beutler, B. (2003). Not 'molecular patterns' but molecules. *Immunity* **19**, 155–156.

Beutler, B. (2004). Innate immunity: An overview. *Mol. Immunol.* **40**, 845–859.

Bidla, G., Lindgren, M., Theopold, U., and Dushay, M. S. (2005). Hemolymph coagulation and phenoloxidase in *Drosophila* larvae. *Dev. Compart. Immunol.* **29**, 669–679.

Brehelin, M., Hoffmann, J. A., Matz, G., and Porte, A. (1975). Encapsulation of implanted foreign bodies by hemocytes in *Locusta migratoria* and *Melolontha melolontha*. *Cell Tissue Res.* **160**, 283–289.

Brivio, M., Mastore, M., and Moro, M. (2004). The role of *Steinernema feltiae* body-surface lipids in host–parasite immunological interactions. *Mol. Biochem. Parasitol.* **135**, 111–121.

Canavoso, L. E., Jouni, Z. E., Karnas, K. J., Pennington, J. E., and Wells, M. A. (2001). Fat metabolism in insects. *Annu. Rev. Nutr.* **21**, 23–46.

Carton, Y., Nappi, A. J., and Poirie, M. (2005). Genetics of anti-parasite resistance in invertebrates. *Dev. Compart. Immunol.* **29**, 9–32.

Castro, V. N., Boman, H. G., and Hammarstroem, S. (1987). Isolation and characterisation of a group of isolectins with galactose-*N*-acetylgalactosamine specificity from hemolymph of the giant moth *Hyalophora cecropia. Insect Biochem.* **17**, 513–523.

Christophides, G. K. (2002). Immunity-related genes and gene families in *Anopheles gambiae. Science* **298**, 159–165.

Christophides, G. K., Vlachou, D., and Kafatos, F. C. (2004). Comparative and functional genomics of the innate immune system in the malaria vector *Anopheles gambiae. Immunol. Rev.* **198**, 127–148.

Clement-Collin, V., Leroy, A., Monteilhet, C., and Aggerbeck, L. P. (1999). Mimicking lipid-binding-induced conformational changes in the human apolipoprotein E *N*-terminal receptor binding domain – Effects of low pH and propanol. *Eur. J. Biochem.* **264**, 358–368.

Cui, L. W., and Webb, B. A. (1996). Isolation and characterization of a member of the cysteine-rich gene family from *Campoletis sonorensis* polydnavirus. *J. Gen. Virol.* **77**, 797–809.

Dahlman, D. L., Rana, R. L., Schepers, E. J., Schepers, T., DiLuna, F. A., and Webb, B. A. (2003). A teratocyte gene from a parasitic wasp that is associated with inhibition of insect growth and development inhibits host protein synthesis. *Insect Mol. Biol.* **12**, 527–534.

Dangl, J. L., and McDowell, J. M. (2006). Two modes of pathogen recognition by plants. *Proc. Natl. Acad. Sci. USA* **103**, 8575–8576.

Dantuma, N. P., Potters, M., De Winther, M. P. J., Tensen, C. P., Kooiman, F. P., Bogerd, J., and Van der Horst, D. J. (1999). An insect homolog of the vertebrate very low density lipoprotein receptor mediates endocytosis of lipophorins. *J. Lip. Res.* **40**, 973–978.

Dean, P., Potter, U., Richards, E. H., Edwards, J. P., Charnley, A. K., and Reynolds, S. E. (2004a). Hyperphagocytic haemocytes in *Manduca sexta. J. Insect Physiol.* **50**, 1027–1036.

Dean, P., Richards, E. H., Edwards, J. P., Reynolds, S. E., and Charnley, K. (2004b). Microbial infection causes the appearance of hemocytes with extreme spreading ability in monolayers of the tobacco hornworm *Manduca sexta. Dev. Compart. Immunol.* **28**, 689–700.

Delotto, Y., and Delotto, R. (1998). Proteolytic processing of the *Drosophila* Spaetzle protein by Easter generates a dimeric NGF-like molecule with ventralising activity. *Mech. Dev.* **72**, 141–148.

Dib, H. S. D., Webb, B. A., and Summers, M. D. (1993). Structure and evolutionary implications of a 'cysteine-rich' *Campoletis sonorensis* polydnavirus gene family. *Proc. Natl. Acad. Sci. USA* **90**, 3765–3769.

Dong, Y., Taylor, H. E., and Dimopoulos, G. (2006). AgDscam, a hypervariable immunoglobulin domain-containing receptor of the *Anopheles gambiae* innate immune system. *PLoS Biol.* **4**, e229.

Dupas, S., Carton, Y., and Poirie, M. (2003). Genetic dimension of the coevolution of virulence-resistance in *Drosophila*–parasitoid wasp relationships. *Heredity* **90**, 84–89.

Duvic, B., and Brehelin, M. (1998). Two major proteins from locust plasma are involved in coagulation and are specifically precipitated by laminarin, a beta-1,3-glucan. *Insect Biochem. Mol. Biol.* **28**, 959–967.

Esmon, C. T. (2004). Interactions between the innate immune and blood coagulation systems. *Trends Immunol.* **25**, 536–542.

Espagne, E., Dupuy, C., Huguet, E., Cattolico, L., Provost, B., Martins, N., Poirie, M., Periquet, G., and Drezen, J. M. (2004). Genome sequence of a polydnavirus: Insights into symbiotic virus evolution. *Science* **306**, 286–289.

Etienne-Manneville, S., and Hall, A. (2002). Rho GTPases in cell biology (review). *Nature* **420**, 629–635.

Federici, B. A., and Bigot, Y. (2003). Origin and evolution of polydnaviruses by symbiogenesis of insect DNA viruses in endoparasitic wasps. *J. Insect Physiol.* **49**, 419–432.

Firrao, S. (1994). Cybernetic interpretation of cell immunological defence mechanism. *Cybernetica* **37**, 125–134.

Glatz, R., Schmidt, O., and Asgari, S. (2003). Characterization of a novel protein with homology to C-type lectins expressed by the *Cotesia rubecula* bracovirus in larvae of the lepidopteran host, *Pieris rapae*. *J. Biol. Chem.* **278**, 19743–19750.

Glatz, R., Roberts, H. L. S., Li, D., Sarjan, M., Theopold, U. H., Asgari, S., and Schmidt, O. (2004). Lectin-induced haemocyte inactivation in insects. *J. Insect Physiol.* **50**, 955–963.

Goetz, P. (1986). Encapsulation in arthropods. In *Immunity in Invertebrates* (M. Brehelin, Ed.), pp. 153–170. Springer Verlag, Berlin.

Haine, E. R., Rolff, J., and Siva-Jothy, M. T. (2007). Functional consequences of blood clotting in insects. *Dev. Compart. Immunol.* **31**, 456–464.

Hall, M., Wang, R., van Antwerpen, R., Sottrup-Jensen, L., and Soderhall, K. (1999). The crayfish plasma clotting protein: A vitellogenin-related protein responsible for clot formation in crustacean blood. *Proc. Natl. Acad. Sci. USA* **96**, 1965–1970.

Harwood, S. H., Grosovsky, A. J., Cowles, E. A., Davis, J. W., and Beckage, N. E. (1994). An abundantly expressed hemolymph glycoprotein isolated from newly parasitized *Manduca sexta* larvae is a polydnavirus gene product. *Virology* **205**, 381–392.

Harwood, S. H., McElfresh, J. S., Nguyen, A., Conlan, C. A., and Beckage, N. E. (1998). Production of early expressed parasitism-specific proteins in alternate sphingid hosts of the braconid wasp *Cotesia congregata*. *J. Invertebr. Pathol.* **71**, 271–279.

Hellers, M., Beck, M., Theopold, U., Kamei, M., and Schmidt, O. (1996). Multiple alleles encoding a virus-like particle protein in the ichneumonid endoparasitoid *Venturia canescens*. *Insect Mol. Biol.* **5**, 239–249.

Hise, A. G., Daehnel, K., Gillette-Ferguson, I., Cho, E., McGarry, H. F., Taylor, M. J., Golenbock, D. T., Fitzgerald, K. A., Kazura, J. W., and Pearlman, E. (2007). Innate immune responses to endosymbiotic *Wolbachia* bacteria in *Brugia malayi* and *Onchocerca volvulus* are dependent on TLR2, TLR6, MyD88, and Mal, but not TLR4, TRIF, or TRAM. *J. Immunol.* **178**, 1068–1076.

Huigens, M. E., Luck, R. F., Klaassen, R. H. G., Maas, M. F. P. M., Timmermans, M. J. T. N., and Stouthamer, R. (2000). Infectious parthenogenesis. *Nature* **405**, 178–179.

Janeway, C. A., Ed. (1989). *Approaching the Asymptode? Evolution and Revolution in Immunology*. Cold Spring Harbor Laboratory Press, Plainview.

Janeway, C. A., and Medzhitov, R. (2002). Innate immune recognition. *Annu. Rev. Immunol.* **20**, 197–216.

Jones, J. D. G., and Dangl, J. L. (2006). The plant immune system. *Nature* **444**, 323–329.

Kanost, M. R., Jiang, H., and Yu, X.-Q. (2004). Innate immune responses of a lepidopteran insect, *Manduca sexta*. *Immunol. Rev.* **198**, 97–105.

Kasza, A., Petersen, H. H., Heegaard, C. W., Oka, K., Christensen, A., Dubin, A., Chan, L., and Andreasen, P. A. (1997). Specificity of serine proteinase serpin complex binding to very-low-density lipoprotein receptor and alpha(2)-macroglobulin receptor low-density-lipoprotein-receptor-related protein. *Eur. J. Biochem.* **248**, 270–281.

Kinuthia, W., Li, D., Schmidt, O., and Theopold, U. (1999). Is the surface of endoparasitic wasp eggs and larvae covered by a limited coagulation reaction? *J. Insect. Physiol.* **45**, 501–506.

Klein, J. (1999). Self-nonself discrimination, histoincompatibility, and the concept of immunology. *Immunogenetics* **50**, 116–123.

Korayem, A. M., Fabbri, M., Takahashi, K., Scherfer, C., Lindgren, M., Schmidt, O., Ueda, R., Dushay, M. S., and Theopold, U. (2004). A *Drosophila* salivary gland mucin is also expressed in immune tissues: Evidence for a function in coagulation and the entrapment of bacteria. *Insect Biochem. Mol. Biol.* **34**, 1297–1304.

Korayem, A. M., Hauling, T., Lesch, C., Fabbri, M., Lindgren, M., Loseva, O., Schmidt, O., Dushay, M. S., and Theopold, U. (2007). Evidence for an immune function of lepidopteran silk proteins. *Biochem. Biophys. Res. Comm.* **352**, 317–322.

Kotani, E., Yamakawa, M., Iwamoto, S., Tashiro, M., Mori, H., Sumida, M., Matsubara, F., Taniai, K., Kadono-Okuda, K., Kato, Y., et al. (1995). Cloning and expression of the gene of hemocytin, an insect humoral lectin which is homologous with the mammalian von Willebrand factor. *Biochim. Biophys. Acta.* **1260**, 245–258.

Kraaijeveld, A. R., and Godfray, H. C. J. (1999). Geographic patterns in the evolution of resistance and virulence in *Drosophila* and its parasitoids. *Am. Nat.* **153**, S61–S74.

Kraaijeveld, A. R., Van Alphen, J. J. M., and Godfray, H. C. J. (1998). The coevolution of host resistance and parasitoid virulence. *Parasitology* **116**, S29–S45.

Krem, M. M., and Cera, E. D. (2002). Evolution of enzyme cascades from embryonic development to blood coagulation. *Trends Biochem. Sci.* **27**, 67–74.

Lavine, M. D., and Strand, M. R. (2001). Surface characteristics of foreign targets that elicit an encapsulation response by the moth *Pseudoplusia includens. J. Insect Physiol.* **47**, 965–974.

Lavine, M. D., and Strand, M. R. (2002). Insect hemocytes and their role in immunity. *Insect Biochem. Mol. Biol.* **32**, 1295–1309.

Le, N. T., Asgari, S., Amaya, K., Tan, F. F., and Beckage, N. E. (2003). Persistence and expression of *Cotesia congregata* polydnavirus in host larvae of the tobacco hornworm, *Manduca sexta. J. Insect Physiol.* **49**, 533–543.

Lemaitre, B., Nicolas, E., Michaut, L., Reichhart, J. M., and Hoffmann, J. A. (1996). The dorsoventral regulatory gene cassette spaetzle/Toll/cactus controls the potent antifungal response in *Drosophila* adults. *Cell* **86**, 973–983.

Li, B., and Paskewitz, S. M. (2006). A role for lysozyme in melanization of Sephadex beads in *Anopheles gambiae. J. Insect Physiol.* **52**, 936–942.

Li, D., Scherfer, C., Korayem, A. M., Zhao, Z., Schmidt, O., and Theopold, U. (2002). Insect hemolymph clotting: Evidence for interaction between the coagulation system and the prophenoloxidase activating cascade. *Insect Biochem. Mol. Biol.* **32**, 919–928.

Li, J. (1994). Egg chorion tanning in *Aedes aegypti* mosquito. *Compart. Biochem. Physiol. A Physiol.* **109**, 835–843.

Lord, E. M., and Russell, S. D. (2002). The mechanisms of pollination and fertilization in plants. *Annu. Rev. Cell Dev. Biol.* **18**, 81–105.

Luckhart, S., and Webb, B. A. (1996). Interaction of a wasp ovarian protein and polydnavirus in host immune suppression. *Dev. Compart. Immunol.* **20**, 1–21.

Ma, G., Hay, D., Li, D., Asgari, S., and Schmidt, O. (2006). Recognition and inactivation of LPS by lipophorin particles. *Dev. Compart. Immunol.* **30**, 619–626.

Marquardt, T., Shirasaki, R., Ghosh, S., Andrews, S. E., Carter, N., Hunter, T., and Pfaff, S. L. (2005). Coexpressed EphA receptors and Ephrin-A ligands mediate opposing actions on growth cone navigation from distinct membrane domains. *Cell* **121**, 127–139.

Meister, M. (2004). Blood cells of *Drosophila*: Cell lineages and role in host defence. *Curr. Opin. Immunol.* **16**, 10–15.

Mellroth, P., Karlsson, J., Hakansson, J., Schultz, N., Goldman, W. E., and Steiner, H. (2005). Ligand-induced dimerization of *Drosophila* peptidoglycan recognition proteins *in vitro. Proc. Natl. Acad. Sci. USA* **102**, 6455–6460.

Murphy-Ullrich, J. E. (2001). The de-adhesive activity of matricellular proteins: Is intermediate cell adhesion an adaptive state? *J. Clin. Invest.* **107**, 785–790.

Mylonakis, E., and Aballay, A. (2005). Worms and flies as genetically tractable animal models to study host–pathogen interactions. *Infect. Immun.* **73**, 3833–3841.

Nappi, A. J., and Christensen, B. M. (2005). Melanogenesis and associated cytotoxic reactions: Applications to insect innate immunity. *Insect Biochem. Mol. Biol.* **35**, 443–459.

Nappi, A. J., Frey, F., and Carton, Y. (2005). *Drosophila* serpin 27A is a likely target for immune suppression of the blood cell-mediated melanotic encapsulation response. *J. Insect Physiol.* **51**, 197–205.

Nardi, J. B., Pilas, B., Bee, C. M., Zhuang, S., Garsha, K., and Kanost, M. R. (2006). Neuroglian-positive plasmatocytes of *Manduca sexta* and the initiation of hemocyte attachment to foreign surfaces. *Dev. Compart. Immunol.* **30**, 447–462.

Niere, M., Dettloff, M., Maier, T., Ziegler, M., and Wiesner, A. (2001). Insect immune activation by apolipophorin III is correlated with the lipid-binding properties of this protein. *Biochem.* **40**, 11502–11508.

Panakova, D., Sprong, H., Marois, E., Thiele, C., and Eaton, S. (2005). Lipoprotein particles are required for *Hedgehog* and *Wingless* signalling. *Nature* **435**, 58–65.

Parkinson, N. M., Conyers, C., Keen, J., MacNicoll, A., Smith, I., Audsley, N., and Weaver, R. (2004). Towards a comprehensive view of the primary structure of venom proteins from the parasitoid wasp *Pimpla hypochondriaca*. *Insect Biochem. Mol. Biol.* **34**, 565–571.

Pech, L. L., and Strand, M. R. (2000). Plasmatocytes from the moth *Pseudoplusia includens* induce apoptosis of granular cells. *J. Insect Physiol.* **46**, 1565–1573.

Pennacchio, F., and Strand, M. R. (2006). Evolution of developmental strategies in parasitic hymenoptera. *Annu. Rev. Entomol.* **51**, 233–258.

Rahman, M. M., Ma, G., Roberts, H. L. S., and Schmidt, O. (2006). Cell-free immune reactions in insects. *J. Insect Physiol.* **52**, 754–762.

Ribeiro, C., Duvic, B., Oliveira, P., Givaudan, A., Palha, F., Simoes, N., and Brehelin, M. (1999). Insect immunity – Effects of factors produced by a nematobacterial complex on immunocompetent cells. *J. Insect Physiol.* **45**, 677–685.

Rizki, R. M., and Rizki, T. M. (1990). Microtubule inhibitors block the morphological changes induced in *Drosophila* blood cells by a parasitoid wasp factor. *Experientia* **46**, 311–315.

Rizki, R. M., and Rizki, T. M. (1991). Effects of lamellolysin from a parasitoid wasp on *Drosophila* blood cells *in vitro*. *J. Exp. Zool.* **257**, 236–244.

Rizki, R. M., Rizki, T. M., and Andrews, C. A. (1975). *Drosophila* cell fusion induced by wheat germ agglutinin. *J. Cell Sci.* **18**, 113–142.

Rizki, T. M., and Rizki, R. M. (1983). Basement membrane polarizes lectin binding sites of *Drosophila* larval fat body cells. *Nature* **303**, 340–342.

Rizki, T. M., and Rizki, R. M. (1992). Lamellocyte differentiation in *Drosophila* larvae parasitized by *Leptopilina*. *Dev. Compart. Immunol.* **16**, 103–110.

Rogers, S. L., Wiedemann, U., Stuurman, N., and Vale, R. D. (2003). Molecular requirements for actin-based lamella formation in *Drosophila* S2 cells. *J. Cell Biol.* **162**, 1079–1088.

Salt, G. (1970). *The Cellular Defence Reactions of Insects*. Cambridge University Press, Cambridge.

Schmidt, O., and Schreiber, A. (2006). Integration of cell adhesion reactions – A balance of forces? *J. Theoret. Biol.* **238**, 608–615.

Schmidt, O., and Schuchmann-Feddersen, I. (1989). Role of virus-like particles in parasitoid–host interaction of insects. In *Subcellular Biochemistry* (J. R. Harris, Ed.), pp. 91–119. Plenum Press, New York.

Schmidt, O., and Theopold, U. (2004). An extracellular driving force of endocytosis and cell-shape changes (hypothesis). *Bioessays* **26**, 1344–1350.

Schmidt, O., Theopold, U., and Strand, M. (2001). Innate immunity and its evasion and suppression by hymenopteran endoparasitoids. *Bioessays* **23**, 344–351.

Schmidt, O., Li, D., Beck, M., Kinuthia, W., Bellati, J., and Roberts, H. L. S. (2005). Phenoloxidase-like activities and the function of virus-like particles in ovaries of the parthenogenetic parasitoid *Venturia canescens*. *J. Insect Physiol.* **51**, 117–125.

Schmidt, O., Söderhäll, K., Theopold, U., and Faye, I. (invited review). The role of adhesion in arthropod immune recognition. *Annu Rev. Entomol.* **54**.

Soderhall, K., and Cerenius, L. (1998). Role of the prophenoloxidase-activating system in invertebrate immunity. *Curr. Opin. Immunol.* **10**, 23–28.

Stasiak, K., Renault, S., Federici, B. A., and Bigot, Y. (2005). Characteristics of pathogenic and mutualistic relationships of ascoviruses in field populations of parasitoid wasps. *J. Insect Physiol.* **51**, 103–115.

Strand, M. R., and Pech, L. L. (1995a). Immunological basis for compatibility in parasitoid host relationships. *Annu. Rev. Entomol.* **40**, 31–56.

Strand, M. R., and Pech, L. L. (1995b). *Microplitis demolitor* polydnavirus induces apoptosis of a specific haemocyte morphotype in *Pseudoplusia includens*. *J. Gen. Virol.* **76**, 283–291.

Strand, M. R., Witherell, R. A., and Trudeau, D. (1997). Two *Microplitis demolitor* polydnavirus mRNAs expressed in hemocytes of *Pseudoplusia includens* contain a common cysteine-rich domain. *J. Virol.* **71**, 2146–2156.

Summers, M. D., and Dib-Hajj, S. D. (1995). Polydnavirus facilitated endoparasitoid protection against host immune defenses. *Proc. Natl. Acad. Sci. USA* **92**, 29–36.

Swanson, J. A., and Baer, S. C. (1995). Phagocytosis by zippers and triggers. *Trends Cell Biol.* **5**, 89–93.

Theopold, U., and Schmidt, O. (1997). *Helix pomatia* lectin and annexin V, molecular markers for hemolymph coagulation. *J. Insect Physiol.* **43**, 667–674.

Theopold, U., Dorian, C., and Schmidt, O. (2001). Changes in glycosylation during *Drosophila* development. The influence of ecdysone on hemomucin isoforms. *Insect Biochem. Mol. Biol.* **31**, 189–197.

Theopold, U., Li, D., Fabbri, M., Scherfer, C., and Schmidt, O. (2002). The coagulation of insect hemolymph. *Cell Mol. Life Sci.* **59**, 363–372.

Theopold, U., Schmidt, O., Soderhall, K., and Dushay, M. S. (2004). Coagulation in arthropods: Defence, wound closure and healing. *Trends Immunol.* **25**, 289–294.

Van der Horst, D. J. (2003). Insect adipokinetic hormones: Release and integration of flight energy metabolism. *Compart. Biochem. Physiol. B Biochem. Mol. Biol.* **136**, 217–226.

Van Hoof, D., Rodenburg, K. W., and Van der Horst, D. J. (2005). Intracellular fate of LDL receptor family members depends on the cooperation between their ligand-binding and EGF domains. *J. Cell Sci.* **118**, 1309–1320.

Vilcinskas, A., Matha, V., and Goetz, P. (1997a). Inhibition of phagocytic activity of plasmatocytes isolated from *Galleria mellonella* by entomogenous fungi and their secondary metabolites. *J. Insect Physiol.* **43**, 475–483.

Vilcinskas, A., Matha, V., and Goetz, P. (1997b). Inhibition of phagocytic activity of plasmatocytes isolated from *Galleria mellonella* by entomogenous fungi and their secondary metabolites. *J. Insect Physiol.* **43**, 475–483.

Vinson, S. B. (1990). How parasitoids deal with the immune system of their hosts: An overview. *Arch. Insect Biochem. Physiol.* **13** (Special Issue), 3–27.

Watson, F. L. (2005). Extensive diversity of Ig-superfamily proteins in the immune system of insects. *Science* **309**, 1874–1878.

Webb, B. A., and Cui, L. W. (1998). Relationships between polydnavirus genomes and viral gene expression. *J. Insect Physiol.* **44**, 785–793.

Webb, B. A., and Luckhart, S. (1994). Evidence for an early immunosuppressive role for related *Campoletis sonorensis* venom and ovarian proteins in *Heliothis virescens*. *Arch. Insect Biochem. Physiol.* **26**, 147–163.

Whitfield, J. B., and Asgari, S. (2003). Virus or not? Phylogenetics of polydnaviruses and their wasp carriers. *J. Insect Physiol.* **49**, 397–405.

Whitten, M. M. A., Tew, I. F., Lee, B. L., and Ratcliffe, N. A. (2004). A novel role for an insect apolipoprotein (apolipophorin III) in {beta}-1,3-glucan pattern recognition and cellular encapsulation reactions. *J. Immunol.* **172**, 2177–2185.

Xu, S. Z., Laccotripe, M., Huang, X. W., Rigotti, A., Zannis, V. I., and Krieger, M. (1997). Apolipoproteins of HDL can directly mediate binding to the scavenger receptor SR-BI, an HDL receptor that mediates selective lipid uptake. *J. Lip. Res.* **38**, 1289–1298.

Yu, X.-Q., and Kanost, M. R. (2002). Binding of hemolin to bacterial lipopolysaccharide and lipoteichoic acid: An immunoglobulin superfamily member from insects as a pattern-recognition receptor. *Eur. J. Biochem.* **269**, 1827–1834.

Zchori-Fein, E., Gottlieb, Y., Kelly, S. E., Brown, J. K., Wilson, J. M., Karr, T. L., and Hunter, M. S. (2001). A newly discovered bacterium associated with parthenogenesis and a change in host selection behavior in parasitoid wasps. *Proc. Natl. Acad. Sci. USA* **98**, 12555–12560.

12

RNAI AND THE INSECT
IMMUNE SYSTEM

STUART E. REYNOLDS* AND IOANNIS
ELEFTHERIANOS**

*Department of Biology and Biochemistry, University of Bath, Claverton Down, Bath, UK

**CNRS-UPR9022, Institut de Biologie Moléculaire et Cellulaire, 15 rue René Descartes,
F-67084 Strasbourg Cedex, France

ABSTRACT: RNA interference (RNAi) is an endogenous, specific gene silencing
mechanism that uses double-stranded RNA (dsRNA) to suppress the expression of

targeted genes. In insects, silencing occurs principally through mRNA degradation, but inhibition of translation may also play an important role. The RNAi machinery of the cell is itself used to protect genome integrity and as an antiviral defense, and some insect viruses have evolved genes that suppress host RNAi. The technology of RNAi can be used experimentally both for gene discovery and also to elucidate gene function. Specific dsRNAs can be used to selectively silence known immune-related genes, even where traditional genetic approaches are unable to achieve this. This can be done through molecular genetics, by transforming the target insect with a hairpin construct that generates dsRNA when transcribed, or pharmacologically, by administering exogenous dsRNA specific to the targeted gene (systemic RNAi). These techniques are well suited to testing hypotheses about the functions of particular immune-related genes, especially interactions between pathogens or parasites and their insect hosts. Silencing can be precisely timed and targeted both *in vivo* and *in vitro*. In the case of insect viruses, RNAi can be used not only to suppress expression of host genes, but also of viral genes. Molecular genetics allows construction of DNA hairpin libraries; these are well suited to screening for genes with novel immune-related phenotypes both *in vivo* and *in vitro*. RNAi is not always effective, and possible reasons for this are discussed. Finally, RNAi treatments may in some cases induce off-target effects, and possible explanations for this are considered. It is well known in vertebrates that dsRNA can itself induce immune responses. This possibility deserves further exploration in insects.

Abbreviations:

2',5'-AS	= 2',5'-oligoAdenylate Synthetase
ADGF	= Adenosine Deaminase-related Growth Factor
ATT	= Attacin
CEC	= Cecropin
DCV	= *Drosophila* C virus
ds	= double-stranded
EIF2	= Eukaryotic Initiation Factor 2
FHV	= Flock House Virus
GFP	= Green Fluorescent Protein
IML-2	= Immulectin-2
IRES	= Internal Ribosome Entry Site
LEB	= Lebocin
LRIM-1	= Leucine Rich–Repeat Immune-1
LYS	= Lysozyme
MHC	= Major Histocompatibility Complex
miRNA	= microRNA
MOR	= Moricin
mRNA	= messenger RNA
NF-κB	= Nuclear Factor kappa B
NK cell	= Natural Killer Cell

nt	= nucleotide
PFV-1	= Primate Foamy Virus Type 1
PGRP	= Peptidoglycan Recognition Protein
PKR	= dsRNA-Activated Protein Kinase
PO	= Phenoloxidase
PPO	= Prophenoloxidase
pre-miRNA	= **miRNA pre**cursor
pri-miRNA	= **pri**mary **miRNA**-encoding transcript
rasiRNA	= **r**epeat-**as**sociated small **i**nterfering RNA
RdRP	= RNA-dependent RNA Polymerase
RISC	= RNA-induced Silencing Complex
RNAi	= RNA interference
RT-PCR	= Reverse-transcription Polymerase Chain Reaction
siRNA	= short interfering RNA
SP	= Serine Proteinase
ss	= single-stranded
TLR	= Toll-like Receptor
TRBP	= TAR RNA-binding Protein
V-ATPase	= Vacuolar ATPase

12.1   INTRODUCTION

Although the immune systems of insects are much simpler than those of mammals, complex recognition, signaling, and effector systems nevertheless underlie insect immune responses. Genetic approaches, especially using *Drosophila*, have led to greatly enhanced understanding of immune-signaling pathways in insects (Lemaitre and Hoffmann, 2007). Additionally, much has been learned about circulating recognition proteins, antimicrobial proteins and peptides, and other humoral defenses such as the phenoloxidase (PO) activation and blood clotting cascades, by studying the biochemistry of immune responses in insects other than *Drosophila* (Cerenius and Soderhall, 2004; Kanost et al., 2004; Theopold et al., 2004). Nevertheless, genetics and biochemistry are not able to reveal everything we need to know about insect immune systems, especially where genes have pleiotropic effects on other essential functions or where full genomic sequence information is not available. Moreover, some important biochemical models of insect immunity, such as the tobacco hornworm *Manduca sexta*, are not easily susceptible to genetic analysis.

The discovery of RNA interference (RNAi) (Fire et al., 1998) has had important consequences for the study of insect immune systems. The cellular machinery of RNAi is itself used in immune defense and is in turn targeted by pathogens. Further, RNAi offers an exceptionally powerful experimental technique, allowing the power of gene deletion to reveal the functional consequences of the loss of a gene product to be used where the deletion cannot be achieved using traditional genetic approaches. The technique of systemic RNAi, in which RNAi reagents are effectively used as pharmacological agents, being injected direct into the recipient animal, is particularly suited for use in insects. It provides a methodology that is attractive to investigators who are interested primarily in biological interactions rather than genetics, and can be used at a stage of the insect's life chosen by the investigator. The technique is not always robust, however, and some limitations, pitfalls, and necessary controls will be discussed at the end of this chapter.

12.2   THE PHENOMENON OF RNAi

RNAi is an endogenous intracellular mechanism of gene expression control that recognizes the presence within a cell of a double-stranded RNA (dsRNA) species, cleaves it into shorter (21–25 nt in length) ds-oligoribonucleotide sequences, and uses these short interfering RNAs (siRNAs) to suppress the expression of proteins encoded by ss (single-stranded) mRNAs sharing the same sequence (Fig. 12.1). It is not our intention to provide a comprehensive account of the mechanism of RNAi, since many recent detailed reviews are already available (e.g. Dillon et al., 2005; Dykxhoorn and Lieberman, 2005; Grishok, 2005; Rana, 2007; Sen

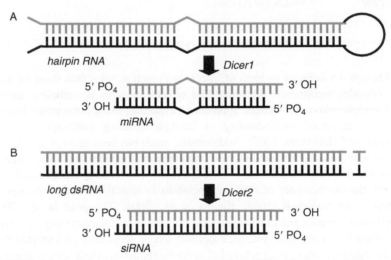

FIGURE 12.1   Processing of double-stranded RNA (dsRNA). (A) Genomic DNA sequences encoding hairpins are transcribed in the nucleus and the inverted repeats fold back on themselves to create the stem-loop structure of a hairpin RNA. The hairpin's ds stem, which may contain mismatches, is then cropped by *Drosha* to form pre-miRNAs, which are capped and polyadenylated. They migrate into the cytoplasm where they are cleaved ('diced') and 5′ phosphorylated by the Dicer1 complex to form 21–23 nt-long miRNAs with 2 nt-long 3′ overhangs. (B) Exogenous long dsRNAs (i.e. from viruses or actively taken up from the extracellular medium) are treated in the same way by the Dicer2 complex to form small interfering RNAs (siRNAs). In each case, the antisense strand, which will ultimately pair with the target mRNA, is called the 'guide strand', while the sense strand is known as the 'passenger strand'. At this stage, however, there is no difference between the two, and the guide strand is only identified by its ability to complement ss mRNA.

and Blau, 2006) and knowledge is rapidly accumulating. A description of the cellular machinery of RNAi will however be helpful in understanding the successes and failures of RNAi as applied to the insect immune system.

The phenomenon of RNAi was originally discovered in plants (Matzke et al., 1989; Napoli et al., 1990) where it was and is still known as gene silencing, and in fungi where it was dubbed 'quelling' (Romano and Macino, 1992) but the key dependence of the effect on dsRNA was elucidated in the model nematode *Caenorhabditis elegans* (Fire et al., 1998; Montgomery et al., 1998) and the fruit fly *Drosophila melanogaster* (Tuschl et al., 1999). RNAi is now recognized to occur in most eukaryotic cells (Dykxhoorn et al., 2003).

This chapter reviews many examples of RNAi in other insects. We focus on RNAi in an experimental context, in which the silencing of target genes is systemically mediated by exogenous dsRNA, and on genetic RNAi screens that engineer DNA sequences encoding inverted tandem repeats into the host genome, which give rise on transcription to hairpin RNAs (loops of ssRNA with a dsRNA stem). In both cases, this gives rise within the cell to siRNAs that silence cognate genes.

12.3   THE EVOLUTIONARY RATIONALE OF RNAi

There appear to be multiple evolutionary 'purposes' of RNAi. One rationale for the ability of cells to recognize the presence of dsRNA and to suppress the expression of cognate mRNAs is itself immunological in nature. RNAi may have arisen early in evolution to protect genomes against invasion by 'foreign' genetic elements. For example, repeat-associated small interfering RNAs (rasiRNAs) that are 24–29 nt-long function in germ-line cells to protect the integrity of the genome through RNAi-mediated silencing of selfish genetic elements (Vagin et al., 2006). This effectively constitutes a component of the immune system that is specialized for germ-line defense. It is becoming clear that the dsRNA-dependent gene silencing machinery underlying this system utilizes different dsRNA processing proteins (Vagin et al., 2006), and has germ-line specific features (O'Donnell et al., 2007). It is thus distinct from 'classical' RNAi (i.e. RNAi that is dependent on exogenous dsRNA). Further, RNAi plays an essential role in the maintenance of heterochromatin (Grewal and Elgin, 2007). Again, this system appears to be distinct in mechanism from that used in 'classical' RNAi; unlike the latter, heterochromatic silencing appears to depend on inhibition of transcription.

It is usually supposed that the principal function of classical RNAi (and therefore its evolutionary origin) is to defend the organism against viruses. As originally discovered in plants (Hamilton and Baulcombe, 1999), the cells of many organisms, including insects, use RNAi to recognize the presence, and prevent the expression of viral dsRNAs (Galiana-Arnoux et al., 2006; Gitlin and Andino, 2003; Li and Ding, 2005; Robalino et al., 2005; Wang et al., 2006). It has been shown that genes encoding the protein components of the RNAi machinery are essential to surviving viral infections in *Anopheles gambiae* (Keene et al., 2004) and *Drosophila melanogaster* (Galiana-Arnoux et al., 2006; van Rij et al., 2006; Wang et al., 2006; Zambon et al., 2006).

It is unsurprising that viruses have themselves evolved counter-measures against such defenses and encode proteins that suppress the host RNAi response (reviewed by Li and Ding, 2005). Examples of insect viruses encoding dsRNA-binding proteins that suppress RNAi include the insect nodavirus flock house virus (FHV) (Li et al., 2002), and the picorna-like insect virus cricket paralysis virus (Wang et al., 2006). It has been shown (Galiana-Arnoux et al., 2006) that the FHV B2 protein inhibits Dicer2, and that loss of Dicer2 function enhances susceptibility to FHV and other viruses. Another mechanism of RNAi suppression seen in human HIV (but not so far in an insect virus) is the sequestration of the Dicer-associated protein TAR RNA-binding protein (TRBP) by the viral protein TAR; this effectively shuts down the host's RNAi response to the virus (Bennasser et al., 2006).

The finding that insect genes encoding proteins of the RNAi machinery have evolved at a particularly high rate indicates the probable existence of antagonistic co-evolutionary arms races between the RNAi genes of insect hosts and viral antiRNAi genes (Obbard et al., 2006).

RNAi based on endogenous dsRNAs also has additional cellular roles, however. MicroRNAs (miRNAs) are a special class of small interfering RNAs of endogenous origin, which are used to control gene expression through RNAi (Ambros and Chen, 2007; Bartel and Chen, 2004; Lim et al., 2005; Nilsen, 2007). miRNAs have conserved sequences and each regulates many target genes. In mammals, some of these miRNAs appear to function specifically in the regulation of expression of immune genes (Chowdhury and Novina, 2005); e.g. specific miRNAs are present in activated leukocytes during inflammatory responses (e.g. O'Connell et al., 2007) and misregulation of miRNA expression is associated with the unregulated proliferation of leukocytes in various leukemias (Calin et al., 2004). It is probable that miRNAs will be shown to have similar roles in regulating the immune systems of insects.

miRNAs may also have antiviral defensive roles, however, perhaps restricting viral host ranges through fortuitous matches with viral RNA genomic sequences. In turn, there will clearly be selective pressure for viruses to evolve so as to subvert miRNA-directed RNAi. Lecellier et al. (2005) have shown that the retrovirus primate foamy virus type 1 (PFV-1) is restricted in this way by human miR-32, and that in turn the virus has acquired a gene *tas* that suppresses miR-32 directed gene silencing.

There will also be selective pressure for viruses to acquire miRNAs similar or identical to those of their hosts, where these can be used to downregulate host immune genes (reviewed by Dykxhoorn, 2007). A nice example is the recent demonstration by Stern-Ginossar et al. (2007) that the human cytomegalovirus genome encodes a miRNA that specifically downregulates a component of the major histocompatibility complex (MHC) in infected cells, thus protecting the virus by preventing infected host cells being killed by NK cells. It seems extremely likely that miRNA-like sequences that abrogate host immunity will also be discovered in insect viruses.

12.4   DICING AND SLICING: THE CELLULAR MACHINERY OF RNAi

The biochemical mechanisms whereby dsRNAs are produced are quite well understood (Fig. 12.1). 'Classical' RNAi is based on exogenous dsRNA supplied by the experimenter. It makes use of the same phenomenon as antiviral RNAi, where the dsRNA is also exogenous, being supplied by an infecting virus. These dsRNAs must first enter the target cell, whether introduced by the virus or actively taken up by the cell (see below). In either case, they enter the cytoplasm rather than the nucleus. By contrast, where the experimenter engineers an expressed tandem repeat into host DNA, initial production of hairpin RNA will take place in the nucleus. This type of RNAi procedure thus takes advantage of endogenous miRNA-based RNAi, where dsRNA sequences are also encoded within the genome. In either of these latter two cases, the primary miRNA-encoding transcripts from genomic DNA hairpins (pri-miRNAs) must first be cleaved by the nuclear RNase III-type endonuclease Drosha to yield 60–80 nt double-stranded miRNA precursors (pre-miRNAs), which are then exported to the cytoplasm (Lee et al., 2003).

Whatever their origin, endogenous or exogenous, once in the cytoplasm, dsRNAs are processed to produce short (~22 nt) dsRNAs. These duplexes, which possess 5'-monophosphates, 3'-hydroxyl groups, and 2-nt 3' overhangs, are primarily responsible for RNAi silencing. siRNAs and miRNAs are both produced by the action of cytoplasmic Dicer proteins, enzymes of the RNase III-type endonuclease class. According to the organism, there may be one or more than one Dicer, each responsible for a different type of short dsRNA product (Meister and Tuschl, 2004). In *Drosophila*, DCR-1 is preferentially used to produce miRNAs, while DCR-2 is responsible for the processing of long dsRNAs into siRNAs (Lee et al., 2004). The rasiRNAs that are used in germ-line defense and heterochromatic gene silencing are probably produced by a different set of cellular machinery, and unlike siRNAs and miRNAs only antisense rasiRNAs are formed (Vagin et al., 2006).

Despite intense attention by researchers, the mechanism by which siRNAs and miRNAs cause silencing of gene expression is currently incompletely understood (Dykxhoorn and Lieberman, 2005; Meister and Tuschl, 2004). Figure 12.2 illustrates current thinking (reviewed by Rana, 2007). 'Classical' RNAi involves siRNA-mediated degradation of target mRNAs (Fire et al., 1998). This requires the siRNA to be incorporated into a multi-subunit ribonucleoprotein RNA-induced silencing complex (RISC) that includes a member of the Argonaute (Ago) family of proteins (Filipowicz, 2005; Peters and Meister, 2007). In the classical model of RNAi, the antisense (guide) strand of an siRNA targets Ago to the complementary (target) sequence of its cognate mRNA and cleaves ('slices') it. Such slicing is promoted by a perfect or near-perfect match between the siRNA or miRNA and its target mRNA sequence. Ago proteins are members of the RNase H family. Classical Slicer activity is due to Ago2, and this protein can direct si/miRNA cleavage alone. But there are other Ago proteins, and in general these do not have a Slicer function, and indeed some lack the necessary catalytic site amino acid

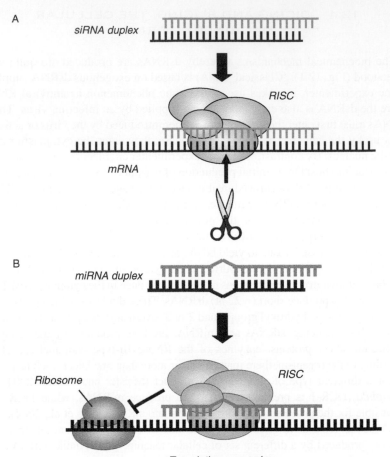

FIGURE 12.2 Silencing of target gene expression by dsRNAs. (A) *Classical RNAi*: siRNA is loaded onto the RISC. The passenger strand is unloaded while the guide strand remains and pairs with its complementary ss mRNA. This leads to the cleavage of the mRNA between residues 10 and 11 of the duplex by the RISC protein Ago2. The guide strand remains associated with RISC and may be recycled for further bouts of target mRNA cleavage. (B) *miRNA-dependent RNAi*: miRNA is loaded onto RISC. Mismatches in the miRNA–mRNA pairing, especially in the central region of the miRNA sequence, prevent cleavage of the cognate mRNA. Instead, the formation of the RISC causes inhibition of translation of the targeted mRNA by an unknown mechanism. This mechanism of gene silencing may also account for some off-target effects of siRNA-mediated (classical) RNAi.

residues to act as an RNase. These proteins are therefore supposed to serve a recognition function that leads to alternative fates for si/miRNAs.

Importantly, it is now clear that non-classical gene silencing mechanisms can also be directed by si/miRNAs. In addition to directed, specific enzymatic cleavage of perfectly complementary target mRNAs, interference with translation (Saxena et al., 2003; Thermann and Hentze, 2007) of the target gene can also occur. This translational gene silencing is characteristic of miRNAs.

In animals, most miRNAs appear to exert their effects on gene expression differently from siRNAs (Nilsen, 2007; Pillai et al., 2007; Valencia-Sanchez et al., 2007), although both siRNAs and miRNAs can function in both RNAi modes (Doench et al., 2003). The difference is probably because animal miRNAs do not in general perfectly match the hybridizing sequence on the target mRNA. Hybridization at residues 2–7 at the 5' end of the sequence (the so-called 'seed sequence') is of greatest importance (Lim et al., 2005). Critically, the sequence match between the miRNA and the target is imperfect in the central part of the miRNA sequence. This local lack of complementarity precludes Ago2-mediated slicing. In some cases, the targeted mRNA may be subject to Slicer-independent decay, perhaps through deadenylation or cap removal (Behm-Ansmant et al., 2006). However, most animal miRNAs appear to function primarily by either preventing or interfering with translation, although mRNA levels may also be affected, perhaps secondarily. Efficient silencing at the translational level appears to require the binding of more than one miRNA to the 3'-untranslated region (3'-UTR) of the targeted transcript. The exact mechanism by which this leads to the suppression of translation is not certain. There is evidence in favor of both inhibition of cap-dependent mRNA loading onto the translational machinery, and premature termination of translation, and indeed both may occur. There is also evidence that mRNAs, which are prevented from being translated, may accumulate in cytoplasmic processing (P) bodies (also called GW bodies after their content of the Ago-binding protein GW182), and there has been speculation that this may be a factor regulating the rate at which these mRNAs are translated. It has recently been shown, however, that this sequestration of miRNA-targeted mRNAs in P-bodies is a consequence, not a cause of silencing (Eulalio et al., 2007).

Yet another mode of gene silencing by dsRNAs is effected through inhibition of transcription. This effect is distinct from both classical and miRNA-mediated RNAi and is associated with heterochromatin maintenance. It has been well established to occur in yeast, plants, and nematodes, but has not been verified to occur in insects or mammals (Lippman and Martienssen, 2004). Nevertheless, the possibility that gene silencing may occur in insects at the transcriptional level in particular cases should not yet be dismissed.

12.5   UPTAKE OF DsRNA INTO TARGET CELLS

In order to be able to induce RNAi systemically (i.e. by injecting dsRNA), exogenous dsRNA must be able to enter target cells. In mammals, systemic RNAi is problematic for a number of reasons, among which are the short half-life of 'naked' siRNAs in nuclease-rich environments like blood plasma, and the poor uptake of exogenous dsRNAs into cells. This has considerably hampered attempts to use RNAi therapeutically (de Fougerolles et al., 2007), and even when using cultured cells, special transfection media are frequently employed to get siRNAs into target cells (e.g. Veldhoen et al., 2006). Goto et al. (2001) used a proprietory transfection reagent ('Effectene') to facilitate entry of dsRNA (all >1 kb in length) into *Drosophila* Kc167 cells, but it is not clear if it was established that this was necessary.

Systemic RNAi, however, has been successfully achieved on many occasions in insects simply by injecting quite long dsRNA molecules directly into the hemolymph (see below). Typically, the injected dsRNAs are 500–1000 bp in length. It is thus evident that insect cells are able to take up dsRNA molecules of considerable size from the hemolymph, presumably converting them into siRNAs by means of Dicer after entry into the cell.

It is not known with certainty how this uptake occurs. It has been shown, however, that the *C. elegans* gene *sid-1* is required for systemic RNAi (Tijsterman et al., 2004; Winston et al., 2002). Sid-1 is a transmembrane protein expressed on cells with direct environmental contact. The demonstration that expression of *C. elegans* Sid-1 protein in a *Drosophila* cell line both enhances uptake of dsRNA and also facilitates RNAi (Feinberg and Hunter, 2003) convincingly shows that this protein is involved in the internalization of dsRNA, although the mechanistic role of this protein is unknown. Significantly, a ubiquitously expressed *sid-1*-like cDNA has been cloned from the locust *Schistocerca americana* (Dong and Friedrich, 2005), and a *sid-1*-like gene *XM_395167.3* is identifiable in the genome of the honey bee, *Apis mellifera*, in which its expression was upregulated during an RNAi response to an unrelated gene (Aronstein et al., 2006). Sid-1-like genes are also present in the genomes of a number of other insects, including the grain beetle *Tribolium castaneum* (two genes: *XM_969743.1 and XM_969161.1*), and the aphids *Aphis gossypii*, *Rhopalosiphum padi*, and *Sitobion avenae* (one EST in each case: *EF533711.1*, *EF533712.1*, and *EF533713.1*).

No obvious Sid-1-like homolog is evident in *Drosophila*, however, suggesting that a different means of uptake of dsRNAs operates in this insect. Saleh et al. (2006) used *Drosophila* S2 cells to show that while long (*ca.* 1 kb) dsRNAs are efficiently taken up into insect cells, siRNAs are only poorly internalized. The uptake process was shown to involve a temperature sensitive (i.e. active) process involving initial binding to the cell surface followed by internalization. An 'RNAi of RNAi' screen revealed a substantial set of genes required for dsRNA-induced inhibition of expression of a reporter gene. Importantly, many of the genes identified in this way are known to be involved in receptor-mediated endocytosis. Among the genes required for efficient dsRNA uptake were components of the vacuolar ATPase (V-ATPase) that is involved in the maturation of early endosomes. Consistent with this, treatment with bafilomycin (a V-ATPase inhibitor) prevented dsRNA internalization. Finally, treatment with known ligands (polyinosine and fucoidin) of scavenger receptors also prevented dsRNA uptake, suggesting that receptors of this type are involved in the process. Additionally, it has been shown that no fewer than nine genes encoding ATP-binding cassette transporters are required for efficient RNAi in *C. elegans*

(Sundaram et al., 2006); the mechanism of this effect is not known, however, nor whether genes of this type are involved in dsRNA uptake in insects.

12.6   RNAi BY FEEDING

In *C. elegans*, dsRNAs of considerable length (1 kb or more) cause RNAi when administered by feeding (Timmons and Fire, 1998). This method of administering dsRNA has obvious advantages, not least among which is the point that it does not involve injuring the animal with an injection needle, which may have unintended consequences on immune function.

Among insects, RNAi has been successfully achieved for two different genes by feeding dsRNA in the lepidopteran *Epiphyas postvittana* (Turner et al., 2006). Gut levels of an mRNA encoding a gut-specific carboxylesterase gene *CXE1* were specifically reduced by 60–80% after feeding 1 µg of dsRNA in a drop of sucrose solution to third stage larvae. The level of mRNA fell during a period of 4–7 days after feeding. A different gene *PBP1*, encoding a pheromone-binding protein, that is only expressed in adult antennae was also examined after feeding dsRNA to larvae. *PBP1* mRNA levels were markedly depressed immediately after adult eclosion, but the effect lasted for only 2–3 days after feeding, when PBP1 levels climbed to levels similar to those seen in controls.

A degree of success with RNAi by feeding dsRNA has also been reported for the bloodsucking bug *Rhodnius prolixus* (Araujo et al., 2006), in which mRNA levels for *nitrophorin 2* were knocked down, although heroic amounts of dsRNA were required to achieve this. Feeding 13 µg dsRNA to second stage larvae achieved a reduction in mRNA level of about half. On the other hand, feeding 80 µg dsRNA to fourth stage larvae had no effect. Successful RNAi by feeding has also been reported in ticks (*Ixodes scapularis*, Soares et al., 2005). Feeding may be an attractive method of achieving tissue-specific RNAi effects (presumably the gut would be more accessible than other tissues), but it not yet clear whether RNAi can be routinely achieved in insects and other arthropods by the oral route. As usual, it is rare for negative results to be published. Rajagopal et al. (2002) noted in passing, however, that they had been unsuccessful in attempts to suppress expression of gut aminopeptidase N in *Spodoptera litura* by soaking insects in dsRNA solutions, or feeding dsRNA in the diet. They did however successfully suppress expression of this gene by injecting dsRNA.

12.7   AMPLIFICATION OF DsRNA

In the nematode *C. elegans*, injected or orally administered dsRNA is replicated by RNA-dependent RNA polymerase (RdRP) and the increased number of copies of the dsRNA greatly facilitates the gene silencing effect of administering dsRNA (Sijen et al., 2001). A similar amplification of exogenous dsRNA has been shown to occur in plants and fungi. This probably accounts for the extraordinary efficacy and long-lasting effects of dsRNA-mediated RNAi in these organisms. Use-dependent dsRNA amplification has been proposed, which would lead to selective amplification of dsRNAs in those tissues where the target mRNA was present. Recent work (Pak and Fire, 2007) has shown that in *C. elegans*, the great majority of siRNAs present during an RNAi response are indeed 'secondary' siRNAs (i.e. they were formed through RdRP action). The mechanism whereby association of 'primary' siRNAs with an mRNA target would lead to production of secondary siRNAs has been shown to depend not on priming of RNA polymerase activity by duplex formation, but on the interaction of the RISC with the target sequence (Pak and Fire, 2007; Sijen et al, 2007).

Unfortunately, however, such RNAi amplification appears not to be universal. RdRP homologs are absent from the *Drosophila* genome and there is no convincing evidence that amplification of exogenous dsRNA occurs in flies (Roignant et al., 2003). Nevertheless, it remains possible that RdRP-dependent amplification of siRNAs may occur in other insects.

Moreover, it should be pointed out that the cleavage of long dsRNAs into siRNAs in effect considerably multiplies the number of RNAi-effecting dsRNA molecules at the site of action. The slicing of a 1 kb dsRNA to 25 bp siRNAs represents an approximately 40-fold increase in molar ratio. Moreover, the efficiency of the silencing process may also be enhanced by the recycling for further use of siRNAs following target mRNA cleavage (Rana, 2007).

12.8   SPREAD OF RNAi WITHIN THE ORGANISM

In plants, it is well established that RNAi-mediated suppression of gene expression is able to spread from the cells in which it was initially evoked to other parts of the plant (reviewed by Lecellier and Voinnet, 2004). This is in part mediated by cell-to-cell movements of primary siRNAs but longer range spread of the RNAi response is mediated by Sid-1 and RdRP-dependent synthesis of secondary siRNAs. RNAi in *C. elegans* is also non-cell autonomous and dependent on dsRNA amplification in a similar way (Winston et al., 2002).

Almost all of the evidence so far available, however, indicates that such cell-to-cell or longer range spread of RNAi does not occur in *Drosophila*. As noted above, evidence for RdRP-dependent amplification of dsRNA is lacking, and when hairpin RNAs are expressed under the control of cell-specific promoters, then the RNAi effect is apparently restricted to the cell type in question (Roignant et al., 2003).

12.9   STAGE AND TISSUE SPECIFICITY OF RNAi

While the cellular machinery that enables RNAi to occur appears to be ubiquitous, for various reasons the efficacy of experimentally introduced dsRNAs to suppress the expression of target genes may vary between experimental organisms, between tissues, and also varies with time and developmental stage. It may be that this variability is due to differential expression of proteins that are required for efficient dsRNA uptake into target cells. Fat body and hemocytes, together with epidermis and gut tissue, are highly represented among those insect tissues that have been successfully targeted by experimental systemic RNAi. These tissues may simply be those that most actively express one or more Sid-1 homologs, or undertake receptor-mediated endocytosis, and are therefore most efficient in taking up exogenous dsRNA. Alternatively, they may be those tissues that are best equipped with the cellular machinery that converts dsRNAs into siRNAs, or the components of the RISC.

Little work has so far been done to explain the selectivity of RNAi in insects. However, understanding is better in *C. elegans* (Grishok, 2005). Genetic screens in this nematode for enhanced RNAi sensitivity (*eri*) have discovered a number of *eri* genes that act to oppose RNAi. The strongest of these effects involves the action of a dsRNA-specific exonuclease, encoded by the gene *eri-1*. Since ERI-1 is most strongly expressed in gonadal and nervous tissue, this may perhaps explain the generally observed low sensitivity of these tissues to RNAi (Kennedy et al., 2004). Recently, it has been noted that mutant strains for the synaptic protein UNC-13 are less susceptible to RNAi. Paradoxically, however, double mutants for *unc-13* and *eri-1* are hypersensitive to RNAi. Thus *unc-13* appears to modulate the effect of *eri-1*. This may also contribute to neuronal insensitivity to RNAi in *C. elegans* (Chapin et al., 2007).

An example of an insect tissue that appears to be much less susceptible to RNAi than others is the salivary gland of the mosquito *Anopheles gambiae* (Boisson et al., 2006). In this insect, injected dsRNA was readily able to suppress a target reporter gene (*green fluorescent protein, GFP*) in the midgut and ovaries, but was effective in salivary gland only when much larger quantities of the dsRNA were used than were required in the other tissues. The mechanism of tissue specificity in this case is unclear. Boisson et al. (2006) suggested that the refractoriness of mosquito salivary tissue was at least in part based on reduced uptake of dsRNA, based on the observation that salivary glands took up less fluorescent labeled siRNA than other tissues. *Anopheles stephensi* and *Aedes aegypti* also behaved in this way. The authors commented that this result might imply that mosquito salivary glands are impervious to exogenous dsRNA, or that they degrade it. In the same study, however, the investigators showed that although mRNAs encoding several known components of the RNAi cellular machinery (*dicer1, dicer2, ago2, ago3*) were present in salivary gland cells, the levels of these transcripts were lower than in other tissues, so that this might also contribute to the insensitivity of this tissue to RNAi. It is not yet clear whether this kind of tissue specificity of RNAi is a generally occurring phenomenon in insects, since descriptions of unsuccessful attempts to effect RNAi are rarely published.

It would be experimentally desirable in some cases to silence a gene in one tissue but not in another. This is now readily achievable in *Drosophila* genetically through *UAS-GAL4* constructs that express an RNA hairpin under the control of a particular promoter, thereby restricting expression to a particular tissue or set of tissues (Dietzl et al., 2007). This is discussed more fully below.

Such targeting is much less readily achieved in other insects that lack the genetic and genomic tools available for *Drosophila*. Attempts are currently being made to target dsRNAs systemically to particular cell types using antibodies, but the technology is not yet well developed (de Fougerolles et al., 2007).

12.10   DIFFERENTIAL SUSCEPTIBILITY OF GENES TO RNAI

It has also sometimes been observed that particular genes expressed within the same tissue are resistant to systemic RNAi when other genes are not. One possible explanation for this is that some genes are regulated by post-translational mechanisms. A good example of a case where it has been proposed that the level of gene regulation accounts for RNAi resistance is provided by two proteins that are expressed in pupal hemocytes of the blowfly *Sarcophaga peregrina* (Nishikawa and Natori, 2001). These genes may have immune-related roles. Expression of the hemocyte-specific scavenger receptor, p120, is regulated at the level of transcription, and p120 mRNA is absent from larval hemocytes but is present in pupal hemocytes. This gene product was efficiently suppressed by injection of specific dsRNA into larvae. By contrast, cathepsin B, also expressed in pupal hemocytes but present in those of larvae at only very low levels, is translationally regulated, and the cathepsin B protein content of hemocytes increases at pupation even though mRNA levels actually fall. Injection of the relevant dsRNA into *Sarcophaga* larvae was found to have no measurable effect on the accumulation of cathepsin B. Nishikawa and Natori concluded that RNAi in this insect (at least at this stage of its development) is effective only when the target mRNA is newly synthesized, and is ineffective in the case of previously synthesized mRNA. It must be conceded, however, that it was not shown directly that this was the case, and the mechanism of the effect is unknown.

Alternative explanations in cases where the efficacy of RNAi appears to depend on the identity of the target gene in question are that the dsRNA reagents or the resulting siRNAs may be subject to sequence-specific degradation, or that the apparent specificity is dependent on the amount and stability of the mRNA in question.

For the experimentalist proposing to make use of RNAi to 'knock down' a gene, it is important to realize that according to the mechanism of the RNAi effect, the dynamics of mRNA synthesis and breakdown may have a profound effect on the outcome of an RNAi experiment. For example, a gene that is transcribed at a high rate, but which has a short-lived mRNA, would be hard to suppress through classical RNAi-mediated transcript degradation, but would be readily knocked down if the effect of RNAi was to inhibit translation through a miRNA-like mechanism. By contrast, a gene that is transcribed at only a low rate, but which has a long-lived mRNA, would be knocked down by either mechanism. Unfortunately, it is rare for the mechanism of RNAi-mediated gene suppression to be verified, so that little can be said at present about this.

12.11   TRANSGENERATIONAL RNAi

Fire et al. (1998) showed that dsRNAs injected into *C. elegans* causes suppression of gene expression not only in the treated generation, but also in their offspring. This transgenerational RNAi effect applies to at least some insect species too. This was shown for *Tribolium castaneum* by Bucher et al. (2002), who used the technique to suppress the expression of two developmentally important transcription factors, *distalless* and *maxillopedia*. It was shown that the procedure led to the suppression of these genes in almost 100% of the insects that hatched from the first batch of eggs laid after treatment. Two limitations were that the proportion of affected offspring (and the intensity of the phenotypic effect) declined in subsequent weeks, and that the dsRNA treatment initially resulted in a markedly reduced number of viable eggs. The authors assumed that the transgenerational RNAi effect (they called it 'parental RNAi') was due to the transfer of dsRNA into the eggs laid by the treated insects, although this was not directly proved.

Transgenerational RNAi effects in insects are not limited to *Tribolium*. The extent of the phenomenon is not yet clear, but at least two other insects can show it. Rajagopal et al. (2002) found that dsRNA injections in fifth stage *Spodoptera litura* larvae could silence expression of aminopeptidase N not only in the gut of the same insects 48 h later, but also in the F1 offspring of the injected insects. An example of transgenerational RNAi applied to a molecule of immune system interest is the study by Bettencourt et al. (2002), in which expression of the immunoglobulin-like protein Hemolin was suppressed by injecting *hemolin*-specific dsRNA into female *Hyalophora cecropia* moths. Hemolin is strongly induced in a number of Lepidoptera following immune challenge. Transgenerational RNAi was successful in preventing expression of Hemolin in eggs, and as a consequence the resulting embryos were malformed and failed to hatch. It is not clear how this relates to the immune function of Hemolin; it seems likely that in this insect the protein is a dual function molecule that is important in cell–cell adhesion during embryogenesis.

12.12   RNA HAIRPINS AS EXPERIMENTAL TOOLS TO INVESTIGATE THE INSECT IMMUNE SYSTEM

Transcription of genetically engineered DNA sequences that encode RNA hairpins can be used to knock down the expression of particular genes to test hypotheses about immune function. The recent availability of an RNAi hairpin library that includes every predicted protein-coding sequence in Release 4.3 of the *Drosophila* genome (Dietzl et al., 2007) is a resource that clearly demands a systematic RNAi test of the function of every putative immune-related gene. Even before the availability of this library, however, researchers were busy; we have chosen several examples of this approach that illustrate its potential.

The first case concerns the specific targeting of a *Drosophila* hemolymph protein, hemolectin (Hml). Goto et al. (2003) showed that Hml is expressed exclusively in a subpopulation of plasmatocytes and crystal cells but not in lamellocytes. The authors used the UAS-GAL4 system to investigate this protein. They created a transgenic line (*hml-GAL4*) that placed expression of the yeast transcriptional regulator GAL4 under the control of 3 kb of upstream *hml* promoter sequence. A second line, *UAS hmlRNAi*, placed an inverted repeat of the *hml* sequence under the control of UAS. Larvae resulting from a cross of these two lines expressed a hairpin *hml* sequence that led to suppression of Hml expression. The nature of the construct means that the dsRNA that interferes with Hml expression is itself only present in those tissues that contain *hml* mRNA, and at the times when the gene is being transcribed. This may be supposed to minimize off-target effects. The *hmlRNAi* larvae were viable and did not display obvious developmental defects, but lost excessive amounts of blood upon injury. Failure to seal the wound was not due to lack of PO or melanin production, and antimicrobial peptides were expressed normally in the *hml* RNAi adults. The conclusion was that Hml is involved in hemostasis and/or coagulation in larvae.

Another example is from work by Bian et al. (2005), in which an immune-related gene was suppressed in a particular physiological situation. *Aedes aegypti* mosquitoes were transformed using a Piggybac vector to express a construct that contained two copies of a large fragment of the gene encoding REL1, the mosquito homolog of the *Drosophila* Toll pathway gene *dorsal*. These sequences were arranged as a head-to-head inverted repeat so that transcription would generate, under the control of a fat body specific vitellogenin (Vg) promoter, a hairpin RNA and ultimately a 626 bp dsRNA. Since Vg is expressed in this insect only in females and only after blood feeding, REL1 expression was suppressed only in these circumstances. Blood fed females were found to be more susceptible than controls to infection by the insect pathogenic fungus *Beauveria bassiana*, and showed reduced expression of two upstream immune regulators, Spaetzle1A (Spz1A) and Serpin-27A, after fungal challenge (Fig. 12.3). As expected, another strain of the mosquito, which had been engineered to overexpress REL1, showed exactly the opposite effect on expression of these two immune regulators. Satisfyingly, a further strain in which the same genetic RNAi technique was used to knock down the mosquito homolog of *cactus*, an IKB inhibitor of the Toll pathway, had the same phenotype as the strain that overexpressed REL1.

A third example is a study on *Drosophila* by Kurucz et al. (2007). These authors used a hairpin construct together with a hemocyte-specific *UAS-GAL4* driver to suppress the expression in hemocytes of the protein encoded by the *nimrod C1* gene. The driver was the upstream DNA sequence for the *hemese* gene (Kurucz et al., 2003). This experiment showed that hemocyte NimC1 is involved in bacterial

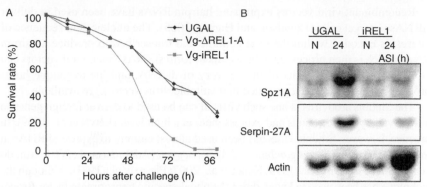

FIGURE 12.3 Increased susceptibility to *Beauveria bassiana* and reduced immune activation of Spz1A and Serpin-27A genes in transgenic mosquitoes with altered expression of *Rel1*. Rel1 is the *Aedes aegypti* homolog of *Drosophila Dorsal*, a regulator of the Toll immune activation pathway. Two different transgenic mosquito strains were constructed, in which the REL1 gene was placed under the transcriptional control of the vitellogenin (Vg) promoter. This meant that one strain (Vg-DREL1-A) overexpressed REL1 following a blood meal, while the second strain (Vg-iREL1), which had an RNA hairpin, expressed dsRNA for REL1 and therefore did not overexpress REL1. The figure shows the second cycle of blood feeding. (A) Transgenic (Vg-DREL1-A and Vg-iREL1) and wild-type (UGAL) mosquitoes were blood fed and after 24 h were challenged with *B. bassiana* spores. (B) mRNA levels of two genes upstream in the Toll activation pathway (Spz1A and Serpin-27A) in fat bodies of transgenic (Vg-iREL1) mosquitoes treated in the same way. Feeding-induced expression of these genes is prevented in the REL1 RNAi strain, indicating that there is positive feedback from the Toll pathway onto the expression of these genes. ASI (h) = hours after septic injury. (Reproduced with permission from Bian et al., 2006.).

species-specific phagocytosis. The number of phagocytosed *S. aureus* (a Gram-positive bacterium) was decreased in the NimC1 knockdown insects, while the number of engulfed *E. coli* (Gram negative) was unaffected. In principle, the *UAS-GAL4* system can be used to target RNAi to any cell or tissue, provided that a suitable driver is available.

12.13 VIRAL RNAi DELIVERY SYSTEMS

The kinds of gene engineering studies described in the previous section are well established in *Drosophila*. In different insect species that lack the necessary genetic tools, however, other methods of delivering RNA hairpins are required. Isobe et al. (2004) used the transposon *piggyBack* to create a transgenic line of silkworms that showed enhanced resistance to infection by the *Bombyx mori* nucleopolyhedrovirus (BmNPV). They introduced an inverted repeat of the BmNPV *lef* gene, known to be essential for viral replication, into the vector and injected it into eggs, eventually recovering transgenic silkworms. The replication of BmNPV in these insects was greatly slowed, but not completely prevented, so that the infected insects eventually died despite their enhanced resistance.

Recombinant viral vectors expressing hairpin RNAs have been used to deliver siRNAs to target cells (Davidson and Harper, 2005). The nucleic acid sequence of interest is encoded within the viral genome and transcribed to produce a hairpin RNA, which is then processed in host cells to form siRNAs. Such viral vectors can be highly efficient in terms of the delivery of dsRNA, and the technique holds promise for use in insects, provided that suitable virus vectors are available.

The Sindbis arbovirus is one such virus. It can be used to deliver foreign genes to a wide variety of arthropods and express them at a high level (Lewis et al., 1999). In practice, however, Sindbis virus has been used almost entirely to express shRNAs in the mosquitoes *Anopheles gambiae* and *Aedes aegypti* (Adelman et al., 2001; Attardo et al., 2003; Johnson et al., 1999; Konet et al., 2007; Pierro et al., 2003). Although the technique has been used to knock down the pupal-specific transcription factor *Broad-Complex* in the silkworm *Bombyx mori* (Uhlirova et al., 2003), the outcome was not entirely satisfactory. The phenotypic effects of virus-mediated *Broad-Complex* RNAi were limited to certain tissues. This was almost certainly due to the fact that the virus enters certain tissues more readily than others. Sindbis virus that expressed GFP was shown first to infect *Bombyx* fat body larval muscles and hemocytes, with nervous system, larval eyes, silk glands, and imaginal disks infected only later. Gonads, Malpighian tubules, and epidermal cells seemed not to be infected at all.

The rather sparse uptake by experimentalists of viral RNAi delivery, other than in mosquitoes (where there are applied reasons to persist with the technique), is presumably because the methodology is relatively complex and simpler techniques are effective. There are also some problems. The Sindbis virus genome is unstable and this can lead to loss of the foreign gene. Further, as noted above, the virus is somewhat selective in its target tissues. It is possible, nevertheless, that other viruses may have better prospects as RNAi vectors in insects. There is a report (Huang et al., 2007) of the construction of baculovirus vector for RNAi using inverted repeat hairpin sequences; it was used to suppress the expression of reporter gene in cultured insect cells.

However, in the present context, there is another problem. Infection of mosquitoes by Sindbis virus results in extensive upregulation of immune-related genes (Sanders et al., 2005). Moreover, viral vectors may themselves interfere with the host cell's RNAi machinery (Li and Ding, 2006). These factors are bound to complicate the interpretation of studies using dsRNAs delivered by viral vectors to investigate immune function.

12.14   RNAi SCREENS

Libraries of genetic constructs expressing RNA hairpins can be created and screened in *Drosophila* and other genetically tractable insects for potential immune-related phenotypes. This can be done *in vivo* or using cultured cells.

A good example of such an *in vivo* RNAi screen is provided by Kambris et al. (2006), who constructed a collection of fly strains carrying inverted repeat sequences for 75 distinct serine proteinase (SP) genes out of the 200 identifiable SPs within the *Drosophila* genome. Each hairpin was placed under the control of the *UAS* promoter sequence, so that when the flies bearing the construct were crossed to flies carrying a ubiquitously expressed GAL4 driver, the F1 flies expressed the encoded hairpin RNA (and thus the derived siRNAs) in all tissues, thus inactivating the cognate SP, allowing its immune function in response to an immune challenge to be assessed. The screen tested the ability of the flies to survive infection with *Enterococcus faecalis* and to express a Drosomycin-GFP reporter. In this way, five novel SPs that appear to function upstream of Spz in the activation of the Toll pathway were identified.

The combination of RNAi and *Drosophila* cell culture is now emerging as an extraordinarily powerful tool for the identification of genes required for particular physiological, immunological, or other phenotypic features (Dietzl et al., 2007; Perrimon and Mathey-Prevot, 2007). Here we give two examples of *in vitro* RNAi screens that have resulted in the discovery of genes with immune functions.

Many viruses, including insect specific viruses like the *Drosophila* C virus (DCV), utilize internal ribosome entry sites (IRESs) for translation. To identify host factors required for IRES-dependent translation and viral replication, Cherry et al. (2005) performed a genome-wide RNAi screen in *Drosophila* cells infected with this virus. Sixty-six ribosomal proteins were identified that were required for DCV growth, but not for that of a different, non-IRES-containing RNA virus.

Stroschein-Stevenson et al. (2006) screened a library containing more than 7000 inverted repeat constructs of *Drosophila* genes that are conserved in other animals, and tested for disruption of phagocytosis of the pathogenic yeast *Candida albicans*. After eliminating housekeeping genes, presumably required for generic processes that ensure cellular responsiveness, 184 *Drosophila* genes were identified as required for efficient phagocytosis. Of particular interest, one of these genes, *macroglobulin complement related* (*mcr*), enables a phagocytic response specifically directed toward *C. albicans* but not toward *E. coli*. Mcr is a secreted protein that appears to function as an opsonin, binding to *C. albicans* and provoking its own phagocytosis. Closely related genes of the *Drosophila* thioester-related protein family encode structurally similar proteins with similar functions but different microbial specificities.

12.15   SYSTEMIC RNAi AS AN EXPERIMENTAL TOOL TO INVESTIGATE THE INSECT IMMUNE SYSTEM

Since its discovery, systemic RNAi has been rapidly adopted as a key experimental technique to investigate the roles of known immune-related proteins in mammalian immunity (Mao et al., 2007). In this section we focus on its use to probe the insect immune system. Space does not permit us to review all published systemic RNAi experiments involving insect immune systems, and we have chosen to give examples of studies that have features of particular interest.

dsRNA can be injected into early stage embryos in order to suppress expression of immune-related proteins later in life. Injection of dsRNA into eggs is a well established technique in *Drosophila*. An example of an immune-related RNAi experiment of this type is provided by Kurucz et al. (2003), who successfully silenced the expression of the *Drosophila* hemocyte-associated glycoprotein Hemese, by injecting specific dsRNA into embryos. The Hemese knockdown insects showed no obvious phenotypic effects until they were exposed to parasitization by the parasitic wasp *Leptopilina boulardi*, when they showed enhanced immune reactivity. Significantly more blood cells were mobilized by Hemese knockdown insects, and more melanotic nodules were formed in response to the parasite. The inference drawn from these results is that the presence of Hemese normally restrains immune responses to foreign materials. In turn this implies that there is a cost associated with immune responses and that such restraint is selectively advantageous.

Injection of dsRNA into eggs of other insects is a less robust technique however. In a non-immune-related example, Quan et al. (2002) injected dsRNA into eggs of *Bombyx mori* to evaluate the method, attempting to silence the silkworm *white* gene. They were able to show dose-related induction of a typical *white egg* phenotype in a high proportion of eggs, and confirmed the loss of *white* mRNA. The hatching success of the eggs after injection (even controls) was quite low (15–30%), however, and of the resulting larvae only a low proportion of those given the highest dsRNA dose displayed the *white* cuticular phenotype. The authors did not report whether the RNAi effect persisted into the adult stage.

The gap between injecting dsRNA and assessing the resultant knockdown phenotype does not have been so long as in the above experiments, however. Levin et al. (2005) used injections of dsRNA into fourth stage *Manduca sexta* larvae to suppress the expression of the hemocyte-specific integrin β1. Unusually, these investigators used siRNAs prepared *in vitro* from longer dsRNA. They showed that this treatment significantly diminished the ability of fifth stage larvae to encapsulate DEAE-Sephadex beads, concluding that integrin β1 must play an important role in this cellular immune defense.

Alternatively, dsRNA injections can be made within the same instar as the stage examined for the effect. This implies that RNAi can work, at least in some cases, very quickly. Eleftherianos et al. (2006a) injected dsRNAs into newly molted final stage larvae of *Manduca sexta* to suppress the expression of three separate hemolymph recognition proteins, Hemolin, Immulectin-2 (IML-2), and Peptidoglycan Recognition Protein (PGRP). Expression of all these proteins in fat body is strongly induced by infection with the pathogen *Photorhabdus luminescens*, but they are expressed at only very low levels in unchallenged insects. Knock down of all these proteins was effective only 6 h after the injection of dsRNA, as was shown when the insects were subsequently challenged with *Photorhabdus* (see Fig. 12.4). In another study, in which injection of dsRNA preceded exposure to the innocuous Gram negative bacterium *E. coli*, Eleftherianos et al. (2006b) showed that the RNAi knockdown was already effective after only 1 h. For all

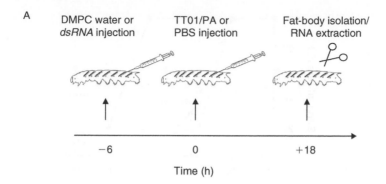

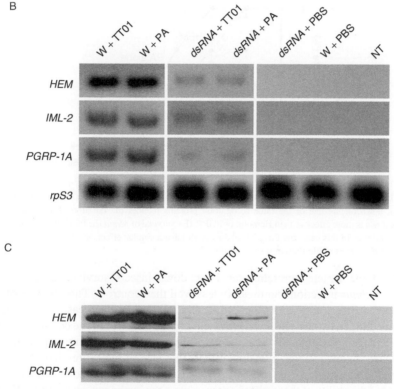

FIGURE 12.4 Systemic RNAi of recognition proteins in *Manduca* reduces survival after bacterial infection. (A) Schematic to show treatment regime. dsRNA specific to one of three different recognition genes was injected 6 h before challenge with one of two different insect pathogenic bacteria, *Photorhabdus luminescens* (TT01) and *P. asymbiotica* (PA). (B) RT-PCR shows that systemic RNAi specifically knocks down the induced expression of mRNAs encoding Hemolin (HEM), Immulectin (IML-2), and peptidoglycan recognition protein (PGRP). PBS, phosphate-buffered saline; NT, no treatment; rpS3, loading control. Sizes of PCR products are indicated. (C) Western blots immunostained to show the three recognition proteins.

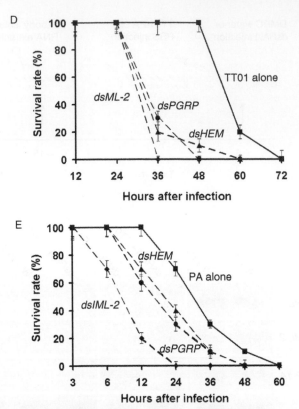

FIGURE 12.4 (Continued)   (D) Survival of *Manduca* following infection with *P. luminescens*. Knock down of any one of the three recognition proteins reduces the duration of survival. IML-2 knock down is more effective than Hemolin or PGRP. (E) Survival of *Manduca* following infection with *P. asymbiotica*. In this case, the three knockdowns all have a similar effect on survival. (Reproduced with permission from Eleftherianos et al., 2006a.)

three of the targeted proteins, the knockdown insects survived a subsequent *Photorhabdus* infection significantly less well than controls. This shows that the induced expression of Hemolin, IML-2 and PGRP is in each case an important, though ultimately unsuccessful component of the insect's immune defenses against this pathogen. Presumably the protective effects of these recognition proteins are mediated by downstream signaling effects that lead to the expression of directly antibacterial effectors. In the case of IML-2, RNAi knockdown significantly impaired the ability of the insect to encapsulate a non-pathogenic bacterium, *E. coli*.

RNAi can also be effective in adult insects. Osta et al. (2004) injected specific dsRNAs into adult mosquitos, *Anopheles gambiae*, to show that knockdown of expression of the *LRIM-1* (leucine rich–repeat immune-1) gene led to a significant increase in the number of surviving *Plasmodium* oocysts in the insect's midgut when the RNAi-treated insects were fed on an infected mouse. In other words, LRIM-1 plays an important role in the ability of the mosquito to resist infection by the parasite. In the same study it was shown that two C-type lectins, *CTL4* and *CTLMA2*, encoded within the mosquito's genome actually enhance the parasite's virulence: RNAi knockdown of these gene products led to a significant reduction in the number of surviving midgut oocysts. There was massive melanization of protozoan cells in the midgets of the *CTL4* and *CTLMA2* knockdown insects.

RNAi can be used to illuminate signaling pathways. Eleftherianos et al. (2006b) showed that the *Manduca* immune system can be effectively primed by prior infection with harmless bacteria so that immunity to infection by a more virulent pathogen is subsequently enhanced. Injection of specific dsRNAs was used to prevent the upregulation in response to prior exposure to *E. coli* of several microbial pattern recognition protein genes (*hemolin, IML-2,* and *peptidoglycan recognition protein*) and antibacterial effector genes (*attacin* (ATT), *cecropin* (CEC), *lebocin* (LEB), *lysozyme* (LYS), and *moricin* (MOR)). The effects of individually knocking down any one recognition protein were greater than knocking down any one effector protein (Fig. 12.5). This was because knock down of any one recognition protein was able to prevent the subsequent appearance of antibacterial activity in hemolymph. The inference is that bacteria are detected by recognition proteins, and that this recognition is subsequently transduced by an unknown signaling pathway into the expression of multiple antimicrobial effectors.

Although the effects of systemic RNAi are not restricted to a particular target tissue, the technique can nevertheless be used to demonstrate tissue-specific effects. For example, in the experiments of Eleftherianos et al. (2006a, b), RNAi-mediated knockdown of Hemolin was shown to reduce survival of *Photorhabdus*-challenged insects. Thus this protein must protect against *Photorhabdus* to some extent. But neither the site of synthesis of the Hemolin in question, nor its effective location, was identified. Similarly, Terenius et al. (2007) used RNAi to suppress expression of Hemolin, in this case in diapausing pupae of the saturniid silkmoth *Hyalophora cecropia*. They found that the dsRNA treatment caused a transient reduction in PO activity. Since prophenoloxidase (PPO) activation takes place in hemolymph plasma, the Hemolin in question could have originated in fat body, hemocytes, some other tissue, or even all of these.

Some light was shed on this question by Eleftherianos et al. (2007a), who showed that injection of dsRNA into newly molted fifth stage *Manduca* larvae suppressed expression of Hemolin and PGRP not only in fat body, but also in hemocytes. By examining *in vitro* the cellular responses of washed hemocytes from dsRNA-treated insects, they were able to show that RNAi suppression of Hemolin, but not PGRP, adversely affected the ability of hemocytes to undertake phagocytosis, and to participate in the formation of melanotic nodules. The implication is that Hemolin synthesized by the hemocytes themselves is required for these activities. Ultimately, however, the only sure proof of the cell autonomous nature of an immune-related gene is tissue or cell-specific RNAi targeting using a hairpin RNA genetic construct that is driven by an appropriately specific driver.

RNAi can be used to prove that a particular host gene is the target of a particular pathogen virulence gene. Eleftherianos et al. (2007b) injected dsRNA into fifth

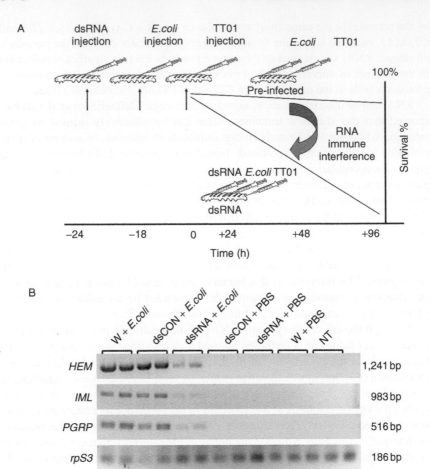

FIGURE 12.5 Injection of dsRNA knocks down *E. coli*-induced overtranscription of *Manduca* immune genes. (A) RT-PCR shows that injection of *E. coli* leads to overexpression of three recognition genes but prior injection of dsRNA specific for each gene greatly reduces mRNA levels (Hemolin, HEM; Immulectin, IML-2, and; peptidoglycan recognition protein, PGRP). (B) Similarly, *E. coli* injection induces five monitored effector genes (Attacin, ATT; Cecropin, CEC; Lebocin, LEB; Lysozyme, LYS; Moricin, MOR), which are all knocked down by prior injection of specific dsRNAs. Control dsRNA (dsCON) has no effect. Note that in both panels all negative controls (either injection of dsRNA alone with no *E. coli* infection) (dsRNA + PBS), injection of PBS alone or no treatment (NT) produce no immune response.

stage *Manduca* larvae in order to knock down the expression of PPO, the inactive precursor to the hemolymph enzyme PO, which is responsible for melanin formation during encapsulation reactions. Although PPO is constitutively expressed at a low level, PPO is induced in insects challenged with bacteria. Eleftherianos et al. showed that activated PO is inhibited by a small molecule inhibitor, ST, synthesized by *Photorhabdus*. Since genetic manipulation of the bacterium to knock out

C

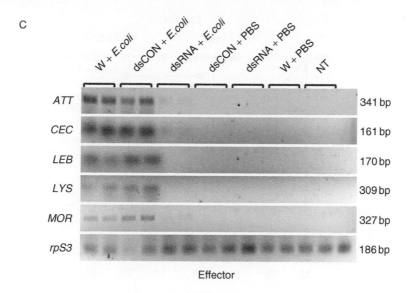

ATT   341 bp

CEC   161 bp

LEB   170 bp

LYS   309 bp

MOR   327 bp

rpS3   186 bp

Effector

D

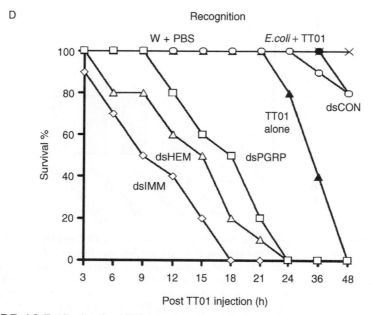

FIGURE 12.5 (Continued) (C) Schematic to show treatment regime. dsRNA specific to the indicated gene was injected 1 h before pre-infection with *E. coli*; 6 h later the insects were challenged with *Photorhabdus luminescens* strain TT01. (D) Survival curve showing that insects experiencing *E. coli* before *Photorhabdus* were protected and survived longer than those given *Photorhabdus* alone. But systemic RNAi knockdown of recognition proteins followed by *E. coli* plus *Photorhabdus* caused insects to die sooner than with *Photorhabdus* alone.

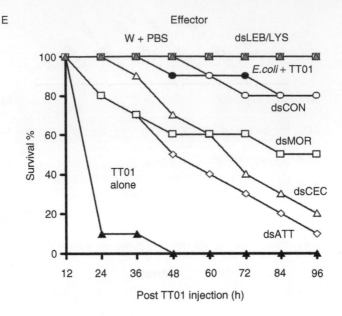

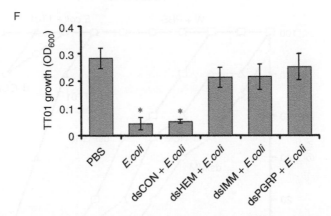

FIGURE 12.5 (Continued)   (E) When the effector proteins ATT, CEC, and MOR were knocked down by systemic RNAi after prior exposure to *E. coli*, survival after *Photorhabdus* challenge was diminished, but much less than in the case of recognition proteins. RNAi of LYS or LEB had no effect on survival, implying that they have little role in defense against *Photorhabdus*. (F) The RNAi-inhibitable immune response is associated with antibacterial activity in cell-free hemolymph plasma. Histogram shows final density (OD600) of *P. luminescens* TT01 after 18 h growth in plasma collected from *Manduca* larvae. Insects were injected with PBS alone or *E. coli* with or without prior injection of dsRNA for Hemolin, IML-2, or PGRP. Note that *Photorhabdus* grows poorly in plasma from insects that had experienced prior infection with *E. coli*, but this growth-inhibiting effect can be overcome by pre-injection of dsRNA from any one of the recognition genes. Pre-injection of the control RNA prior to *E. coli* infection has no effect on *Photorhabdus* growth. (Reproduced with permission from Eleftherianos et al., 2006b.)

a key ST synthesizing enzyme led both to the failure of the normally observed inhibition of PO and also to reduced bacterial virulence, the supposition was that production of ST during infection is a specific adaptation of the bacterial pathogen to the insect host's defense using activated PO. This was confirmed when RNAi knockdown of *Manduca* PPO restored the virulence of bacteria lacking ST to the extent that there was no longer any difference between the mutant and wild-type *Photorhabdus*. Evidently in this case, there appears to be a gene-for-gene interaction between ST production in the bacterium and PPO synthesis in the host insect.

Conversely, RNAi can also be used to target virulence genes of the pathogen itself during infection, in order to test hypotheses about function. An example is provided by the work of Ikeda et al. (2004) on the *Hyphantria cunea* nucleopoly-hedrovirus (HycuNPV). Like all other known lepidopteran baculoviruses, HycuNPV encodes a number of *inhibitor of apoptosis* (*iap*) genes. Proteins encoded by genes of this family have the potential to downregulate host apoptosis, a defensive response to viral infection. RNAi was used to knock down each of the Iap genes during *in vitro* infection of Sf9 cells, and showed that only Iap3 functions to suppress caspase activation and host cell apoptosis.

RNAi can be achieved in cultured insect cells by simply adding dsRNA to the medium. In mammalian cells this would require the use of siRNAs but in insect cells, long dsRNAs are evidently effective. By adding dsRNA to cell culture medium, Tanaka et al. (2006) were able to suppress the expression of insect derived growth factor (IDGF) a member of the adenosine deaminase-related growth factor (ADGF) family that promotes hemocyte growth. Even with very high concentrations (20 µg/ml) of dsRNA, however, they could not reduce protein levels to zero. The authors suggested that this may have been due to the inability of RNAi to suppress expression from pre-existing mRNA, but it is also possible that there was pre-existing protein in the cells or that the dsRNA did not persist over the whole of the 7 days culture period.

12.16   NON-SPECIFIC IMMUNE EFFECTS
OF DsRNAs

The potential of RNAi to cause 'off-target' effects is a legitimate concern. While the classical RNAi gene silencing effects of siRNAs are highly specific, it should be noted that miRNAs have many (probably hundreds of) targets, and because miRNAs match their targets imperfectly it is likely that dsRNA sequences designed to suppress particular genes through classical RNAi might inadvertantly have miRNA-like effects. Jackson et al. (2006a) found that in human cultured cells, the off-target effects of siRNAs are widespread and are probably the consequence of miRNA-like effects. This is something that should be taken into account in designing RNAi reagents. Sequences of siRNAs can be checked against known miRNAs using miRBase (<http://microrna.sanger.ac.uk/sequences/>). Moreover, Jackson et al. (2006b) have shown that incorporation of a 2'-*O*-methyl group at position 2 of the guide strand can alleviate such off-target gene silencing effects.

This is perhaps a suitable solution to the problem when designing siRNAs for therapeutic use, but is of little use when using long dsRNAs to achieve systemic, classical RNAi. Fortunately, however, the off-target effects seen when using relatively long dsRNAs are likely to be much less serious than when using an siRNA, since the off-target miRNA-like effects will be specific to only a few of the many siRNAs generated from the long dsRNA, while the 'classical' RNAi effects of all these siRNAs will be primarily directed toward the selected target.

In the context of using RNAi to investigate immunity, it is relevant to note that in mammals, it has long been known that long (>30 bp) dsRNA sequences, usually of viral origin, are potent triggers for a wide variety of inflammatory and antiviral effects (reviewed by Wang and Carmichael, 2004). Among these is the production and secretion of interferons. These inducible cytokines have both autocrine and paracrine antiviral effects, and also regulate inflammatory cellular responses. Insects are not known to produce interferons, but the possibility of undiscovered analogous responses must be borne in mind. Interferon effects in mammals are diverse, but include the upregulation of several genes encoding dsRNA-activated cytoplasmic enzymes.

The production of interferons by mammalian cells in reponse to viral infection is mediated by a specific Toll-like receptor (TLR3) that specifically recognizes dsRNA. Like other TLRs this is a cell surface protein although it may be localized to endosomal vesicles. Interferon responses may not be exclusively due to TLR3, and a cytoplasmic RNA helicase RIG-1 may also be involved (Schroder and Bowie, 2005).

These enzymes are themselves concerns in the context of possible off-target effects. One such enzyme is the dsRNA-activated protein kinase, PKR (Garcia et al., 2006; Wang and Carmichael, 2004). This enzyme has several effects, but notably results in a general downregulation of protein synthesis through phosphorylation of the eukaryotic initiation factor 2, eIF2, which regulates translation. On the other hand, PKR also phosphorylates the immune-related transcription factor inhibitor, IKB, well known to be a central player in innate immune activation. This releases IKB from its association with cytoplasmic NF-κB, which then migrates to the nucleus where it activates the transcription of a large number of proinflammatory genes. PKR may also activate proapoptotic pathways.

Another important dsRNA-responsive pathway in this context is 2′,5′-oligoadenylate synthetase (2′,5′-AS)/RNase L (Wang and Carmichael, 2004). dsRNAs of >70 bp activate 2′,5′-AS, which is itself upregulated by interferons. 2′,5′-AS activates RNase L, which degrades both mRNAs and rRNAs, thus negatively regulating protein synthesis. To add to the complexity of this story, activation of this pathway may specifically regulate PKR mRNA levels, while dsRNA may also induce an RNase L inhibitor. The genome of the herpes simplex virus type 1 encodes Us11, an inhibitor of 2′,5′-AS, a protein with a dsRNA-binding domain (Sanchez and Mohr, 2007).

Finally, and more speculatively, it should be noted that long dsRNAs (optimally >100 bp) are subject to the action of the RNA editing enzyme RNA-specific adenosine deaminase (ADAR) (Wang and Carmichael, 2004). This enzyme

promiscuously converts adenosine residues to inosine, with the probable consequence that the edited duplexes will extensively unwind, thus rendering them subject
to degradation. There is evidence that ADAR may function specifically to edit and
thus suppress certain miRNAs (Yang et al., 2006). Nevetheless, the extensive conversion of A to I raises the possibility that the specificity of edited dsRNAs that
escape destruction may be altered, leading to off-target actions. ADAR is normally
and ubiquitously expressed in the nucleus, but importantly, a cytoplasmic form is
upregulated in mammals as a result of interferon activation. This leads to the possibility that RNAi experiments in the context of immune challenged individuals
may be particularly prone to such mistargeting.

    The extent to which off-target responses of this kind to dsRNA may also occur
in insects is as yet relatively unexplored. There is certainly some evidence that
RNAi in insects may sometimes have unintended effects. One immune-related
example concerns the lepidopteran immunoglobulin-related protein Hemolin. It is
normally found at only very low levels, but is very highly expressed in a number of
Lepidoptera following immune challenge with bacteria (Faye et al., 1975). In an
attempt to knock down its expression, Hirai et al. (2004) injected a dsRNA specific to
the Hemolin gene into diapausing pupae of the saturniid silkmoth *Antheraea pernyi*,
using a GFP dsRNA as a control. Surprisingly they found that both dsRNAs induced
increased levels of *Hemolin* mRNA. This was accompanied by inhibition of baculovirus replication within the treated pupae. The explanation advanced by the authors
is that dsRNAs cause enhanced expression of Hemolin, regardless of sequence.

    The antiviral effects seen in *A. pernyi* were supposed by Hirai et al. (2004) to be
due to upregulation of Hemolin, but it seems likely that this sequence-independent
dsRNA response extends beyond the enhanced expression of Hemolin to include
upregulation of other proteins. These other proteins may also have contributed to, or
even been entirely responsible for the observed effects on baculovirus replication.
Prominent among the candidates for antiviral activity in induced hemolymph proteins
would be PPO and proteins associated with its activation (Shelby and Popham, 2006).
Hemolin itself may have effects on PPO activation (Terenius et al., 2007). A similar
antiviral response to dsRNA regardless of sequence has also been made in a non-
insect arthropod, the penaid shrimp *Penaeus monodon* (Robalino et al., 2007).

    Unanticipated effects of RNAi could also arise through indirect effects on the
interaction of one protein with another. For example, a reduction in the level of a
target protein that is normally present in the cell only as part of protein complex
might lead to changes in the sizes of the pools of the other proteins. This appears
to be what happened when Goto et al. (2001) used RNAi to silence the expression
of laminins in *Drosophila* KC cells. These proteins are major constituents of basement membranes. α, β, and γ laminins are post-translationally assembled into a
mature complex of all three proteins. Although dsRNA silencing of all three
laminins was in each case specific to the targeted mRNA, knock down of α laminin
resulted in a marked increase in the intracellular protein pools of β and γ laminins.
The authors speculated that this was the result of the requirement for the association of preformed β–γ dimers with α laminin before the final complex is exported.

Although no immune-related effects of this kind have been reported in insects, the possibility for such functional interactions at the protein level should be considered in designing any RNAi experiment.

12.17   ESSENTIAL CONTROLS

In view of the limitations and problems mentioned above, it is clear that neither the success nor the specificity of RNAi can be taken for granted. Thus experimentalists using RNAi must be aware of the need to do appropriate controls. First, it is vital to verify that the RNAi treatment administered to an insect has achieved the expected reduction in the level of the targeted gene's mRNA. This should be accompanied by measurement of the levels of other non-target mRNAs. By the same token it is necessary to check that the same RNAi knockdown effect is not produced by an irrelevant control dsRNA.

Unfortunately, it is never possible to be sure that some completely unexpected off-target effect has not occurred, and this limitation of the RNAi technique has to be accepted. One way of doing this would be to check on the expression levels of a large panel of other genes, say using a microarray, but this is very rarely done.

It is important to remember that knock down of the targeted gene's mRNA level is not the same thing as reducing the level of protein. This is particularly the case where the expression of a particular protein is controlled at the translational level. It is normally considered important in RNAi experiments to check the level of the protein in question by doing a Western blot, but this depends on the availability of antibodies.

The possibility that injection of dsRNA or even the injury associated with it could actually induce the expression of the protein that is being targeted, instead of suppressing it (as was the case for Hirai et al., 2004), means that it is desirable to check that the systemic RNAi treatment does not itself cause upregulation of the gene of interest at the mRNA and protein levels. Controls of all these types are incorporated in Figs 12.4 and 12.5.

12.18   CONCLUSION

RNAi occupies a position at the heart of insect immunity to viruses, and also offers an extraordinarily powerful experimental technique that is only just beginning to be exploited in the study of insect immune responses. It will be fascinating to see how future research reveals further roles for RNAi and makes use of these techniques.

REFERENCES

Adelman, Z. N., Blair, C. D., Carlson, J. O., Beaty, B. J., and Olson, K. E. (2001). Sindbis virus-induced silencing of dengue viruses in mosquitoes. *Insect Mol. Biol.* **10**, 265–273.

Ambros, V., and Chen, X. M. (2007). The regulation of genes and genomes by small RNAs. *Development* **134**, 1635–1641.

Araujo, R. N., Santos, A., Pinto, F. S., Gontijo, N. F., Lehane, M. J., and Pereira, M. H. (2006). RNA interference of the salivary gland nitrophorin 2 in the triatomine bug *Rhodnius prolixus* (Hemiptera: Reduviidae) by dsRNA ingestion or injection. *Insect Biochem. Mol. Biol.* **36**, 683–693.

Aronstein, K., Pankiw, T., and Saldivar, E. (2006). SID-I is implicated in systemic gene silencing in the honey bee. *J. Apicul. Res.* **45**, 20–24.

Attardo, G. M., Higgs, S., Klingler, K. A., Vanlandingham, D. L., and Raikhel, A. S. (2003). RNA interference-mediated knockdown of a GATA factor reveals a link to anautogeny in the mosquito *Aedes aegypti*. *Proc. Natl. Acad. Sci. USA* **100**, 13374–13379.

Bartel, D. P., and Chen, C. Z. (2004). Micromanagers of gene expression: The potentially widespread influence of metazoan microRNAs. *Nat. Rev. Genet.* **5**, 396–400.

Behm-Ansmant, I., Rehwinkel, J., Doerks, T., Stark, A., Bork, P., and Izaurralde, E. (2006). mRNA degradation by miRNAs and GW182 requires both CCR4:NOT deadenylase and DCP1:DCP2 decapping complexes. *Gene. Dev.* **20**, 1885–1898.

Bennasser, Y., Yeung, M. L., and Jeang, K.-T. (2006). HIV-1 TAR RNA subverts RNA interference in transfected cells through sequestration of TAR RNA-binding protein, TRBP. *J. Biol. Chem.* **281**, 27674–27678.

Bettencourt, R., Terenius, O., and Faye, I. (2002). Hemolin gene silencing by ds-RNA injected into cecropia pupae is lethal to next generation embryos. *Insect Mol. Biol.* **11**, 267–271.

Bian, G., Shin, S. W., Cheon, H. M., Kokoza, V., and Raikhel, A. S. (2005). Transgenic alteration of Toll immune pathway in the female mosquito *Aedes aegypti*. *Proc. Natl. Acad. Sci. USA* **102**, 13568–13573.

Boisson, B., Jacques, J. C., Choumet, V., Martin, E., Xu, J. N., Vernick, K., and Bourgouin, C. (2006). Gene silencing in mosquito salivary glands by RNAi. *FEBS Lett.* **580**, 1988–1992.

Bucher, G., Scholten, J., and Klingler, M. (2002). Parental RNAi in *Tribolium* (Coleoptera). *Curr. Biol.* **12**, R85–R86.

Calin, G. A., Liu, C. G., Sevignani, C., Ferracin, M., Felli, N., Dumitru, C. D., Shimizu, M., Cimmino, A., Zupo, S., Dono, M., Dell'Aquila, M. L., Alder, H, Rassenti, L., Kipps, T. J., Bullrich, F., Negrini, M., and Croce, C. M. (2004). MicroRNA profiling reveals distinct signatures in B cell chronic lymphocytic leukemias. *Proc. Natl. Acad. Sci. USA* **101**, 11755–11760.

Cerenius, L., and Soderhall, K. (2004). The prophenoloxidase-activating system in invertebrates. *Immunol. Rev.* **198**, 116–126.

Chapin, A., Correa, P., Maguire, M., and Kohn, R. (2007). Synaptic neurotransmission protein UNC-13 affects RNA interference in neurons. *Biochem. Biophys. Res. Commun.* **354**, 1040–1044.

Cherry, S., Doukas, T., Armknecht, S., Whelan, S., Wang, H., Sarnow, P., and Perrimon, N. (2005). Genome-wide RNAi screen reveals a specific sensitivity of IRES-containing RNA viruses to host translation inhibition. *Gene. Dev.* **19**, 445–452.

Chowdhury, D., and Novina, C. D. (2005). RNAi and RNA-based regulation of immune system function. *Adv. Immunol.* **88**, 267–292.

Davidson, B. L., and Harper, S. Q. (2005). Viral delivery of recombinant short hairpin RNAs. *Meth. Enzymol.* **392**, 145–173.

de Fougerolles, A., Vornlocher, H. P., Maraganore, J., and Lieberman, J. (2007). Interfering with disease: A progress report on siRNA-based therapeutics. *Nat. Rev. Drug Discov.* **6**, 443–453.

Dietzl, G., Chen, D., Schnorrer, F., Su, K. C., Barinova, Y., Fellner, M., Gasser, B., Kinsey, K., Oppel, S., Scheiblauer, S., Couto, A., Marra, V., Keleman, K., and Dickson, B. J. (2007). A genome-wide transgenic RNAi library for conditional gene inactivation in *Drosophila*. *Nature* **448**, 151–157.

Dillon, C. P., Sandy, P., Nencioni, A., Kissler, S., Rubinson, D. A., and Van Parijs, L. (2005). RNAI as an experimental and therapeutic tool to study and regulate physiological and disease processes. *Annu. Rev. Physiol.* **67**, 147–173.

Doench, J. G., Petersen, C. P., and Sharp, P. A. (2003). siRNAs can function as miRNAs. *Gene. Dev.* **17**, 438–442.

Dong, Y., and Friedrich, M. (2005). Nymphal RNAi: Systemic RNAi mediated gene-knockdown in juvenile grasshopper. *BMC Biotechnol.* **5**, article 25.

Dykxhoorn, D. M. (2007). MicroRNAs in viral replication and pathogenesis. *DNA Cell Biol.* **26**, 239–249.

Dykxhoorn, D. M., and Lieberman, J. (2005). The silent revolution: RNA interference as basic biology, research tool, and therapeutic. *Annu. Rev. Med.* **56**, 401–423.

Dykxhoorn, D. M., Novina, C. D., and Sharp, P. A. (2003). Killing the messenger: Short RNAs that silence gene expression. *Nat. Rev. Mol. Cell Biol.* **4**, 457–467.

Eleftherianos, I., Millichap, P. J., ffrench-Constant, R. H., and Reynolds, S. E. (2006a). RNAi suppression of recognition protein mediated immune responses in the tobacco hornworm *Manduca sexta* causes increased susceptibility to the insect pathogen *Photorhabdus. Dev. Compart. Immunol.* **30**, 1099–1107.

Eleftherianos, I., Marokhazi, J., Millichap, P. J., Hodgkinson, A. J., Sriboonlert, A., ffrench-Constant, R. H., and Reynolds, S. E. (2006b). Prior infection of *Manduca sexta* with non-pathogenic *Escherichia coli* elicits immunity to pathogenic *Photorhabdus luminescens*: Roles of immune-related proteins shown by RNA interference. *Insect Biochem. Mol. Biol.* **36**, 517–525.

Eleftherianos, I., Gokcen, F., Felfoldi, G., Millichap, P. J., Trenczek, T. E., ffrench-Constant, R. H., and Reynolds, S. E. (2007a). The immunoglobulin family protein Hemolin mediates cellular immune responses to bacteria in the insect *Manduca sexta. Cell. Microbiol.* **9**, 1137–1147.

Eleftherianos, I., Boundy, S., Joyce, S. A., Aslam, S., Marshall, J. W., Cox, R. J. Simpson, T. J. Clarke, D. J. ffrench-Constant, R. H., and Reynolds, S. E. (2007b). An antibiotic produced by an insect-pathogenic bacterium suppresses host defenses through phenoloxidase inhibition. *Proc. Natl. Acad. Sci. USA* **104**, 2419–2424.

Eulalio, A., Behm-Ansmant, I., Schweizer, D., and Izaurralde, E. (2007). P-body formation is a consequence, not the cause, of RNA-mediated gene silencing. *Mol. Cell. Biol.* **27**, 3970–3981.

Faye, I., Pye, A., Rasmuson, T., Boman, H. G., and Boman, I. A. (1975). Insect immunity II. Simultaneous induction of antibacterial antibacterial activity and selective synthesis of some hemolymph proteins in diapausing pupae of *Hyalophora cecropia* and *Samia cynthia. Infect. Immun.* **12**, 1426–1438.

Feinberg, E. H., and Hunter, C. P. (2003). Transport of dsRNA into cells by the transmembrane protein SID-1. *Science* **301**, 1545–1547.

Filipowicz, W. (2005). RNAi: The nuts and bolts of the RISC machine. *Cell* **122**, 17–20.

Fire, A., Xu, S. Q., Montgomery, M. K., Kostas, S. A., Driver, S. E., and Mello, C. C. (1998). Potent and specific genetic interference by double-stranded RNA in *Caenorhabditis elegans. Nature* **391**, 806–811.

Galiana-Arnoux, D., Dostert, C., Schneemann, A., Hoffmann, J. A., and Imler, J. L. (2006). Essential function *in vivo* for Dicer-2 in host defense against RNA viruses in *Drosophila. Nat. Immunol.* **7**, 590–597.

Garcia, M. A., Gil, J., Ventoso, I., Guerra, S., Domingo, E., Rivas, C., and Esteban, M. (2006). Impact of protein kinase PKR in cell biology: From antiviral to antiproliferative action. *Microbiol. Mol. Biol. Rev.* **70**, 1032–1060.

Gitlin, L., and Andino, R. (2003). Nucleic acid-based immune system: The antiviral potential of mammalian RNA silencing. *J. Virol.* **77**, 7159–7165.

Goto, A., Aoki, M., Ichihara, S., and Kitagawa, Y. (2001). alpha- beta- or gamma-chain-specific RNA interference of laminin assembly in *Drosophila* Kc167 cells. *Biochem. J.* **360**, 167–172.

Goto, A., Kadowaki, T., and Kitagawa, Y. (2003). *Drosophila* hemolectin gene is expressed in embryonic and larval hemocytes and its knock down causes bleeding defects. *Dev. Biol.* **264**, 582–591.

Grewal, S. I. S., and Elgin, S. C. R. (2007). Transcription and RNA interference in the formation of heterochromatin. *Nature* **447**, 399–406.

Grishok, A. (2005). RNAi mechanisms in *Caenorhabditis elegans. FEBS Lett.* **579**, 5932–5939.

Hamilton, A. J., and Baulcombe, D. C. (1999). A species of small antisense RNA in posttranscriptional gene silencing in plants. *Science* **286**, 950–952.

Hirai, M., Terenius, O., Li, W., and Faye, I. (2004). Baculovirus and dsRNA induce Hemolin, but no antibacterial activity, in *Antheraea pernyi*. *Insect Mol. Biol.* **13**, 399–405.

Huang, Y., Deng, F., Hu, Z. H., Vlak, J. M., and Wang, H. Z. (2007). Baculovirus-mediated gene silencing in insect cells using intracellularly produced long double-stranded RNA. *J. Biotechnol.* **128**, 226–236.

Ikeda, M., Yanagimoto, K., and Kobayashi, M. (2004). Identification and functional analysis of *Hyphantria cunea* nucleopolyhedrovirus *iap* genes. *Virology* **321**, 359–371.

Isobe, R., Kojima, K., Matsuyama, T., Quan, G. X., Kanda, T., Tamura, T., Sahara, K., Asano, S. I., and Bando, H. (2004). Use of RNAi technology to confer enhanced resistance to MmNPV on transgenic silkworms. *Arch. Virol.* **149**, 1931–1940.

Jackson, A. L., Burchard, J., Schelter, J., Chau, B. N., Cleary, M., Lim, L., and Linsley, P. S. (2006a). Widespread siRNA 'off-target' transcript silencing mediated by seed region sequence complementarity. *RNA* **12**, 1179–1187.

Jackson, A. L., Burchard, J., Leake, D, Reynolds, A., Schelter, J., Guo, J., Johnson, J. M., Lim, L., Karpilow, J., Nichols, K., Marshall, W., Khvorova, A., and Linsley, P. S. (2006b). Position-specific chemical modification of siRNAs reduces 'off-target' transcript silencing. *RNA* **12**, 1197–1205.

Johnson, B. W., Olson, K. E., Allen-Miura, T., Rayms-Keller, A., Carlson, J. O., Coates, C. J., Jasinskiene, N., James, A. A., Beaty, B. J., and Higgs, S. (1999). Inhibition of luciferase expression in transgenic *Aedes aegypti* mosquitoes by Sindbis virus expression of antisense luciferase RNA. *Proc. Natl. Acad. Sci. USA* **96**, 13399–13403.

Kambris, Z., Brun, S., Jang, I.-H., Nam, H.-J., Takahashi, K., Lee, W.-J., Ueda, R., and Lemaitre, B. (2006). *Drosophila* immunity: A large scale *in vivo* RNAi screen identifies five serine proteases required for Toll activation. *Curr. Biol.* **16**, 808–813.

Kanost, M. R., Jiang, H. B., and Yu, X. Q. (2004). Innate immune responses of a lepidopteran insect, *Manduca sexta*. *Immunol. Rev.* **198**, 97–105.

Keene, K. M., Foy, B. D., Sanchez-Vargas, I., Beaty, B. J., Blair, C. D., and Olson, K. E. (2004). RNA interference acts as a natural antiviral response to O'nyong-nyong virus (Alphavirus; Togaviridae) infection of *Anopheles gambiae*. *Proc. Natl. Acad. Sci. USA* **101**, 17240–17245.

Kennedy, S., Wang, D., and Ruvkun, G. (2004). A conserved siRNA-degrading RNase negatively regulates RNA interference in *C. elegans*. *Nature* **427**, 645–649.

Konet, D. S., Anderson, J., Piper, J., Akkina, R., Suchman, E., and Carlson, J. (2007). Short-hairpin RNA expressed from polymerase III promoters mediates RNA interference in mosquito cells. *Insect Mol. Biol.* **16**, 199–206.

Kurucz, E., Zettervall, C. J., Sinka, R., Vilmos, P., Pivarcsi, A., Ekengren, S., Hegedus, Z., Ando, I., and Hultmark, D. (2003). Hemese, a hemocyte-specific transmembrane protein, affects the cellular immune response in *Drosophila*. *Proc. Natl. Acad. Sci. USA* **100**, 2622–2627.

Kurucz, E., Markus, R., Zsamboki, J., Folkl-Medzihradszky, K., Darula, Z., Vilmos, P., Udvardy, A., Krausz, I., Lukacsovich, T., Gateff, E., Zettervall, C. J., Hultmark, D., and Ando, I. (2007). Nimrod, a putative phagocytosis receptor with EGF repeats in *Drosophila* plasmatocytes. *Curr. Biol.* **17**, 649–654.

Lecellier, C. H., and Voinnet, O. (2004). RNA silencing: no mercy for viruses? *Immunol. Rev.* **198**, 285–303.

Lecellier, C. H., Dunoyer, P., Arar, K., Lehmann-Che, J., Eyquem, S., Himber, C., Saib, A., and Voinnet, O. (2005). A cellular microRNA mediates antiviral defense in human cells. *Science* **308**, 557–560.

Lee, Y., Ahn, C., Han, J. J., Choi, H., Kim, J., Yim, J., Lee, J., Provost, P., Radmark, O., Kim, S., and Kim, V. N. (2003). The nuclear RNase III Drosha initiates microRNA processing. *Nature* **425**, 415–419.

Lee, Y., Kim, M., Han, J. J., Yeom, K. H., Lee, S., Baek, S. H., and Kim, V. N. (2004). MicroRNA genes are transcribed by RNA polymerase II. *EMBO J.* **23**, 4051–4060.

Lemaitre, B., and Hoffmann, J. (2007). The host defense of *Drosophila melanogaster*. *Annu. Rev. Immunol.* **25**, 697–743.

Levin, D. M., Breuer, L. N., Zhuang, S. F., Anderson, S. A., Nardi, J. B., and Kanost, M. R. (2005). A hemocyte-specific integrin required for hemocytic encapsulation in the tobacco hornworm, *Manduca sexta*. *Insect Biochem. Mol. Biol.* **35**, 369–380.

Lewis, D. L., DeCamillis, M. A., Brunetti, C. R., Halder, G., Kassner, V. A., Selegue, J. E., Higgs. S., and Carroll, S. B. (1999). Ectopic gene expression and homeotic transformations in arthropods using recombinant Sindbis viruses. *Curr. Biol.* **9**, 1279–1287.

Li, H. W., and Ding, S. W. (2005). Antiviral silencing in animals. *FEBS Lett.* **579** (Special Issue), 5965–5973.

Li, H., Li, W. X., and Ding, S. W. (2002). Induction and suppression of RNA silencing by an animal virus. *Science* **296**, 1319–1321.

Lim, L. P., Lau, N. C., Garrett-Engele, P., Grimson, A., Schelter, J. M., Castle, J., Bartel, D. P., Linsley, P. S., and Johnson, J. M. (2005). Microarray analysis shows that some microRNAs down-regulate large numbers of target mRNAs. *Nature* **433**, 769–773.

Lippman, Z., and Martienssen, R. (2004). The role of RNA interference in heterochromatic silencing. *Nature* **431**, 364–370.

Mao, C. P., Lin, Y. Y., Hung, C. F., and Wu, T. C. (2007). Immunological research using RNA interference technology. *Immunology* **121**, 295–307.

Matzke, M. A., Primig, M., Trnovsky, J., and Matzke, A. J. M. (1989). Reversible methylation and inactivation of marker genes in sequentially transformed tobacco plants. *EMBO J.* **8**, 643–649.

Meister, G., and Tuschl, T. (2004). Mechanisms of gene silencing by double-stranded RNA. *Nature* **431**, 343–349.

Montgomery, M. K., Xu, S. Q., and Fire, A. (1998). RNA as a target of double-stranded RNA-mediated genetic interference in *Caenorhabditis elegans*. *Proc. Natl. Acad. Sci. USA* **95**, 15502–15507.

Nilsen, T. W. (2007). Mechanisms of microRNA-mediated gene regulation in animal cells. *Trends Genet.* **23**, 243–249.

Nishikawa, T., and Natori, S. (2001). Targeted disruption of a pupal hemocyte protein of *Sarcophaga* by RNA interference. *Eur. J. Biochem.* **268**, 5295–5299.

Obbard, D. J., Jiggins, F. M., Halligan, D. L., and Little, T. J. (2006). Natural selection drives extremely rapid evolution in antiviral RNAi genes. *Curr. Biol.* **16**, 580–585.

O'Connell, R. M., Taganov, K. D., Boldin, M. P., Boldin, M. P., Cheng, G., and Baltimore, D. (2007). MicroRNA-155 is induced during the macrophage inflammatory response. *Proc. Natl. Acad. Sci. USA* **104**, 1604–1609.

O'Donnell, K. A., and Boekel, J. D. (2007). Mighty piwis defend the germline against genome intruders. *Cell* **129**, 37–44.

Osta, M. A., Christophides, G. K., and Kafatos, F. C. (2004). Effects of mosquito genes on *Plasmodium* development. *Science* **303**, 2030–2032.

Pak, J., and Fire, A. (2007). Distinct populations of primary and secondary effectors during RNAi in *C. elegans*. *Science* **315**, 241–244.

Perrimon, N., and Mathey-Prevot, B. (2007). Applications of high-throughput RNA interference screens to problems in cell and developmental biology. *Genetics* **175**, 7–16.

Peters, L., and Meister, G. (2007). Argonaute proteins: Mediators of RNA silencing. *Mol. Cell* **26**, 611–623.

Pierro, D. J., Myles, K. M., Foy, B. D., Beaty, B. J., and Olson, K. E. (2003). Development of an orally infectious Sindbis virus transducing system that efficiently disseminates and expresses green fluorescent protein in *Aedes aegypti*. *Insect Mol. Biol.* **12**, 107–116.

Quan, G. X., Kanda, T., and Tamura, T. (2002). Induction of the *white egg 3* mutant phenotype by injection of the double-stranded RNA of the silkworm *white* gene. *Insect Mol. Biol.* **11**, 217–222.

Rajagopal, R., Sivakumar, S., Agrawal, N., Malhotra, P., and Bhatnagar, R. K. (2002). Silencing of midgut aminopeptidase N of *Spodoptera litura* by double-stranded RNA establishes its role as *Bacillus thuringiensis* receptor. *J. Biol. Chem.* **277**, 46849–46851.

Rana, T. M. (2007). Illuminating the silence: Understanding the structure and function of small RNAs. *Nat. Rev. Mol. Cell Biol.* **8**, 23–36.

Robalino, J., Bartlett, T., Shepard, E., Prior, S., Jaramillo, G., Scura, E., Chapman, R. W., Gross, P. S., Browdy, C. L., and Warr, G. W. (2005). Double-stranded RNA induces sequence-specific antiviral silencing in addition to nonspecific immunity in a marine shrimp: Convergence of RNA interference and innate immunity in the invertebrate antiviral response? *J. Virol.* **79**, 13561–13571.

Roignant, J. Y., Carre, C., Mugai, B., Szymczak, D., Lepesant, J. A., and Antoniewski, C. (2003). Absence of transitive and systemic pathways allows cell-specific and isoform-specific RNAi in *Drosophila. RNA* **9**, 299–308.

Romano, N., and Macino, G. (1992). Quelling – Transient inactivation of gene-expression in *Neurospora crassa* by transformation with homologous sequences. *Mol. Microbiol.* **6**, 3343–3353.

Saleh, M. C., van Rij, R. P., Hekele, A., Gillis, A., Foley, E., O'Farrell, P. H., and Andino, R. (2006). The endocytic pathway mediates cell entry of dsRNA to induce RNAi silencing. *Nat. Cell Biol.* **8**, 793–819.

Sanchez, R., and Mohr, I. (2007). Inhibition of cellular 2′ -5′ oligoadenylate synthetase by the herpes simplex virus type 1 Us11 protein. *J. Virol.* **81**, 3455–3464.

Sanders, H. R., Foy, B. D., Evans, A. M., Ross, L. S., Beaty, B. J., Olson, K. E., and Gill, S. S. (2005). Sindbis virus induces transport processes and alters expression of innate immunity pathway genes in the midgut of the disease vector, *Aedes aegypti. Insect Biochem. Mol. Biol.* **35**, 1293–1307.

Saxena, S., Jonsson, Z. O., and Dutta, A. (2003). Small RNAs with imperfect match to endogenous mRNA repress translation – Implications for off-target activity of small inhibitory RNA in mammalian cells. *J. Biol. Chem.* **278**, 44312–44319.

Schroder, M., and Bowie, A. G. (2005). TLR3 in antiviral immunity: Key player or bystander? *Trends Immunol.* **26**, 462–468.

Sen, G. L., and Blau, H. M. (2006). A brief history of RNAi: The silence of the genes. *FASEB J.* **20**, 1293–1299.

Shelby, K. S., and Popham, H. J. R. (2006). Plasma phenoloxidase of the larval tobacco budworm, *Heliothis virescens*, is virucidal. *J. Insect Sci.* **6**, article 13.

Sijen, T., Fleenor, J., Simmer, F., Thijssen, K. L., Timmons, L., Plasterk, R. H. A., and Fire, A. (2001). On the role of RNA amplification in dsRNA-triggered gene silencing. *Cell* **107**, 297–307.

Sijen, T., Steiner, F. A., Thijssen, K. L., and Plasterk, R. H. A. (2007). Secondary si RNAs result from unprimed RNA synthesis and form a distinct class. *Science* **315**, 244–247.

Soares, C. A. G., Lima, C. M. R., Dolan, M. C., et al. (2005). Capillary feeding of specific dsRNA induces silencing of the *isac* gene in nymphal *Ixodes scapularis* ticks. *Insect Mol. Biol.* **14**, 443–452.

Stern-Ginossar, N., Elefant, N., Zimmermann, A., Wolf, D. G., Saleh, N., Biton, M., Horwitz, E., Prokocimer, Z., Prichard, M., Hahn, G., Goldman-Wohl, D., Greenfield, C., Yagel, S., Hengel, H., Altuvia, Y., Margalit, H., and Mandelboim, O. (2007). Host immune system gene targeting by a viral miRNA. *Science* **317**, 376–381.

Sundaram, P., Echalier, B., Han, W., Hull, D., and Timmons, L. (2006). ATP-binding cassette transporters are required for efficient RNA interference in *Caenorhabditis elegans. Mol. Biol. Cell* **17**, 3678–3688.

Tanaka, Y., Yamaguchi, S., Fujii-Taira, I., Iijima, R., Natori, S., and Homma, K. J. (2006). Involvement of insect-derived growth factor (IDGF) in the cell growth of an embryonic cell line of flesh fly. *Biochem. Biophys. Res. Commun.* **350**, 334–338.

Terenius, O., Bettencourt, R., Lee, S. Y., Li, W. L., Soderhall, K., and Faye, I. (2007). RNA interference of Hemolin causes depletion of phenoloxidase activity in *Hyalophora cecropia. Dev. Compart. Immunol.* **31**, 571–575.

Theopold, U., Schmidt, O., Soderhall, K., and Dushay, M. S. (2004). Coagulation in arthropods: Defence, wound closure and healing. *Trends Immunol.* **25**, 289–294.

Thermann, R., and Hentze, M. W. (2007). *Drosophila* miR2 induces pseudo-polysomes and inhibits translation initiation. *Nature* **447**, 875–878.

Timmons, L., and Fire, A. (1998). Specific interference by ingested dsRNA. *Nature* **395**, 854.

Tijsterman, M., May, R. C., Simmer, F., Okihara, K. L., and Plasterk, R. H. A. (2004). Genes required for systemic RNA interference in *Caenorhabditis elegans. Curr. Biol.* **14**, 111–116.

Turner, C. T., Davy, M. W., MacDiarmid, R. M., Plummer, K. M., Birch, N. P., and Newcomb, R. D. (2006). RNA interference in the light brown apple moth, *Epiphyas postvittana* (Walker) induced by double-stranded RNA feeding. *Insect Mol. Biol.* **15**, 383–391.

Tuschl, T., Zamore, P. D., Lehmann, R., Bartel, D. P., and Sharp, P. A. (1999). Targeted mRNA degradation by double-stranded RNA *in vitro*. *Gene. Dev.* **13**, 3191–3197.

Uhlirova, M., Foy, B. D., Beaty, B. J., Olson, K. E., Riddiford, L. M., and Jindra, M. (2003). Use of Sindbis virus-mediated RNA interference to demonstrate a conserved role of Broad-Complex in insect metamorphosis. *Proc. Natl. Acad. Sci. USA* **100**, 15607–15612.

Vagin, V. V., Sigova, A., Li, C. J., Seitz, H., Gvozdev, V., and Zamore, P. D. (2006). A distinct small RNA pathway silences selfish genetic elements in the germline. *Science* **313**, 320–324.

van Rij, R. P., Saleh, M. C., Berry, B., Foo, C., Houk, A., Antoniewski, C., and Andino, R. (2006). The RNA silencing endonuclease Argonaute 2 mediates specific antiviral immunity in *Drosophila melanogaster. Gene. Dev.* **20**, 2985–2995.

Veldhoen, S., Laufer, S. D., Trampe, A., and Restle, T. (2006). Cellular delivery of small interfering RNA by a non-covalently attached cell-penetrating peptide: Quantitative analysis of uptake and biological effect. *Nucleic Acids Res.* **34**, 6561–6573.

Quan, G. X., Kanda, T., and Tamura, T. (2002). Induction of the *white egg 3* mutant phenotype by injection of the double-stranded RNA of the silkworm *white* gene. *Insect. Mol. Biol.* **11**, 217–222.

Wang, Q., and Carmichael, G. G. (2004). Effects of length and location on the cellular response to double-stranded RNA. *Microbiol. Mol. Biol. Rev.* **68**, 432–452.

Wang, X. H., Aliyari, R., Li, W. X., Li, H. W., Kim, K., et al., (2006). RNA interference directs innate immunity against viruses in adult *Drosophila. Science* **312**, 452–454.

Winston, W. M., Molodowitch, C., and Hunter, C. P. (2002). Systemic RNAi in C-elegans requires the putative transmembrane protein SID-1. *Science* **295**, 2456–2459.

Yang, W. D., Chendrimada, T. P., Wang, Q. D., Higuchi, M., Seeburg, P. H., Shiekhattar, R., and Nishikura, K. (2006). Modulation of microRNA processing and expression through RNA editing by ADAR deaminases. *Nat. Struct. Mol. Biol.* **13**, 13–21.

Zambon, R. A., Vakharia, V. N., and Wu, L. P. (2006). RNAi is an antiviral immune response against a dsRNA virus in *Drosophila melanogaster. Cell. Microbiol.* **8**, 880–889.

13

EPILOGUE: PATHWAYS INTO THE FUTURE OF INSECT IMMUNOLOGY

NANCY E. BECKAGE

*Departments of Entomology and Cell Biology and Neuroscience and
Center for Disease Vector Research, University of California-Riverside,
Riverside, CA 92521, USA*

Insect immunology is a critical subject relevant to insect pathology, parasitology, insect pest control, and human health that interfaces with many areas of insect biology and ecology. Potentially fruitful areas of future research in insect immunology have been identified throughout the preceding chapters of this book. Identification of signaling pathways critical to non-self recognition of parasites and pathogens represents an especially promising field that challenges future investigators, as these pathways are critical to determining whether the biotic agent is virulent or avirulent to the host. The dynamic interplay between virulence genes of the pathogen and gene(s) encoding factors responsible for susceptibility or resistance of the host provides evidence for the evolution of complex strategies of molecular host–pathogen and host–parasite interaction in many insect systems. While such relationships are often described as representative of a co-evolutionary arms race taking place between the host and its invader, the evidence accumulated thus far suggests that pathogens and parasites can manipulate their relationship with the host to establish a balanced equilibrium between pathogen virulence characteristics and host resistance mechanisms which culminates in successful infection of susceptible hosts.

Recent studies of the social Amoeba *Dictyostelium discoideum* have revealed that these evolutionarily primitive organisms have cells capable of non-self recognition and defense. The presence of functional immune 'sentinel cells' in these species, representing some of the first multicellular organisms of ancient origin, argues that immunity is a basic property of organismal biology. Hence, many features of insect immunity likely had their origins within simple multicellular organisms. Continued focus on comparative immunological analyses are very likely to reveal how these components, and the regulatory pathways they mediate, co-evolved on a temporal scale in both invertebrate and vertebrate phyla.

The evolutionary relationships and physiological interface between insect and mammalian immunity are especially complex in insect vectors that feed on vertebrate blood. Moreover, the bloodfeeding behavior of insect vectors of human and animal diseases offers a wealth of potential opportunities for controlling rates of disease transmission among vertebrate hosts. To achieve this goal, a better integration of the study of mammalian and arthropod immunity is not only logical, given the contribution of insects to our understanding of mammalian innate immunity, but also an absolute necessity, given the immunological interface of bloodfeeding. The physiological juxtaposition of vertebrate and insect immune systems in such relationships provides strong evidence for the evolution of sophisticated strategies of physiological interaction between two evolutionarily distinct taxonomic animal groups. To reduce rates of disease transmission, there are two opportunities to manipulate this bloodfeeding interface: at the point of feeding via injection of vector saliva into the vertebrate host, and within the gut of the arthropod vector to block the movement of the parasite/pathogen into the hemocoel and its ultimate transfer to the salivary glands. The conserved signaling pathways and effectors in vector species provide tremendous insight into the evolution of immunity, a fact that receives little attention in most immunology courses. However, this conservation also facilitates a surprisingly complex immunological communication between these biologically disparate systems. Dissection of this biology to identify the 'master switches' will probably require a computational or systems biology approach. From an applied perspective, manipulation of these interfaces provides a basis for strategies to disrupt pathogen transmission at multiple levels. Thus, to study the immunology of arthropods and their mammalian hosts in isolation overlooks the basic fact that they are intricately connected through bloodfeeding.

Additionally, a survey of vertebrate host immunological responses to insect vector feeding activity provides ample opportunities to curb rates of disease transmission among susceptible individuals via enhancement of immunological reactivity in the vertebrate host. As described above, manipulation of the vector's immunological responses to parasitic or pathogenic infection can result in enhancement in levels of vector refractoriness to the disease agent, thereby reducing rates of disease transmission. Identification of genetic components contributing to vector refractoriness will have a similar beneficial effect on reducing vertebrate infection rates as these critical genes provide avenues for manipulation of vector competence.

How pathways of insect immunity are regulated by cross-talk from other regulatory systems in the host can form the foci for new areas of research that will also potentially impact agriculture and disease transmission. For example, interactions between the immune and nervous systems of insects are only now just beginning to be addressed as a new subfield of insect immunology. At the moment, we lack a clear understanding of the details of immune-neural connections at all levels in insect systems. While psychoneuroimmunology is already an established field in vertebrate (particularly mammalian) immunology, such links have proved difficult to decipher in insects. The chemical identity of the cytokines that are the major players in altering neuronal function in insects, and assessment of whether they cross the insect blood–brain barrier, remain open questions. Isolation of the neuronal receptors for these cytokines, as well as characterization of the subset(s) of neurons responding to specific molecules, in the insect host also remain subjects for future study.

In the other direction (neural to immune), we lack any direct evidence that the nervous system has a direct effect on insect immune function (e.g. via neural connections to immunologically responsive tissues such as fat body). Integration of information from both insect neurobiology and immunology will be critical to understand the biological significance of immune function changes induced by neural substances such as octopamine. Thus, our base of knowledge about insect neuroimmunological interactions is still in its infancy and a variety of new experimental approaches will be required to tease these complex but crucial relationships apart.

Similarly, the impact of putative hormonal factors on host immunocompetence has been a subject of much scientific speculation but minimal experimentation thus far in insect biology, although such influences are well-established in vertebrates (e.g. the critical roles of sex hormones in regulating vertebrate host immunity). In holometabolous insects, older instars of insect larvae often display enhanced resistance to viral and bacterial pathogens, as well as parasitoids, compared to younger developmental stages, suggestive of a possible hormone-mediated regulation of immunocompetence. Both juvenile hormone and ecdysteroids are candidate modulators of immunity, as hemolymph levels of these hormones fluctuate in amount in correlation with the insect's stage of development and the temporal progression of the molt cycle. Whether peptide hormones also play regulatory roles in insect humoral and/or cellular immunity is another question, which merits experimentation by future physiologists.

Studies of relationships between insect host behavior and immunity have given rise to the newly emerging field of evolutionary immunology, which frames insect immunity in an ecological context. This new area poses novel challenges for future investigators that have not been presented in this book, which focuses on physiological and molecular aspects of immunity. The fitness costs incurred by mobilization of immune responses to pathogens and parasites have seminal relevance to ongoing efforts to develop strategies of immunologically based manipulation of vector refractoriness to reduce rates of vertebrate disease transmission. While the expression of refractoriness-related gene products in transgenic insects has yet to be implemented successfully in field populations of vectors to control disease agents, we are possibly not far in time from ultimately achieving this goal. The ecological impacts of different aspects of insect immunity alone could form the focus of a separate book on evolutionary immunology to draw new investigators with both basic and applied perspectives to weigh the balances between the fitness trade-off costs and benefits of mobilizing effective immune responses to fight infection which impact fitness traits, such as longevity and reproductive output.

While the initial conception of a gene-for-gene co-evolutionary arms race operating between a host and its invaders had its origins in the field of plant pathology, the dynamics of the juxtaposition of virulence characteristics of pathogens *vis a vis* the resistance traits of the host are similar on molecular as well as organismal levels in both insects and plants. Many features of the virulence traits shown by insect pathogens and parasites have parallels in plant pathology, even though the molecular mediators of virulence and resistance are quite different in these two diverse groups of organisms.

The genetic contributors to determining whether a given insect species or strain is susceptible to a particular parasite or pathogen can frequently be identified by studying genes and gene products expressed in semi- or non-permissive host species that successfully resist infection. While this has been done superbly in *Drosophila* due to the accumulated wealth of genetic and genomic studies focusing on this species for decades, the time is now ripe for characterization of these elements in other agriculturally important host species due to the rapid expansion of genomic and proteomic information for a taxonomically diverse range of insect species. Additionally, while the malaria parasite, *Plasmodium falciparum*, was the first for which the complete genomic sequences of the parasite, vector, and human host were first made available to researchers, the expected completion of full genome sequences for other vector species will facilitate future identification of genes contributing to susceptibility and resistance traits of vectors that transmit parasites and pathogens that threaten human health.

While immunological memory and specificity are usually considered to be two defining characteristics of mammalian immunity, recent studies have pinpointed similar processes operating in insects as described in this book. The phrase 'immunological memory' is often interpreted as synonymous with 'antibody production' in the biological literature, but we now know that insects have evolved different strategies to cope with repeated infections. Although insects lack antibody-mediated responses *per se* they have evolved the capacity to exhibit 'immune priming' and increased tolerance following repeated exposure to an infectious agent, analogous to effects induced by 'vaccination' of the host. Future emphasis on deciphering how this process operates in insects on a mechanistic level will yield more comprehensive understanding of the evolutionary development of host defense pathways throughout the animal kingdom. The use of microarrays and even newer technologies will facilitate immunological specificity studies, to generate surveys of anti-parasite, anti-parasitoid, and anti-pathogenic immune responses of insects and make comparisons amongst the different classes of defenses mobilized against a particular type of infection. While the molecular aspects of memory and specificity of immunity are not yet clearly defined in insects, in contrast to mammals, the similarities reported thus far yield additional insights into the analogous strategies of molecular interaction evolved in insects and vertebrates.

While the utility and relevance of using insect models to study vertebrate immune mechanisms has not yet been fully appreciated by immunologists focusing on disease etiology in humans, many parallels certainly exist. For example, many developmental signaling molecules and pathways play dual immunological roles in insects, and homologous molecules and regulatory processes occur in many vertebrates including humans. The use of insect systems as mammalian disease models offers a plethora of opportunities for unraveling complex modes of physiological and molecular interaction among parasite, pathogen, and host.

Future biochemical studies of insect immunity will also uncover new classes of molecules not previously thought to play immunological roles. While insect lipids and lipid-carrying molecules were previously hypothesized to play a primarily metabolic role, their tandem role in immunity as demonstrated by recent experimental work is illustrated in this book. Our knowledge and identification of potentially important insect cytokines is a field still in its infancy compared to the status of this subfield of vertebrate immunology, and future research emphasis on identification of insect cytokines is likely to prove especially fruitful in isolation of biologically active molecules that can influence host susceptibility versus resistance to parasitism and disease.

Insect parasitoids are unusual parasites in that their presence within the host invariably causes premature host death, thus preventing host reproduction and proliferation of potentially highly susceptible hosts. A broader understanding of the molecular interactions responsible for successful parasitism of a particular host species or strain, as well as the physiological components contributing to definition of a parasitoid's host range, will dramatically increase the potential for successful implementation of biological control programs based upon the exploitation of parasitoids to control expanding populations of pest species. Additional insights into molecular strategies involved in evolution of host resistance to a given insect parasitoid species will also enhance such programs by pinpointing potentially critical host–parasitoid immunological elements and processes that facilitate successful parasitism of pest species. Continued advances in utilization of genomic and proteomic approaches to study the roles of polydnaviruses produced by hymenopteran parasitoids in manipulating lepidopteran host immunity to infection is but one example of how future research directions can impact biotechnological developments in insect pest control. These goals can be attained via expression of immunologically active polydnavirus gene products targeted to the host with implications for their potential exploitation in insect pest associations with pathogens, parasites, and plants.

Deciphering which genes and gene expression products confer disease susceptibility versus resistance also offers exciting opportunities for development of novel biopesticides and a range of biologically based strategies to reduce populations of pest insects. Disease resistance is especially critical to the health of beneficial insects including many parasitoids, predators, and pollinators that play crucial beneficial roles in agriculture. Development of transgenic plants and generation of recombinant biopesticides expressing genes encoding insect-active virulence factors to target growth and development of insect pests is in progress in many countries around the globe, with beneficial economic impacts. Genetic modification of foods or other economically important products remains controversial in several countries due to their as yet not fully documented effects on environmental and human health. Nevertheless, implementation of well-designed bioengineering technologies will continue to expand the approaches available in our arsenal to increase the efficiency of agricultural production and output, to ultimately beneficially impact human and animal nutrition and health. Similarly, expression of immunologically active molecules in transgenic insect vectors offers fresh opportunities to eradicate global disease transmission by short-circuiting the life cycle of the pathogen in the insect, although this strategy has not been widely implemented as yet.

As the chapters in this book have revealed, many new pathways into the future of insect immunology are now challenging us as we focus on both near and far horizons in this field. Pattern recognition processes, the molecules involved in nonself recognition, and signaling pathways that mobilize cellular and humoral defenses, are three topics of current focus that will continue to engage insect physiologists and molecular immunologists for decades to come. The formidable power of genomic and bioinformatic approaches has produced massive assemblages of immunological information about insect systems in just the past 5 years. As we project forward into the future, this rapid pace of progress is anticipated to continue as we introduce many new generations of investigators to the exciting and dynamic discipline of insect immunology.

INDEX

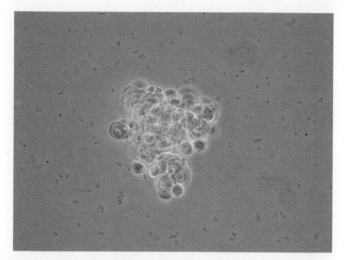

PLATE 3.3   A photomicrograph (400 × magnification) of a hemocytic microaggregate, taken 1 h after injecting a tobacco hornworm, *Manduca sexta*, with the bacterium *Serratia marcescens*. The cells in this image are about 10–12 microns in diameter. (Prepared and photographed by Jon S. Miller.)

PLATE 3.4   A photomicrograph (40 × magnification) of nodules formed in a tobacco hornworm, *M. sexta*, 4 h after injection with the bacterium *S. marcescens*. Nodules are typically darkened by a final melanization reaction and attached to body and organ surfaces. The nodules seen here are attached to the alimentary canal which is framed by Malpighian tubules. The muscles surrounding the alimentary canal feature large tracheae, clearly visible in this image (from Stanley and Miller, 2006, with permission of Wiley-Blackwell).

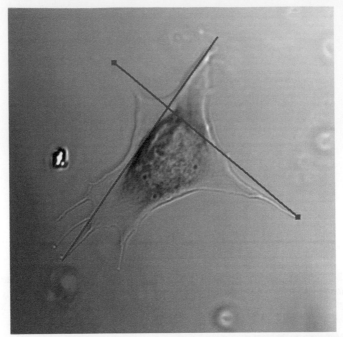

PLATE 3.5 A photomicrograph of a plasmatocyte from an untreated tobacco hornworm, *Manduca sexta*, after spreading on a glass cover slip for 1 h. The red lines represent digital measurements of the cell dimensions. This photograph was taken through confocal optics at 400×. (Prepared and photographed by Jon S. Miller.)

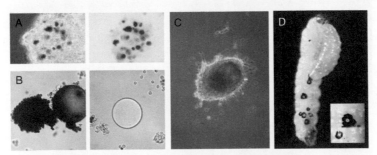

PLATE 4.1 Melanin formation in response to infection or wounding: (A) A hemocyte nodule dissected from a *M. sexta* larva 1 day after injection of *Micrococcus luteus*. The left image was obtained with phase contrast optics and highlights the presence of multiple layers of hemocytes. The right image was obtained with bright field conditions and shows the presence of spots of melanin within the nodule. (B) Encapsulation and melanization of two nickel–agarose chromatography beads coated with a 6-His tagged domain from *M. sexta* immulectin-2 and incubated with hemolymph *in vitro* (left panel). Control beads coated with a different recombinant protein (*M. sexta* cuticular protein CP36) were not encapsulated or melanized (right panel). (C) Melanization and encapsulation *in vivo* of a *Cotesia congregata* egg in an unsuitable host, the sphingid lepidopteran *Pachysphinx occidentalis*. (D) Melanized wounds on a *Plodia interpunctella* larva injured by the feeding of an ectoparasitoid larva, *Habrobracon hebetor* (Baker and Fabrick, 2002). The insert shows such melanized regions at higher magnification. (B is from Xiao-Qiang Yu, University of Missouri, Kansas City, MO; C is from Nancy Beckage, University of California, Riverside, CA; D is from Jeffrey Fabrick, USDA-ARS Arid-Land Agricultural Research Center, Maricopa, AZ.)

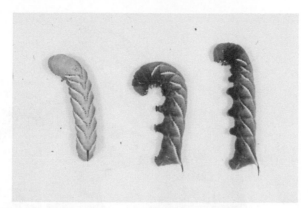

PLATE 4.3   Systemic melanization resulting from uncontrolled phenoloxidase activation. *Manduca sexta* larvae were injected with water (left), 200 μg of *Micrococcus luteus* (middle), or with 2 mg of bovine chymotrypsin (right) and photographed 24 h later.

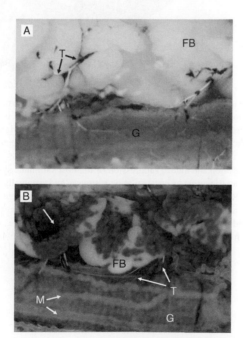

PLATE 9.2   Impact of genetics on virus propagation within an insect host: Larvae of the semi-permissive host *Ostrinia nubilalis* (A) and the permissive host *Heliothis virescens* (B) dissected 72 h after infection with *Rachiplusia ou* multiple nucleopolyhedrovirus expressing β-galactosidase under the control of the hsp70 promoter. Infection of *H. virescens* is significantly more advanced with extensive lacZ blue staining indicative of infection in the tracheoles (T), fat body (FB) and gut epithelium (G). Infection of *O. nubilalis* is restricted to foci of infection in these tissues (B). The Malpighian tubules (M) are also shown.

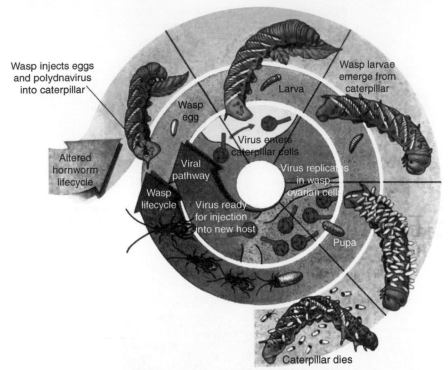

Wasp injects eggs
and polydnavirus
into caterpillar

Wasp
egg

Larva

Wasp larvae
emerge from
caterpillar

Virus enters
caterpillar cells

Altered
hornworm
lifecycle

Viral
pathway

Virus replicates
in wasp
ovarian cells

Wasp
lifecycle

Virus ready
for injection
into new host

Pupa

Caterpillar dies

PLATE 10.1 The interrelationships of the life cycles of insect hosts, parasitoids, and PDVs as illustrated by tobacco hornworm, *Manduca sexta*, larvae parasitized by the braconid wasp *Cotesia congregata*. The disrupted life cycle of the parasitized larva is shown to the right. The *C. congregata* bracovirus (CcBV) replicates in wasp ovarian calyx cells and virions are injected into the caterpillar host along with the eggs during wasp oviposition. Viral genes are expressed in host *M. sexta* larvae in the absence of viral replication and induce suppression of the host's immune system. The parasitoids develop in the host without triggering an encapsulation response, and eventually emerge from the host to spin pupal cocoons. Replication of the PDV begins during the wasp's late pupal stage, and the female wasp ecloses as an adult carrying mature virions to parasitize new hosts. The tobacco hornworm larva with emerged wasps is developmentally arrested in the larval stage. The cycle then begins again. Figure reprinted from Beckage (1997) with permission from *Scientific American* and Roberto Osti Illustrations.

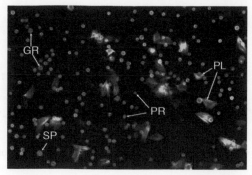

PLATE 10.7 Light micrograph showing appearance of prohemocytes (PR), plasmatocytes (PL), granulocytes (GR), and spherule cells (SP) in hemolymph from a non-parasitized fourth instar *Manduca sexta* larva. Note that the plasmatocytes show filopodial extensions resulting from actin polymerization in the cells that spread. The granulocytes are round with accumulations of dense granules. The smallest cells are prohemocytes (see chapter 2).

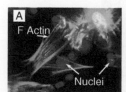 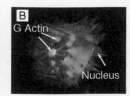 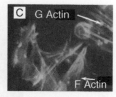

PLATE 10.9 Appearance of hemocytes from (A) non-parasitized control fourth instar *Manduca sexta* larva compared with (B) and (C) the disrupted appearance of cells collected from parasitized larvae 24 h post-oviposition by *Cotesia congregata*. Actin was labeled with phalloidin (green) and the nuclei were stained with DAPI (red). In non-parasitized larvae (A) note the preponderance of polymerized filamentous actin (F actin) in long filopodial extensions of plasmatocytes which adhere to the slide. In non-parasitized larvae, hemocyte nuclei are round, dense, and centrally located in the cells. In contrast, as seen in (B) and (C), the hemocytes of newly parasitized larvae have disrupted cytoskeletal networks. Note the presence of depolymerized globular actin (G actin) and disrupted ectopic nuclei in hemocytes of newly parasitized larvae. The hemocytes of parasitized larvae are destined to undergo apoptosis (see text). Photograph modified from Dumpit et al. (submitted).

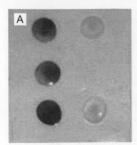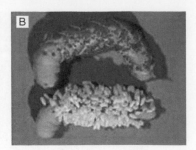

PLATE 10.10   Panel (A) compares hemolymph melanization rates in non-parasitized control larvae (left) versus parasitized (right) fifth instar *Manduca sexta* larvae. The hemolymph from the parasitized larva was collected on the day *Cotesia congregata* larvae emerged from the host. Hemolymph from the unparasitized larva was collected on day 3 of the fifth instar. Photograph was taken 1.5 h following spotting of the hemolymph samples on Parafilm. Panel (B) shows that when the parasitoid cocoons are stripped from the host (upper larva), localized wound responses and melanization can be seen encircling the sites of parasitoid exit through the integument despite the reduced rates of melanization seen in host hemolymph at this time (A). See text for further details.

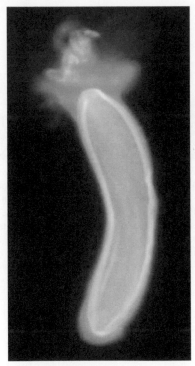

PLATE 11.1   Mature oocyte of the parthenogenetic wasp *Venturia canescens* stained with a Gal-specific lectin, peanut agglutinin (PNA). The oocyte (or non-fertilized egg) was dissected from the wasp egg reservoir, incubated with FITC-conjugated PNA and inspected under indirect UV light on a confocal microscope. Note the bushy chorion protrusions at the anterior end of the egg shell, which are translucent in the non-stained egg. Encapsulation of the egg usually starts at these heavily stained protrusions (Schmidt and Schuchmann-Feddersen, 1989).

Printed and bound by CPI Group (UK) Ltd, Croydon, CR0 4YY

03/10/2024

01040415-0009